Advances in Intelligent Systems and Computing

Volume 851

Series editor

Janusz Kacprzyk, Polish Academy of Sciences, Warsaw, Poland
e-mail: kacprzyk@ibspan.waw.pl

The series "Advances in Intelligent Systems and Computing" contains publications on theory, applications, and design methods of Intelligent Systems and Intelligent Computing. Virtually all disciplines such as engineering, natural sciences, computer and information science, ICT, economics, business, e-commerce, environment, healthcare, life science are covered. The list of topics spans all the areas of modern intelligent systems and computing such as: computational intelligence, soft computing including neural networks, fuzzy systems, evolutionary computing and the fusion of these paradigms, social intelligence, ambient intelligence, computational neuroscience, artificial life, virtual worlds and society, cognitive science and systems, Perception and Vision, DNA and immune based systems, self-organizing and adaptive systems, e-Learning and teaching, human-centered and human-centric computing, recommender systems, intelligent control, robotics and mechatronics including human-machine teaming, knowledge-based paradigms, learning paradigms, machine ethics, intelligent data analysis, knowledge management, intelligent agents, intelligent decision making and support, intelligent network security, trust management, interactive entertainment, Web intelligence and multimedia.

The publications within "Advances in Intelligent Systems and Computing" are primarily proceedings of important conferences, symposia and congresses. They cover significant recent developments in the field, both of a foundational and applicable character. An important characteristic feature of the series is the short publication time and world-wide distribution. This permits a rapid and broad dissemination of research results.

More information about this series at http://www.springer.com/series/11156

Shailesh Tiwari · Munesh C. Trivedi
Krishn K. Mishra · A. K. Misra
Khedo Kavi Kumar

Editors

Smart Innovations in Communication and Computational Sciences

Proceedings of ICSICCS-2018

 Springer

Editors
Shailesh Tiwari
Computer Science and Engineering
  Department
ABES Engineering College
Ghaziabad, Uttar Pradesh, India

Munesh C. Trivedi
Department of Information Technology
Rajkiya Engineering College
Azamgarh, Uttar Pradesh, India

Krishn K. Mishra
Department of Computer Science
  and Engineering
Motilal Nehru National Institute
  of Technology
Allahabad, Uttar Pradesh, India

A. K. Misra
Department of Computer Science
  and Engineering
Motilal Nehru National Institute
  of Technology
Allahabad, Uttar Pradesh, India

Khedo Kavi Kumar
Department of Computer Science
  and Engineering
University of Mauritius
Moka, Mauritius

ISSN 2194-5357   ISSN 2194-5365 (electronic)
Advances in Intelligent Systems and Computing
ISBN 978-981-13-2413-0   ISBN 978-981-13-2414-7 (eBook)
https://doi.org/10.1007/978-981-13-2414-7

Library of Congress Control Number: 2018937328

This Springer imprint is published by the registered company Springer Nature Singapore Pte Ltd.
The registered company address is: 152 Beach Road, #21-01/04 Gateway East, Singapore 189721, Singapore

Preface

The 2nd International Conference on *Smart Innovations in Communications and Computational Sciences (ICSICCS-2018)* was held at Indore, MP, India, **during April 28–29, 2018**. ICSICCS-2018 was organized and supported by **Devi Ahilya Vishwavidyalaya, Indore, MP, India**. The main purpose of ICSICCS-2018 is to provide a forum for researchers, educators, engineers, and government officials involved in the general areas of communication, computational sciences and technology to disseminate their latest research results and exchange views on the future research directions of these fields; exchange computer science and integrate its practice, and application of the academic ideas; improve the academic depth of computer science and its application; and provide an international communication platform for educational technology and scientific research for the world's universities, business intelligence engineering field experts, professionals, and business executives.

The field of communication and computational sciences always deals with finding the innovative solutions for problems by proposing different techniques, methods, and tools. Generally, innovation refers to finding new ways of doing usual things or doing new things in a different manner, but due to increasingly growing technological advances with a speedy pace, *smart innovations* are needed. Smart refers to '*how intelligent the innovation is?*' Nowadays, there is a massive need to develop new '*intelligent*' '*ideas, methods, techniques, devices, tools.*' The proceedings cover those systems, paradigms, techniques, and technical reviews that employ knowledge and intelligence in a broad spectrum.

ICSICCS-2018 received around 250 submissions from 603 authors of 8 different countries such as Taiwan, Sweden, Italy, Saudi Arabia, China and many more. Each submission has gone through the plagiarism check. On the basis of a plagiarism report, each submission was rigorously reviewed by at least two reviewers. Even some submissions have more than two reviews. On the basis of these reviews, 47 high-quality papers were selected for publication in this proceedings volume, with an acceptance rate of 18.6%.

We are thankful to the speakers: Prof. Srikanta Patnaik, SOA University, Bhubaneswar, India, and Mr. Vinit Goenka, Member—Governing Council—CRIS, Min. of Railways, and Member—IT. We are thankful to the delegates and the authors for their participation and their interest in ICSICCS as a platform to share their ideas and innovation. We are also thankful to Prof. Dr. Janusz Kacprzyk, Series Editor, AISC, Springer, and Mr. Aninda Bose, Senior Editor, Hard Sciences, Springer, India, for providing continuous guidance and support. Also, we extend our heartfelt gratitude and thanks to the reviewers and technical program committee members for showing their concern and effort during the review process. We are indeed thankful to everyone directly or indirectly associated with the conference organizing team leading it toward the success.

We hope you enjoy the conference proceedings and wish you all the best.

Indore, MP, India   Organizing Committee
                                                                                      ICSICCS-2018

Contents

Part I Smart Computing Techniques

Design of Eye Template Matching Method for Head Gesture Recognition System   3
Rushikesh T. Bankar and Suresh S. Salankar

Real-Time Meta Learning Approach for Mobile Healthcare   11
Dipti Durgesh Patil and Vijay M. Wadhai

Firearm Detection from Surveillance Cameras Using Image Processing and Machine Learning Techniques   25
Fraol Gelana and Arvind Yadav

Segmentation of Optic Disc by Localized Active Contour Model in Retinal Fundus Image   35
Shreenidhi H. Bhat and Preetham Kumar

A Novel Genetic Algorithm Based Scheduling for Multi-core Systems   45
Aditi Bose, Tarun Biswas and Pratyay Kuila

Various Preprocessing Methods for Neural Network Based Heart Disease Prediction   55
Kavita Burse, Vishnu Pratap Singh Kirar, Abhishek Burse
and Rashmi Burse

Intelligent Passenger Information System Using IoT for Smart Cities   67
Polamarasetty Anudeep and N. Krishna Prakash

Continuous Monitoring and Detection of Epileptic Seizures Using Wearable Device   77
Darshan Mehta, Tanay Deshmukh, Yokesh Babu Sundaresan
and P. Kumaresan

**Brain Tumor Detection Using Cuckoo Search Algorithm
and Histogram Thresholding for MR Images**   85
Sudeshna Bhakat and Sivagami Periannan

**eMDPM: Efficient Multidimensional Pattern Matching Algorithm
for GPU**   97
Supragya Raj, Siddha Prabhu Chodnekar, T. Harish
and Harini Sriraman

**Computer Vision-Based Fruit Disease Detection
and Classification** 105
Abhay Agarwal, Adrija Sarkar and Ashwani Kumar Dubey

**Automated Brain Tumor Detection Using Discriminative
Clustering Based MRI Segmentation** 117
Abhilash Panda, Tusar Kanti Mishra and Vishnu Ganesh Phaniharam

Cancer Prediction Based on Fuzzy Inference System 127
Soumi Dutta, Sujata Ghatak, Abhijit Sarkar, Rechik Pal, Rohit Pal
and Rohit Roy

**Energy Consumption Data Analysis and Operation Evaluation
of Green Buildings** 137
Weiyan Li, Dong-Lin Wang, Chen-Fei Qu, Xuantao Zhang, Wen-Jing Wu
and Pengcheng Zhao

**The Method of Random Generation of Electronic Patrol Path
Based on Artificial Intelligence** 147
Wen-Xia Liu, Dong-Lin Wang, Wen-Jing Wu and Chen-Fei Qu

Part II   Intelligent Communications & Networking

**Channel Power Estimation for DVBRCS to DVBS2 Onboard
DSP Payload** 161
Krishna K. Wadiwala, Neeraj Mishra, Deepak Mishra and Hetal Patel

**A Slotted Microstrip Patch Antenna for 5G Mobile Phone
Applications** 173
Kausar Parveen, Mohammad Sabir, Manju Kumari and Vishal Goar

**Reliability Study of Sensor Node Monitoring Unattended
Environment** 179
V. Mahima, G. R. Kanagachidambaresan, M. Balaji and Jagannath Das

RSOM-Based Clustering and Routing in WSNs 189
G. R. Asha and Gowrishankar Subrahmanyam

**Evaluation of Received Signal Strength Indicator (RSSI)
for Relay-Based Communication in WBAN** 201
Pulkit Pandey, Arthav S. Patial and Sindhu Hak Gupta

Side-Channel Attacks on Cryptographic Devices and Their Countermeasures—A Review 209
M. M. Sravani and S. Ananiah Durai

Implementation and Analysis of Different Path Loss Models for Cooperative Communication in a Wireless Sensor Network 227
Niveditha Devarajan and Sindhu Hak Gupta

Parallel Approach for Sub-graph Isomorphism on Multicore System Using OpenMP 237
Rachna Somkunwar and Vinod M. Vaze

Optimized Solution for Employee Transportation Problem Using Linear Programming 247
Alind and Himanshu Sahu

Spectrum Prediction Using Time Delay Neural Network in Cognitive Radio Network 257
Sweta Jain, Apurva Goel and Prachi Arora

Reliability Factor Based AODV Protocol: Prevention of Black Hole Attack in MANET 271
Prakhar Gupta, Pratyaksh Goel, Pranjali Varshney and Nitin Tyagi

Part III   Web & Informatics

A Survey of Lightweight Cryptographic Algorithms for IoT-Based Applications 283
Ankit Shah and Margi Engineer

Fruit Disease Detection Using Rule-Based Classification 295
Vippon Preet Kour and Sakshi Arora

Fault Tolerance Through Energy Balanced Cluster Formation (EBCF) in WSN 313
Hitesh Mohapatra and Amiya Kumar Rath

Mining Social Networks: Tollywood Reviews for Analyzing UPC by Using Big Data Framework 323
V. Kakulapati and S. Mahender Reddy

Cloud-Based E-Learning: Using Cloud Computing Platform for an Effective E-Learning 335
Shams Tabrez Siddiqui, Shadab Alam, Zaki Ahmad Khan and Ashok Gupta

Deadline-Aware Scheduling for Scientific Workflows in IaaS Cloud 347
Mainak Adhikari and Tarachand Amgoth

Table Detection and Metadata Extraction in Document Images 361
Anand Gupta, Devendra Tiwari, Tarasha Khurana and Sagorika Das

**Interactive Mobile Application to Determine and Enhance
User's Skills in Their Respective Field of Interest** 373
Akshay Talke, Rohit Kr. Singh, Sanyam Raj, Virendra Patil,
Ameya Jawalgekar and Ajitkumar Shitole

**Back-Propagated Neural Network on MapReduce Frameworks:
A Survey** 381
Jenish Dhanani, Rupa Mehta, Dipti Rana and Bharat Tidke

TWEESENT: A Web Application on Sentiment Analysis 393
Sweta Swain and K. R. Seeja

Personalized Secured API for Application Developer 401
R. Maheswari, S. Sheeba Rani, P. Sharmila and S. Rajarao

**Classification of Query Graph Using Maximum Connected
Component** 413
Parnika Paranjape, Meera Dhabu, Rushikesh Pathak, Nitesh Funde
and Parag Deshpande

**All Domain Hidden Web Exposer Ontologies: A Unified Approach
for Excavating the Web to Unhide Deep Web** 423
Manpreet Singh Sehgal and Jay Shankar Prasad

**Data Stream Classification Using Dynamic Model for Labeling
Strategy** 433
Nitesh Funde, Meera Dhabu and Parnika Paranjape

**Regulatory Framework for Standardization of Online
Transactions Using Cryptocurrencies** 443
Astitva Narayan Pandey and Himanshu Gupta

**Autonomics of Self-management for Service Composition
in Cyber Physical Systems** 455
Swati Nikam and Rajesh Ingle

Ranking-Based Sentence Retrieval for Text Summarization 465
Abhishek Mahajani, Vinay Pandya, Isaac Maria and Deepak Sharma

**Design of Dmey Wavelet Gaussian Filter (DWGF) for De-noising
of Skin Lesion Images** 475
Ginni Arora, Ashwani Kumar Dubey and Zainul Abdin Jaffery

Medical Image Watermarking in Transform Domain 485
Harsh Vikram Singh and Ankur Rai

**Denoising of Brain MRI Images Using a Hybrid Filter Method
of Sylvester-Lyapunov Equation and Non Local Means** 495
Krishna Kumar Sharma, Dheeraj Gurjar, Monika Jyotyana
and Vinod Kumari

**Dynamical Simulation of TT&C Based on STKX Components
and MATLAB** 507
Hu Mengzhong

Author Index 519

About the Editors

Dr. Shailesh Tiwari currently works as Professor in Computer Science and Engineering Department, ABES Engineering College, Ghaziabad, India. He is an alumnus of Motilal Nehru National Institute of Technology Allahabad, India. His primary areas of research are software testing, implementation of optimization algorithms, and machine learning techniques in various problems. He has published more than 50 publications in international journals and proceedings of international conferences of repute. He has edited *Scopus, SCI, and E-SCI-indexed journals*. He has also edited several books published by Springer. He has organized several international conferences under the banner of IEEE and Springer. He is Senior Member of IEEE, Member of IEEE Computer Society, and Fellow of Institution of Engineers (FIE).

Dr. Munesh C. Trivedi currently works as Professor in Computer Science and Engineering Department, ABES Engineering College, Ghaziabad, India. He has published 20 textbooks and 80 research publications in different international journals and proceedings of international conferences of repute. He has received Young Scientist and numerous awards from different national as well as international forums. He has organized several international conferences technically sponsored by IEEE, ACM, and Springer. He is on the review panel of IEEE Computer Society, *International Journal of Network Security, Pattern Recognition Letter and Computer & Education* (Elsevier's journal). He is Executive Committee Member of IEEE UP Section, IEEE India Council, and also IEEE Asia Pacific Region 10.

Dr. Krishn K. Mishra is currently working as Visiting Faculty, Department of Mathematics and Computer Science, University of Missouri, St. Louis, USA. He is an alumnus of Motilal Nehru National Institute of Technology Allahabad, India, which is also his base working institute. His primary areas of research include evolutionary algorithms, optimization techniques, and design and analysis of algorithms. He has also published more than 50 publications in international journals and proceedings of internal conferences of repute. He is serving as a program committee member of several conferences and also editing *Scopus and SCI-indexed journals.*

Prof. A. K. Misra retired from the post of Professor in Computer Science and Engineering Department, Motilal Nehru National Institute of Technology Allahabad, India. Presently, he is associated with SPMIT, Allahabad, India, as Advisor. He has more than 45 years of experience in teaching, research, and administration. His areas of specialization are software engineering and nature-inspired algorithms. He has fetched grants as Coordinator, Co-investigator, and Chief Investigator for several research projects such as Indo-UK: REC Project, Development of a Framework for Knowledge Acquisition and Machine Learning for Construction of Ontology for Traditional Knowledge Digital Library (TKDL), Semantic Web Portal for Tribal Medicine and completed them successfully. He has guided 148 PG and 20 doctorate students. He has published more than 90 research articles in international journals and proceedings of international conferences of repute. He is Fellow of Institution of Engineers (FIE), Member of ISTE, and Member of IEEE and CSI. He has organized several national and international conferences in the capacity of general chair under the flagship of ACM and IEEE.

Dr. Khedo Kavi Kumar is Associate Professor in the Department of Computer Science and Engineering, University of Mauritius, Mauritius. His research interests are directed toward wireless sensor networks, mobile ad hoc networks, context-awareness, ubiquitous computing, and Internet of things. He has published research papers in renowned international conferences/high impact journals and has presented his research works in reputed international conferences around the world. He has also served on numerous editorial boards of distinguished international journals and on technical program committees of popular international conferences (EEE Africon 2013, IEEE ICIT 2013, InSITE 2011, IEEE WCNC 2012, ICIC 2007, WCSN 2007, COGNITIVE 2010, *WILEY International Journal on Communication Systems, International Journal of Sensor Networks, International Journal of Computer Applications*). He has also served as Head of Department in the Department of Computer Science and Engineering, Faculty of Engineering, University of Mauritius. He was awarded the UoM Research Excellence Award in February 2010 and ICT Personality of the Year 2013 (runner-up) which recognizes the outstanding young academic who has contributed significantly to promote research at the university and to an individual that has demonstrated exemplary growth and performance in the ICT industry in the year 2013 in Mauritius.

Part I
Smart Computing Techniques

Design of Eye Template Matching Method for Head Gesture Recognition System

Rushikesh T. Bankar and Suresh S. Salankar

1 Introduction

Today's electric wheelchair is based on hand gesture based, eye based, voice based, and joystick based used by the handicapped peoples. The recognition of the gestures, hand or head generates the control signals. In the process of recognition of the user's head gestures, the control signals generated by the user's body parts mean hand or head is recognized by the receiver. The gestures are expressive, having meaningful user's body motions which involve the physical movements of the human body parts such as hands, arms, fingers, and head which conveys the meaningful information or doing some needful actions. In any gesture recognition, the environmental conditions may play an important role. The applications of the recognition of the user's head gestures have various applications namely security purpose to recognize the sign language means hand gesture recognition, for the medical applications, such as handicapped peoples, monitoring automobile driver's alertness/drowsiness levels, etc.

In head gesture recognition based interface (HGI) process, the face detection, in the particular environmental condition or the user using the system or wheelchair in indoor or outdoor environmental conditions, is very useful. The face tracking is also important to track under different environmental conditions. Some difficulties deal with traditional electric wheelchairs controlled by eye, voice based, hand control based, and head controlled based wheelchair. These difficulties are when the user's

R. T. Bankar
Department of Electronics Engineering, G H Raisoni College of Engineering, Nagpur, India
e-mail: rushikesh.bankar@raisoni.net

S. S. Salankar (✉)
Department of Electronics & Telecommunication Engineering, G H Raisoni
College of Engineering, Nagpur, India
e-mail: suresh.salankar@raisoni.net

© Springer Nature Singapore Pte Ltd. 2019
S. Tiwari et al. (eds.), *Smart Innovations in Communication and Computational Sciences*, Advances in Intelligent Systems and Computing 851,
https://doi.org/10.1007/978-981-13-2414-7_1

head may be the out of the image, in various changing illumination conditions, the user's face color may change, the users have different facial appearances, etc.

These limitations can be solved in the paper [1] where P. Jia et al. present the hand's free control system of an intelligent wheelchair using the gestures of the head. This system requires the detection of the face as well as tracking of the user's face for generating the control signals which is used by the wheelchair. This system is used as the human-friendly interface to operate an intelligent wheelchair using the head gestures for the handicapped people whose limb movements not worked. The system uses camshift algorithm for the face tracking of the user, but the face tracking method/algorithm has limitations. These are: it is unable to precisely track the face when the illumination condition changes and it cannot work well under the cluttered environment.

2 Existing Methods

The challenging task for recognizing the gestures under indoor as well as outdoor environmental conditions is the recognition of the head gestures. The recognition of head gestures is the interface system which is based on the human gestures that are head. The head gesture based system includes the detection of user's face in real as well as real-time face tracking and gesture recognition processes. It is used as the human–robot interface for the intelligent wheelchair. There are various existing methods for recognizing the head gestures. They are the head gesture recognition using

lips position detection method, optical flow based head gesture recognition system, head gesture recognition system using depth sensor, head gesture recognition system using accelerometer and magnetometer, etc. These head gesture recognizing methods are used only for indoor environmental conditions. Also in these methods, the concept of environmental conditions is not considered.

Rathore et al. [2] proposed an intelligent system which can assist physically handicapped, visually impaired as well as elderly people. It consists of a navigation system. The system uses the accelerometer and magnetometer. The system also consists of a navigation pad which can be tied to the head for navigating the intelligent wheelchair. The obstacle detection/avoidance are one of the most important parameters for the intelligent wheelchair used by the user. For the obstacle avoidance system, the four ultrasonic sensors are used.

Manju Davy and Deepa [3] proposed an intelligent wheelchair based on the accelerometer sensor to recognize the head movements of the user. The project works for the handicapped peoples who cannot perform their controlled movement. The system or an intelligent wheelchair will be used for the patients who are particularly suffering from the diseases. To generate the motion control commands for the controller, the recognized gestures of the user are used. So that wheelchair can control the motion of intelligent wheelchair according to the interaction of the user. The position controlled by the head is the gesture which can be performed by the patients having quadriplegic. So, the movement of the user's head is the gesture for such patients.

Martin et al. [4] proposed the system in which movements of the head are based on optical flow and gesture analysis used in the automotive surroundings. The head gesture recognition is a fundamental part of look inside a vehicle when manipulative an intelligent driver support system. The authors in this paper used the optical flow method for recognizing the head gestures of the user. The optical flow based head movement and the gesture analyzer are a user autonomous, very vigorous to the occlusions from the head turns of the user and the changing illumination conditions. The system segments the head gestures into the motion states of head or no head motion states. Sarika and Das [5] proposed an intelligent system of background subtraction based head gesture recognition. The system also used optical flow based classification method for recognizing the user's head gestures. This paper presents a technique in which a real-time head gesture is to be recognized. The method uses a Gaussian mixture model which is accompanied by the optical flow algorithm. The optical flow algorithm provides the required information about the movement of the user's head. This intelligent system can be implemented in the various motion control system. The limitation of this system is that the system considered the gesture only from real time. The system can be further improved by giving order and control the robots.

Kawarazaki and Diaz [6] present an intelligent system for the wheelchair based on a depth sensor. In traditional/conventional wheelchair, the user controls the wheelchair using the joystick. However, when the user holds an object by both the hands then the user cannot control the joystick. In this paper, the wheelchair used moves according to the position hand of the user. For recognizing gestures of the hand quickly, a depth sensor is used.

Rehman et al. [7] present an intelligent system in which the controlling of an electric wheelchair is based on the vibrotactile rendering of head gestures. The system is very useful for the persons with severe disabilities. In this system, a stereo camera is used to obtain the real-time range information. This information is used to locate and segment the user's face images in the real time. In this paper, for the head pose estimation, an isomap based nonlinear manifold learning map of user's facial textures is used. To command the wheelchair, the user required to gesture his head. The system uses vibrotactile rendering of the head gestures as a feedback.

3 Experimental Setup as a Prototype

The experimental setup for an intelligent wheelchair as a prototype is shown in the figure.

Figure 1 shows a prototype for an intelligent wheelchair as an intelligent wheelchair with the user. The BOT uses two CC2500RF modules as a transmitter and receiver. For the supply voltage, a 6 V battery with 7805 voltage regulator is required. The microcontroller used is the ATMEGA16 microcontroller. For moving

Fig. 1 Experimental setup

Fig. 2 Flowchart for
proposed BOT as a prototype
of an intelligent wheelchair

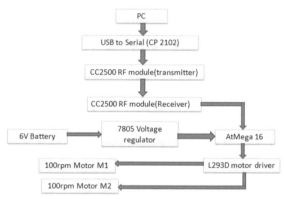

the BOT, two 100 RPM motors are used. The head gesture recognition is done in the image processing tool that is the MATLAB R2013a. In this paper, a novel integrated approach is used named as head gesture based interface that is the HGI. The head gesture based interface consists of three stages namely the detection of the user's face in real time and real-time tracking of the user's face.

Figure 2 shows flowchart for our proposed BOT as a prototype of an intelligent wheelchair. The input of the BOT is connected to the PC/Laptop. The connections are via USB to serial port that is CP2102. For the signals transmission and receiving purpose, we have to use CC2500 RF Module that is the transmitter as well as receiver. The output of the receiver goes to the ATMEGA 16 Microcontroller. The complete BOT/Module requires 6 V power supply. For that, we have to use 6 V battery. For regulating the 6 V supply voltage, we require 7805 voltage regulator. The output of the ATMEGA 16 Microcontroller goes to the L293D motor driver module. The two motors M1 and M2 are connected to the motor driver module. The speed of the motors is 100 RPM.

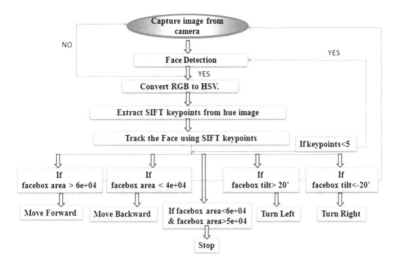

Fig. 3 Proposed architecture for the system

Figure 3 shows the complete proposed architecture of the system. First, we take the video of the user in various illumination conditions. From the video, we capture or grab the images. After capturing the images, we detect the user's face using face detection method. If detection of the user's face gets completed then we use the RGB to HSV conversion process. From the hue images, we have to extract the SIFT keypoints. The face tracking of the user is based on the parameter that is SIFT keypoints extraction. If the condition keypoints <5 is satisfied then the process goes to the detection of the user's face. After tracking the user's face sited on wheelchair, movement of the wheelchair is to be decided.

4 Face Tracking by Improved Camshift

The conventional camshift algorithm works very fast. This algorithm is used for the face tracking of different environmental conditional images. However, face tracking of the user in different environmental conditions using camshift face tracker is not proper in outdoor environmental conditions. Therefore, we used the improved camshift with SIFT for face tracking for improving the results in the outdoor environmental conditions. The face detection is done by using the face detection method. The face detection method of the user's face in different environmental conditions is highly used to detect face.

First, the image is grabbed from the video using webcam then the face is detected using Viola–Jones algorithm. After it, conversion of RGB to HSV takes place. As this proposed algorithm is improved camshift, only hue image is taken into consideration because camshift works on hue image. So SIFT key points are extracted from hue channel data and face tracking takes place using these SIFT key points. Tracking of key points is obtained by applying point tracker from frame to frame. In this way improved camshift algorithm works and it tracks the face properly in outdoor environments under various conditions.

5 Method Used for Recognition of Head Gestures

The existing head gesture recognition method used nose template matching method for the recognition of the head gestures. The profile left, profile right, profile up, and profile down head gestures are recognized using the nose template matching method. As per the nose template directions of the user, the movement of the head is to be decided. Accordingly, the control signal goes to the wheelchair system and the wheelchair moves as per the directions.

But the system deals with some uncertainty in the practical application of the electric wheelchair. If any person calls the user and the user seated on the wheelchair is in the opposite direction the person, then in this situation, we normally move the head. So the user moves his head and accordingly the wheelchair moves. But the users do not want to move the wheelchair. So such miss-operation occurred while the system uses nose template matching method with head movement.

To overcome this limitation, we have to propose a system with head tilting. For the head tilting of the user, the system requires an eye template matching method. For the movement of the wheelchair, the user uses the head tilting method. In this situation, if anybody calls the user then the user move his head and not tilt the head. So the mis-operation of the wheelchair is not performed by the user. For the movement of the electric wheelchair, the head tilting method is the best method. For performing the head tilting method, we require the method of template matching. So the eye template matching method is the best method to recognize the head gestures of the user.

6 Eye Template Matching Results

Profile Face Downward

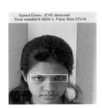

```
>> new_x

new_x =

        173

>> new_y

new_y =

        84
```

New 'y'=84

Ref 'y'=118

➢ The output shows the Upward image
  If 'y' difference < speed threshold

➢ 'y' diff.=(new 'y'-ref 'y')
  'y' diff.=(84-118)

        = -34

➢ Speed threshold=8

➢ Hence -34 < 8

Profile Face Upward

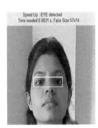

```
>> new_x

new_x =

        158

>> new_y

new_y =

        101
```

New 'y'=118

Ref 'y'=61

➢ The output shows the Downward image
  If 'y' difference > speed threshold

➢ 'y' diff.=(new 'y'-ref 'y')
  'y' diff.=(118-61)

        =57

➢ Speed threshold=8

➢ Hence 57>8

Profile Face Right

```
>> new_x

new_x =

        173

>> new_y

new_y =

        118
```

New 'x'=158

Ref 'x'=173

➢ The output shows the Right turn
  If 'x' difference < Turn threshold

➢ 'x' diff.=(new 'x'-ref 'x')
  'x' diff.=(158-173)

=-15

➢ Turn threshold=5

➢ Hence -15 < 5

Profile Face Left

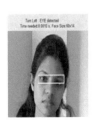

```
>> new_x

new_x =

        173

>> new_y

new_y =

        77
```

New 'x'=173

Ref 'x'=158

➢ The output shows the Left turn
  If 'x' difference > Turn threshold

➢ 'x' diff.=(new 'x'-ref 'x')
  'x' diff.=(173-158)

        = 15

➢ Turn threshold=5

➢ Hence 15 > 5

7 Conclusion

The system presents implementation of intelligent wheelchair for the handicapped peoples having no limb movements due to the diseases like parkingsons. The system designs a prototype of an intelligent wheelchair. The system fulfills the need of handicapped peoples having very restricted limb movements or ruthless handicap, for elderly and the disabled peoples. The gestures which are recognized are used for generating the motion control commands for the ATmega 16 microcontroller. So that it can control the movement of an intelligent wheelchair prototype according to the intention of the user.

Acknowledgements The authors express grateful thanks to the anonymous referees for their useful comments and suggestions to improve the presentation of this paper. The pictures in the figures of all the simulation work were carried out with Dolly Junghare and Neha Sakure with her permissions.

References

1. Jia, P., Hu, H., Lu, T., Yuan, K.: Head gesture recognition for hands free control of an intelligent wheelchair. In: 2010 IEEE International Conference on Computer Application and System Modeling (ICCASM 2010)
2. Rathore, D.K., Srivastava, P., Pandey, S., Jaiswal, S.: A novel multipurpose smart wheelchair. In: 2014 IEEE Student's Conference on Electrical, Electronics and Computer Science
3. Davy, M., Deepa, R.: Hardware implementation based on head movement using accelerometer sensor. Int. J. Appl. Sci. Eng. Res. **03**(01) (2014)
4. Martin, S., Tran, C., Tawari, A., Kwan, J., Trivedi, M.: Optical flow based head movement and gesture analysis in automotive environment. In: 2012 15th International IEEE Conference on Intelligent Transportation Systems Anchorage, Alaska, USA, 16–19 September, 2012
5. Saikia, P., Das, K.: Head gesture recognition using optical flow based classification with reinforcement of GMM based background subtraction. Int. J. Comput. Appl. (0975–8887), **65**(25) (2013)
6. Kawarazaki, N., Diaz, A.I.B.: Gesture recognition system for wheelchair control using a depth sensor. In: 2013 IEEE International Conference
7. Rehman, S.U.R., Raytchev, B., Yoda, I., Liu, L.: Vibrotactile rendering of head gestures for controlling electric wheelchair. In: 2009 IEEE International Conference on Systems, Man and Cybernetics San Antonio, TX, USA, October 2009

Real-Time Meta Learning Approach for Mobile Healthcare

Dipti Durgesh Patil and Vijay M. Wadhai

1 Introduction

Vital signals like heart rate, blood pressure, etc. of human body may change instanta-neously. This may affect human health suddenly. As such situations are unpredictable; it is infeasible to keep any patient waiting in hospital to record such changes. The designed Wireless Body Sensor Network (WBSN) based Intelligent Mobile Real-Time Health Monitoring framework (WIMRHM) [1] provides the ease of allowing a patient to be mobile and still be monitored continuously [2]. Different wireless body sensors are deployed on human body and the sensed vital signals are then transmitted to a smartphone for real-time analysis through any wireless technology like Bluetooth^{TM}. Some of the healthcare systems developed have applied only pat-tern matching techniques to predict the health risk where the rules are mined from patient's historical data and then these rules are used to predict the health risk of another patient. As the body tendency changes from person to person and time to time, medical profile of one person having same medical background may not be always applicable to another person, hence there is a need to build either patient or disease-specific model based on both historical as well as current behavior of vital signals of the patient. Our research has come up with a novel meta learning approach where meta learning means learning from learned knowledge. The historical rule base is formed statically using patient's prerecorded vital signals to form a base classifier and another base classifier is formed on the fly using continuously arriving real-time vital signal data. Then, these two base classifiers are combined to form a meta classifier to predict the level of health risk of monitored patient.

D. D. Patil (✉)
MKSSS's Cummins College of Engineering for Women, Pune, India
e-mail: diptivt@gmail.com

V. M. Wadhai
Trinity Academy of Engineering, Pune, India
e-mail: wadhai.vijay@gmail.com

© Springer Nature Singapore Pte Ltd. 2019
S. Tiwari et al. (eds.), *Smart Innovations in Communication and Computational Sciences*, Advances in Intelligent Systems and Computing 851,
https://doi.org/10.1007/978-981-13-2414-7_2

The paper is organized as follows. Section 2 discusses previous research work related to different real-time health monitoring systems and the online data stream mining algorithms for real-time learning. Section 3 describes the new real-time healthcare system framework and its various modules in detail. Section 4 discusses Partition based Adaptive Real-Time Clustering Stream (PARC-Stream) data stream mining algorithm to form real-time clusters. The results of the experiments evaluating the effectiveness of the meta learning approach to predict health risk are presented in Sect. 5. Finally, Sect. 6 draws conclusions and presents the future scope of the proposed approach.

2 Related Work

Agent-based meta learning concept has been used in many of the application areas [3–5] and proved effective in evaluated results. From the literature many of the features of meta learning are pointed out, they are listed below:

1. Meta learning executes base learners in parallel on partitioned data to improve efficiency and support scalability. This has two benefits, first, the same serial code can directly be used as base learner without the tedious task of parallelizing it and second, learning from partitioned data makes in-memory operations possible.
2. Meta learning boosts predictive accuracy by combining results of different base learners, each having different ways of representation, search heuristics or search space. These separately learned concepts increase the chances to derive a higher level learned model which in turn expresses a large database more accurately than any of the individual learners.
3. Scalability feature of meta learning has the ability to support generalized hierarchical multi-level meta learning.
4. The meta learning strategy is cost effective and an efficient way to apply in big data analytics on cloud-based architecture.

There is a lot of ongoing research to develop real-time healthcare system (RTHS). But our real-time meta learning approach for health risk prediction gives the highest accuracy as compare to other RTHSs. It is a fact that the global population is both growing and aging. The total number of people suffering from some type of disability will continue to rise. The first example in this category is mHealth system [6, 7] developed to extract automatic rules. This system does offline learning of the OSA events detected from ECG sensors and measures the HRV parameters. Agent-based meta learning concept has been used in many of the application areas and proved effective in evaluated results. WIMRHM developed by us uses this meta learning approach in real time for clustering and classification of vital signals to increase the prediction accuracy. WE-CARE [8] is a smartphone based wearable platform, where ECG signal is monitored continuously sensed using wireless ECG sensor. ECG signal is analyzed in real time to detect abnormal trends related to cardiovascular disease. Some of the other cell phone based ECG analysis solutions are proposed by ([9–12]). Most of them make analysis and classification of signals offline instead of real-time feedback.

Issue of network link breakage in WSN based u-healthcare system is [13], where researchers have designed fast link exchange minimum cost forwarding routing protocol. The healthcare framework discussed [14] does only offline learning of the patient's recorded signals and pattern matching is done in real time. IBM [15] has discussed a system that can monitor data streams from ICU and make a prediction. This system needs an offline system to analyze the medical database using cluster and Locally Supervised Metric Learning (LSML). Real-time clinical decision support system is discussed which uses supervised learning for analysis of vital signals [16]. Due to the supervised learning approach, this system needs initial training before formal use which is avoided in our developed system with the help of unsupervised approach.

UniMiner is a unified framework of data mining suggested to handle big data on smart devices [17]. There are various stream mining algorithms available for processing real-time data [1, 18, 19]. RA-VFKM (Resource-Aware Very Fast K-Means) [19], whose design is based on the effective stream clustering technique from the VFKM (Very Fast K-Means) algorithm, utilizes the AOG resource-aware technique to solve the problem of mining failure with constrained resources in VFKM. It increases the allowable error and the value of probability to decrease the number of runs and the number of amples, when the available free memory reaches a critical stage. VFKM and RA-VFKM also have the compulsion of keeping fix number of clusters and also involves a lot of redundant distance calculations which is reduced in PARC-Stream using the principle of Triangle inequality [21].

3 WBSN-Based Intelligent Mobile Real-Time Health Monitoring (WIMRHM) Framework

In healthcare domain, the physiological signals are continuous and time varying. Also, it generates huge amount of data in seconds. For predicting health risk and monitoring, the behavior of vital signals in the correlation of each other may improve the accuracy of diagnosis.

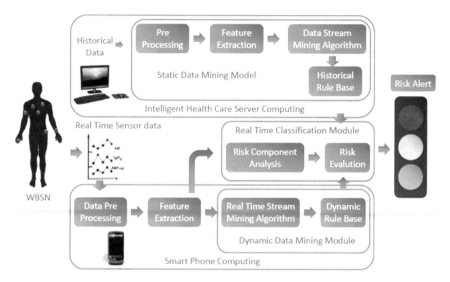

Fig. 1 WIMRHM framework [1]

Figure 1 shows the architecture of WIMRHM. Physiological vital signals are captured through Wireless Body Sensor Network and transferred on smartphone using Bluetooth™ or WPAN technology. Missing values and noisy data are rectified in data preprocessing phase. Null values are substituted to handle missing values which may be caused by sensor failure.

3.1 Preprocessing, Feature Extraction, and Risk Component Analysis Phase

For the implementation of this health monitoring system, we have considered three features namely slope, distance from normality thresholds and offset which are calculated in feature extraction phase [1]. *Slope* indicates short-term trends. Abrupt variations in vital signals are noted with this feature. This slope function evaluates the rate of signal change. *Distance from normality threshold* describes the deviation of current vital signal value from standard normality thresholds provided in medical literature [20]. The third feature *offset* calculates the difference between average conditions of vital signals in recent past and current value. Using these three features, three risk components are inferred namely sharp changes (Z1), long-term trends (Z2), and distance from normal behavior (Z3).

3.2 Static and Dynamic Data Mining Module

As the sensed signals are continuous and time dependent, stream mining algorithms are needed to analyze this data. The architecture consists of two modules, viz. Static data mining module (SDM) and dynamic data mining (DDM) module which make use of data stream mining algorithms. In SDM, after extracting the relevant features, the mining algorithms are used to cluster these features into three types of risk levels, i.e., normal, moderate and high; this forms the historical rule base. The same procedure is followed in DDM module with the only difference being the usage of real-time Stream Mining algorithms like PARC-Stream. These algorithms dynamically update the rule base if any concept drift [19] occurs.

3.3 Meta Classification

In this module, risk components are classified into their appropriate risk level based on the historical as well as the dynamic rule base computed using meta learning process.

The steps involved in meta learning process are:

1. Partition the input data (in this case Historical and Real-time patient data).
2. Build a base classifier for each partition.
3. Combine the outputs of the base classifiers.
4. Apply the rules from combined decision table for risk prediction.
5. Get the final classification.

WIMRHM is designed in such a way that the meta learning and classification phase is carried out in parallel as shown in Fig. 2. Base learners form the historical and dynamic meta model using prerecorded and real-time vital signals' extracted risk components Z1, Z2, and Z3. This process of forming combined classifier is called meta learning. In parallel to meta learning, the incoming vital signals are also classified as one of the three health risk levels using the meta model named as TT() as represented in Fig. 2.

4 Implementation

To work efficiently with ubiquitous environment, new PARC-Stream clustering algorithm is designed.

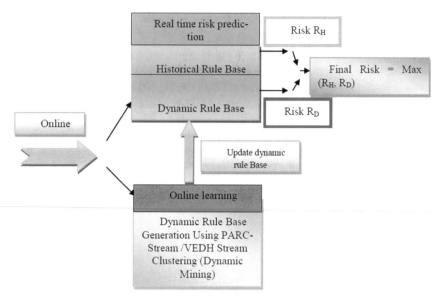

Fig. 2 Meta learning and risk prediction process

4.1 PARC-Stream Clustering Algorithm

The PARC-Stream algorithm is adaptive, efficient, resource conscious with high prediction accuracy. The algorithm uses a partition based approach which is described in Figs. 3, 4a and b. The novelty points that are introduced in PARC-Stream which makes it faster than K-Means and VFKM algorithms are novel lower threshold which helps to skip the inner loop of K-means most of the time [21]. PARC-Stream is simple to implement and easy to optimize. Its latency time is less due to the absence of spatial index construction. The algorithm is iterative in nature hence it avoids costly recursive calls. The memory requirement of the algorithm is also very low, which is a prerequisite for ubiquitous computing.

4.2 Description of Algorithm

PARC-Stream algorithm calculates available memory, performs initialization, and then processes sample X from the stream indefinitely. As each sample x arrives, it is added to the window until the window is full and then passed to the PARC-K-Means procedure for cluster formation. PARC-K-Means provides support to the main algorithm. After initializing the higher and lower limits, the algorithm iterates until the centers converge. The higher limit h limits the distance between x and its nearest center. Rather than keeping many k lower limits for each data point, PARC-

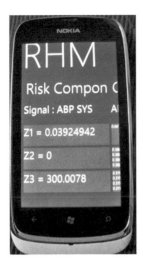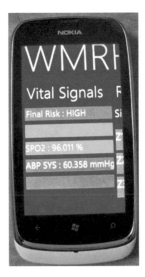

Fig. 3 Dynamic rule base formation of risk components and risk prediction using meta classification

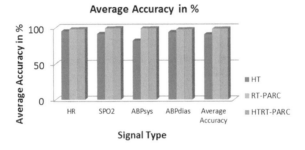

Fig. 4 Comparison of average accuracy with offline and online clustering algorithms for Z3 risk component

Stream maintains only one lower limit for each sample point, i.e., lo. PARC-Stream guarantees the following:

i. Higher limit h(i) will always be greater than equal to distance between sample value x(i) and corresponding cluster centroid c(a(i))

ii. Lower limit $lo(i) \geq min_{j \neq a(i)}|x(i) - c(j)|$ where, i is index of the current example.

Check on these bounds LL_m, MU_m, and threshold T gives the greatest upper limit from history centers and help to skip the inner loop over all cluster centers. If this condition fails, the algorithm recalculates LL_m if the limit of the threshold was not strong enough, and tests the condition again. If both the test fails then it calls the inner loop for all centers which updates LL_m, MU_m and threshold T. Thus, PARC-Stream has the potential to avoid the inner loop entirely. PARC-Stream also checks for outliers or any change in concept by comparing whether the data point is very far from the threshold, if yes then stores it separately for new cluster formation. When the formed clusters are not very well separated, the algorithm supports merging of nearest clusters and reforms the clusters with better statistics. This increases the adaptiveness of this algorithm by never restricting on lower limit of number of clusters.

PARC-Stream Algorithm

Input Parameters: **X**: set of vital signals
    W: size of sliding window   **K**: No. of clusters : in this case K=3 for three risk levels.
    T: maximum threshold limit on data to center distance
    LL_m: Lowest limit of memory usage MU_m: Memory in use
    A_m: Remaining rate of memory T_m:Total memory size
    TH : maximum threshold limit on center to center distance
    c: Cluster centers
Output Parameters: Cluster centroids c = $\{c_1, c_2, \ldots c_k\}$
Procedure PARC-Stream(X, W, K, c)

1. Compute
2. Set the size of window W.
3. Initialize no. of initial clusters K, set threshold value T, TH.
4. Read the W no. of data points.
5. select k initial centroids randomly and store it in c.
6. **repeat**
7.   PARC-K-Means(X, c, T_m,LL_m ,A_m , T, TH,MU_m)
8.   Read the W no. of data points
9. **Until** EOF

PARC-K-Means Algorithm

Input parameters :
d (.;.) : distance matrix for point to centroid and centroid to centroid pairs.

c' : sum of all sample points in respective cluster
q : Count of sample points assigned to each cluster,
p : amount of distance by which cluster centroid c last moved,
s : Distance from cluster centroid c to its nearest other centroid
a : centroid index to which x is assigned
h : higher limit on the distance between sample point x and centroid c to which x is as-
      signed

lo: lower limit on the distance between the sample point x and its second nearest centroid-that is, the nearest center to x that is not c.

Output: Cluster centroids $c = \{c_1, c_2, \dots c_k\}$ for current window W

Procedure PARC-K-Means(X, c, T_m, LL_m, A_m, T, TH, MU_m)

1. Initialize higher limit h, and lower limit lo
2. Assign X to its nearest cluster centroid
3. Update vectors c, c', q, a of cluster statistics
4. **While** not converged do
5. **for** each cluster centroid
6.   Find distance from current centroid to its nearest other centroid and store in s
7.   **for** each example x(i)
8.   **if** the upper bound u(i) on x(i) doesn't satisfy triangle inequality m and violets threshold constraint T
       /* tighten the upper bound*/
9.   Update higher limit h(i) with current nearest cluster centroid
10. **if** h(i) > m and h(i) < T then   /* Second limit test */
11.   store currently assigned cluster centroid a(i) of x(i) in a'
12.   recalculate closest a(i) to point x(i), higher limit h(i) and lower limit lo(i)
13. **if** $a' \neq a(i)$ and $d(x(i), a(i)) < T$ then
14.   Assign x(i) to its new nearest cluster center
15.   Update cluster statistics of both previous and new clusters of x(i)
16. **else if** x(i) seems to be an outlier violating threshold constraints
17.   Store such example in separate array m' and keep the count cnt of such examples
18. Compute current memory usage MU_m
19. Compute remaining memory A_m
20. **if** $A_m < LL_m$ then

    /* Reduce Memory Consumption */
21.   Decrease threshold value T and window size W
22. **if** remaining memory A_m greater than 80%
23.   Increase threshold value T and Size of window W

24. **if** no. of unassigned data points in $m' \geq \left\lceil \dfrac{W}{4} \right\rceil$

    /* handling outliers*/
25.   Select new centroids cc from unassigned data items, update c as, $c = cc \cup c$
          /* Increase inter cluster dissimilarity */
26.   **if** distance between two cluster centroids $d(c_i, c_j) < TH \ \forall c_i \in c$ where i, j =1...n
          and $i \neq j$
27.   Merge such clusters into one
28.   Reassign their data points
29.   Update the cluster statistics
30.   Recalculate cluster centers based on current example assignments
31.   Update bounds for these newly calculated cluster center
32. **Return** clusters c

5   Experimental Design and Results

The data used in this study are four biomedical signals, viz. SPO2, HR, ABP systolic, ABP diastolic which are taken from publically available clinical database named MIMIC (Multi-parameter Intelligent Monitoring for Intensive Care) [22]. The database contains around 200 patient days of real-time signals. Records from MIMIC numeric section are used and sampled at 1 Hz. Signals are processed in two phases: (a) to form historical rules statically (b) to form dynamic rules in real time. The vital signal data read for processing are taken in simple text format. Experiments are carried out to study the effectiveness of meta learning approach over traditional approach of classification for health risk prediction. WBSN environment is simulated for RPMS. PC acts as sensor which transfers vital signal samples through Bluetooth connection to the smart phone. The mined historical rules on PC are uploaded on the smart phone. Online learning and prediction phase is then carried out on smart phone in parallel. The smart phone receives the vital signal values through Bluetooth^{TM} in text format, simulated as if they were coming from real sensors, and analysis is carried out in real time. The framework has been implemented using Microsoft .NET framework. Microsoft SQL Server 2010 is used for maintaining historical rules.

Performance of both model building and mobile classification are analyzed in terms of time required for processing and resources requested.

5.1   Clustering Results

K-Means and PARC-Stream clustering algorithms are used to form two base classifiers. The window size of 18 samples per window is set and then it varies depending on the memory usage. The parameters setting are like $k = 3$, $LLm = 30\%$, $T = 10$ and $TH = 30$. The samples are divided for online and offline training phase.

The historical rule base is formed after applying K-Means algorithm on the input samples of different vital signals. These rules are then uploaded on smartphone for real-time classification of incoming vital signals.

Figure 3 shows the snapshot of the real-time patient monitoring system deployed on smartphone. Red color indicates high risk, orange indicates moderate risk, and green color indicates normal status. Global risk is declared based on overall risk status of four signals using meta classification techniques and displayed on mobile, denoted as Final Risk. If there is high risk the alert is given to patient and concerned doctor to take preventive action. Dynamic rule base forms on the fly on Smartphone using PARC-Stream algorithm.

Fig. 5 Comparison of average TPR with offline and online clustering algorithms for Z3 risk component

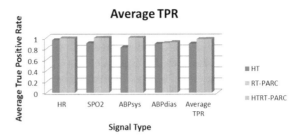

5.2 Functional Accuracy and Evaluation Results of Real-Time Patient Monitoring System

Functional accuracy of RPMS risk prediction is evaluated for each risk component with and without meta learning approach. Three types of rule base combinations tested are:

a. Risk prediction using only Historical Rule base (HT)
b. Risk Prediction using only Dynamic rule base trained with PARC-Stream algorithm (RT-PARC)
c. Risk Prediction using meta classifier using combination of Historical and Dynamic rule base.

For evaluating system accuracy, although three risk levels are considered, moderate, and high-risk prediction are considered as abnormal situation or True positives whereas normal signal value and deviation by plus or minus five of moderate risk are considered as normal or true negatives. As a moderate situation cannot be directly considered as high risk or no risk, deviation of five is introduced.

If incoming signals are only clustered using historical rule base by pattern matching technique, the results are not as admirable as those processed by the support of dynamic rules. This happens because there may be a possibility that some of the patterns may not be present in historical database and rules are not available which can be dug out during online learning phase. The accuracy achieved by various combination of rule base for Z3 risk component and their comparisons are shown in Fig. 4. The accuracy of prediction is evaluated by comparing system predicted risk with the normality thresholds provided by medical experts. After evaluating overall average accuracy, TPR and FPR for Z3 component of all four vital signals, it has been observed that HTRT-PARC system gave highest accuracy of 99.04%, highest TPR of 0.98 and FPR of 0.01 as shown in Figs. 4 and 5.

There are no direct measures to compare the sharp changes and long-term trends accuracy like absolute threshold values given by medical literature for Z3 components. But the two risk components Z1 and Z2 are also very important to find abnormality in the functioning of human body. To check whether the changes in these risk components are correctly detected by indicating appropriate risk levels, we set some threshold values for the normal and abnormality range by consulting domain experts.

With that base combination of Z1, Z2, and Z3 (Z1Z2Z3), the accuracy of the system is evaluated. HTRT system gave the highest accuracy of 99%.

Most of the times risk prediction with only historical rule base gave lower accuracy, which may be due to the absence of events while training rule base which occurred in real time. Same performance degradation may occur during prediction using only dynamic rule base. This proves our meta learning approach is really effective to give accurate risk prediction of health in real time.

6   Conclusion and Further Work

An innovative combinatorial approach for learning and adapting vital signal behavior of human body and predicting health risk level in real time on smart phone, based on both patient history and current health status. Real-time patient monitoring system is modeled and tested using newly designed real-time clustering algorithm named PARC-Stream.

Designed PARC-Stream algorithm is resource-aware as it handles memory constraints in real time and suits any kind of ubiquitous computing. The algorithm is adaptive to changes called as concept drift, occurring in incoming signals. Number of clusters may vary accordingly. The PARC-Stream algorithm is designed on the principle of triangle inequality using which loop required for distance calculations among data to centers is skipped and performance gets accelerated.

This research work is now looking towards developing the various disease and patient-specific healthcare systems.

References

1. Patil, D., Wadhai, V.: Dynamic data mining approach to WMRHM. In: Proceedings of 7th IEEE Conference on Industrial Electronics and Applications (ICIEA), Singapore, pp. 1978–1983 (2012)
2. Chen, S.-L., Lee, H.-Y., Chen, C.-A., Huang, H.-Y., Luo, C.-H.: Wireless body sensor network with adaptive low-power design for biometrics and healthcare applications. Syst. J. IEEE 3(4), 398–409 (2009)
3. Rong, W.-H.L., Jin, R., Hauptmann, A.: Meta-classification of multimedia classifiers. In: International Workshop on Knowledge Discovery in Multimedia and Complex Data, Taipei, Taiwan, 6 May (2002)
4. Gorodetsky, V., Karsaeyv, O., Samoilov, V.: Multi-agent technology for distributed data mining and classification. In: Proceedings of the IEEE/WIC International Conference on Intelligent Agent Technology (IAT'03) (2003)
5. Bernstein A., Provost, F., Hill, S.: Toward intelligent assistance for a data mining process: an ontology-based approach for cost-sensitive classification. IEEE Trans. Knowl. Data Eng. 17(4) (2005)
6. Sannino, G., De Falco, I., De Pietro, G.: An automatic rules extraction approach to support OSA events detection in an mHealth system. IEEE J. Biomed. Health Inf. 18(5), 1518, 1524 (2014)

7. Van Halteren, A., Konstantas, D., Bults, R., Wac, K., Dokovsky, N., Koprinkov, G., Jones, V., Widya, I.: Mobihealth: ambulant patient monitoring over next generation public wireless networks. Stud. Health Technol. Inf. **106**, 107–122 (2004)
8. Huang, A., et al.: WE-CARE: an intelligent mobile telecardiology system to enable mHealth applications. IEEE J. Biomed. Health Inf. **18**(2), 693–702 (2014)
9. Rodrıguez, A.R., Rodríguez, G.M., Almeida, R., Pina, N., Oca, G.M.: Design and evaluation of an ECG holter analysis system. In: Proceedings of the IEEE Computing in Cardiology, pp. 521–523 (2010)
10. Oresko, J.J., Jin, Z., Cheng, J., Huang, S., Sun, Y., Duschl, H., Cheng, A.C.: A wearable smartphone-based platform for real-time cardiovascular disease detection via electrocardiogram processing. IEEE Trans. Inf Technol. Biomed. **14**(3), 734–740 (2010)
11. Kang, K., Park, K.J., Song, J.J., Yoon, C.H., Sha, L.: A medical-grade wireless architecture for remote electrocardiography. IEEE Trans. Inf. Technol. Biomed. **15**(2), 260–267 (2011)
12. Liou, S.H., Wu, Y.H., Syu, Y.S., Gong, Y.L., Chen, H.C., Pan, S.T.: Real-time remote ECG signal monitor and emergency warning/positioning system on cellular phone. Intelligent Information and Database Systems, pp. 336–345. Springer-Verlag, New York, NY, USA (2012)
13. Lee, S.-C., Chung, W.-Y.: A robust wearable u-healthcare platform in wireless sensor network. J. Commun. Netw. **16**(4), 465–474 (2014)
14. Apiletti, D., Baralis, E., Bruno, G., Cerquitelli T.: Real-time analysis of physiological data to support medical applications. IEEE Trans. Inf. Technol. Biomed. **13**(3), 313–321 (2009)
15. Sun, J.M., Sow, D., Hu J.Y., Ebadollahi, S.: A system for mining temporal physiological data streams for advanced prognostic decision support. In: Proceedings of the 10th IEEE International Conference on Data Mining (ICDM '10), pp. 1061–1066 (2010)
16. Zhang, Y., Fong, S., Fiaidhi, J., Mohammed, S: Real-time clinical decision support system with data stream mining. J. Biomed. Biotechnol. **2012**, Article ID 580186, 1–8 (2012)
17. Ur Rehman, M.H., Liew, C.S., Wah, T.Y.: UniMiner: towards a unified framework for data mining. In: Information and Communication Technologies (WICT), 2014 Fourth World Congress, pp. 134–139 (2014)
18. Chao, C.-M., Chao, G.-L.: Resource-aware density and grid based clustering in ubiquitous data streams. In: 26th International Conference on Advanced Information Networking and applications Workshops (WAINA), pp. 1203–1208 (2012)
19. Shah, R., Krishnaswamy, S., Gaber, M.M.: Resource-aware very fast K-means for ubiquitous data stream mining. In: Proceedings of the 2nd International Workshop on Knowledge Discovery in Data Streams, pp. 40–50 (2005)
20. Kasper, D.L., Braunwald, E., Fauci, A., Hauser, S., Longo, D., Jameson, J.L.: Harrison's, Principles of Internal Medicine, pp. 315, 1349. McGraw-Hill, New York (1992)
21. Hamerly, G.: Making k-means even faster. In: Proceedings of the 2010 SIAM International Conference on Data Mining (2010)
22. The MIMIC and MIT-BIH database on PhysioBank. http://www.physionet.org/physiobank/database. 27 Feb (2018)

Firearm Detection from Surveillance Cameras Using Image Processing and Machine Learning Techniques

Fraol Gelana and Arvind Yadav

1 Introduction

In this modern time of surveillance and security, the number of Closed-Circuit Television Systems (CCTV) deployed in public and private places such as Cinemas, Malls, and Hotels have increased exponentially. Currently, there are millions of CCTV cameras in operation in India. Therefore the increasing density of surveillance camera footage makes it a challenge for a human operator to inspect, analyze, and decide whether a potentially dangerous situation is about to happen. One such dangerous scenario is the Active Shooter Event such as the Colorado Theater Shooting (USA), Oslo (Norway), and Paris (France) shows that rapid detection and identification of Armed Shooter is essential in reducing the number of casualties.

2 Literature Review

The image processing based approached used for firearm detection is as follows: Various background subtraction algorithms proposed by researchers has been discussed in this work. Among this Zivkovic et al. [1] proposed the Gaussian Mixture Model (GMM), the GMM algorithm creates a background model by computing the probability density distribution of the values observed at each pixel over time by a weighted mixture of Gaussians and the values are constantly updated by performing

F. Gelana (✉) · A. Yadav
Department of Electronics and Communication Engineering, Parul Institute
of Engineering and Technology, Vadodara, Gujarat, India
e-mail: fraolgela@gmail.com

A. Yadav
e-mail: arvind.yadav@paruluniversity.ac.in

© Springer Nature Singapore Pte Ltd. 2019
S. Tiwari et al. (eds.), *Smart Innovations in Communication and Computational Sciences*, Advances in Intelligent Systems and Computing 851,
https://doi.org/10.1007/978-981-13-2414-7_3

recursive equations. The GMM algorithm performs well in outdoor situations where the image consists of low-frequency objects such as waving tree branches.

Later on, Olivier and Droogenbroeck [2] proposed the Visual Background Extractor algorithm. This algorithm assigns a value for each pixel in the current frame that is taken in the previous frames at the exact pixel neighborhood. Then it compares these computed values to the current pixel value to determine whether this pixel belongs to the background frame and changes the background model by selecting randomly the values which should be replaced first. This method tries to classify a pixel as a background or foreground with respect to its immediate neighbors rather than with an explicit model of the background thus it is more reliable and efficient.

A real-time situational recognition and awareness from CCTV image analysis has been proposed in [3]. This system supports the CCTV by automatically detecting a "Dangerous Object" and raising an alarm. Detection is performed by a pre trained neural network. The proposed algorithm utilizes an MPEG-7 classifier in cascade with the artificial neural network in order to decrease the number of false positives, this algorithm has a very low detection sensitivity while giving a 100% specificity.

A system for the automatic detection of suspicious and dangerous objects in a cluttered Three-Dimensional (3D) computed tomography (CT) imagery of baggage had been proposed by Megherbi et al. [4]. The proposed system utilizes a 3D CT imagery rather than a most commonly used 2D X-ray imagery technique, the proposed system uses a linear Support Vector Machine (SVM) as a classifier.

A method for classification of objects of an X-ray image of airport baggage using transfer learning and a deep convolutional network has been proposed by Akcay et al. [5]. This system uses transfer learning paradigm to overcome the need for large amounts of training data. Mery et al. [6] have proposed a system for detecting prohibited objects with predefined shapes and sizes from X-ray images of airport baggage. A system for detecting concealed weapons in an X-ray image has been proposed by Roomi and Rajashankarii [7]. This system uses the fuzzy KNN (K-Nearest Neighbors) to classify objects as a threat or non-threat. Multiple objects are extracted by using shape-based image segmentation method and feature extraction methods. A method using CNN Tensor flow-based implementation for detecting and classifying weapons in images has been proposed by Lai and Maples [8]. The researchers used over 1.3 million images with approximately 3000 weapon based images. A large amount of training data is selected so that to include every situation and orientation a weapon might show up on CCTV footage. A method for detecting concealed weapon using active millimeter wave radar had been proposed in [9]. The system uses a three-layered Artificial Neural Network with 160 input nodes and 1 output nodes. In the year 2017, Olmos et al. [10] proposed an automatic system for detecting handguns from CCTV videos. The researchers attempted to reduce the number of false detections by using a Deep Convolutional Neural Network (CNN) classifier.

The objective of this research is to implement an automated and real-time firearm detection algorithm that assists a human operator by raising an alarm whenever a potentially threatening object (gun) is detected. To get the desired outcome the algorithm must be fully automatic, operate in real-time, accurate, and run on a standard personal computer.

3 Methodology for the Detection of Firearm

This section describes the overall working of the proposed firearm detection methodology.

3.1 Description of the Firearm Detection Methodology

The method for the detection of a firearm is implemented as shown in Fig. 1. Elements of the algorithm will be discussed in the subsequent Sections.

1. **RGB to Grayscale Conversion**.

RGB to Grayscale Conversion is performed in order to simplify the complexity of each frame and speed up the operation of the subsequent Background subtraction

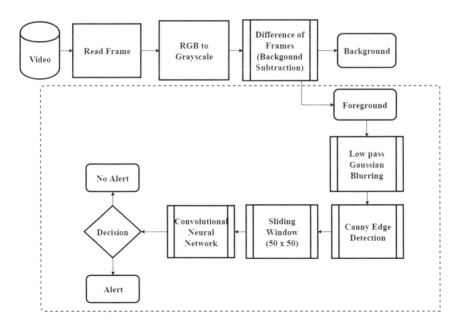

Fig. 1 Block diagram of the firearm detection methodology

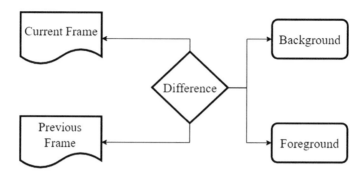

Fig. 2 Difference of frames algorithm

and segmentation stages. Grayscale images are computed much faster compared to RGB images when performing segmentation operations such Canny [11].

2. **Background Subtraction**.

Background subtraction is one of the key techniques for automatic video analysis, especially in the domain of video surveillance [12]. In this work, three different approaches towards background subtraction and segmentation were tested.

The Visual Background Extractor [2] and Improved Gaussian Mixture Model [1] methods and Difference of frame background subtraction algorithm (Fig. 2).

3. **Filtering Operation**.

Due to varying lighting conditions and other interferences, the extracted foreground object is highly noisy. This affects the performance of consequent operations that will take later on and demand high computational requirements by creating false regions of interest which carry minimal information. Dilation and Erosion operations were performed on the extracted foreground object in order to remove small white noises that occur due to lighting changes and joining disparate elements in an image. The kernel (structuring element) size that yields the desired output is selected after successive experiments.

4. **Segmentation/Edge Detection**.

In order for the next stages of the algorithm to perform as desired, edges must be detected. For this purpose, the famous Canny [12] edge detection algorithm is used. The Canny algorithm takes the filtered foreground object as input and returns the edges as output.

5. **Sliding Window**.

As the dangerous object can be at any location in the foreground frame a sliding window technique is used. A sliding window is a rectangular region of fixed width and height that slides across an image. The sliding window technique significantly minimizes the area to be inspected by the learning algorithm, the size and slide step is

Fig. 3 Frame with a sliding windows

selected after numerous experiments and is subject to change in the future. Figure 3 shows a frame with multiple windows on the foreground object.

6. **Classification**.

Classification of an object as either a treat (gun) or non-treat is done by a Tensorflow-based implementation of Convolutional Neural Network (CNN) algorithm. The input to the CNN is an image of either fixed or variable dimension that is generated by the sliding window operation in the previous stage. Then the image is resized to a predetermined size and shape into the CNN for classification.

4   Results and Discussion

A set of videos used to test the proposed firearm detection method were obtained at [13] which were used by researchers in [3]. The CNN training\testing dataset consists of 4000 negative and 1869 positive images with 30% split. To reduce the training time and the computational complexity of the learning algorithm, the dataset contains only the edge information and the color information is discarded. Experimental results have shown that the information carried by the edge is more than enough to describe the gun object thus the color is discarded to reduce the dimensionality.

From the experimental results and observations made it was clear that the predictive models like ViBe and IAGMM give a much more detailed result and are also much better resistant to noise and sudden intensity changes that occur especially when

Fig. 4 Result of implementing ViBe algorithm on test video

Fig. 5 Result of implementing IAGMM algorithm on test video

operating in an outdoor environment. Out of the two methods, the ViBe algorithm is substantially faster and gives a better result. But both the algorithms are comparatively much slower, require high computational power and memory. The Frame Difference method, on the other hand, is much faster and is less computationally intensive but gives a less detailed result which is affected by noise significantly.

Since the Automatic detection of firearms is mostly meant to be from security cameras found inside an indoor environment (Theaters, Mall, Classrooms, etc.) that have a controlled scenery with less frequent intensity changes and high frequency moving objects (trees). Thus the Frame Difference approach can be used without the loss of much detail available for the next stages of the system (Figs. 4, 5 and 6).

Performing morphological operations on the extracted foreground object has shown effect of reducing small flickering white noises that occur due to sudden lighting changes and camera autofocus, but this operations have undesired effect on the performance of the canny edge detection because the foreground objects edges get significantly diminished those causing discontinuities in the output of the canny operation. Therefore in this work morphological operations are taken as optional only applied if the scenery is very noisy, otherwise, results without morphological operation appear to be desirable.

Fig. 6 Result of implementing the difference of frames algorithm on test video

Table 1 Confusion matrix for a 1758 testing set data

	Gun	Not gun	Accuracy (%)
Gun	549	36	93.84
Not gun	3	1170	99.73
Total accuracy (%)	99.45	97.01	97.78

Table 2 Results for images containing firearm (gun) object

Number of images	585
Sensitivity	93.84%

Table 3 Results for images not containing firearm (gun) object

Number of images	1173
Specificity	99.73%

The result of the classification algorithm on a testing set of 1758 images is given in Table 1 (Tables 2 and 3).

Out of the 1758 images used for testing the algorithm 585 contain positive and 1173 negative images. In the context of firearm detection the most important factor is minimizing the number of false positives without affecting the detection sensitivity. The method proposed in this paper achieved a specificity of 99.73% for images that contain objects that are not gun and detection accuracy of 93.84% for images that contain objects that are gun (Fig. 7).

The method proposed in this paper tries to minimize the complexity and computational requirements to a minimum when compared to firearm detection algorithms such as [3] and [10] This is achieved by removing computationally exhaustive stages such as PCA and secondary classifier algorithms like MPEG-7 classifier used by researchers at [3]. Instead in this paper, we found out that the edge information alone is more than enough for the learning algorithm to accurately describe the gun object.

Fig. 7 A frame from test results movie

By using only the edges it was possible to reduce the dimensionality of the input significantly thus eliminating the need for PCA. Also implementing a multilayered Convolutional Neural Network trained on a bigger dataset was found to give a much faster detection rate while keeping the number of false positives as low as possible. As of now, a further work is being performed to increase the sensitivity and specificity of the method; this includes preparing an even larger training dataset and improving the CNN learning algorithm.

The major limitations of the proposed algorithm are as follows:

- The algorithm is only capable of detecting a single type of firearm. Due to the complexity and performance requirements of implementing an algorithm that is capable of detecting multiple types of firearms such as rifles, shotguns, pistols, revolvers, etc. It is decided in this research to implement an algorithm that detects a single type of firearm. Therefore this algorithm is limited on detecting only pistols from CCTV images.
- The algorithm is only capable of detecting a firearm that is not hidden (concealed) by the perpetrator.
- The algorithm requires good quality CCTV video to achieve high detection accuracy.

5 Conclusion

In this paper, we presented an automatic real-time firearm detection methodology that assists a human operator by alerting him/her when a potentially suspicious object is detected. The method uses a convolutional neural network to classify parst of a CCTV video frame as either containing a firearm or not. The method proposed in this paper uses only the edge information as a feature in order to increase accuracy and reduce complexity. Based on our experimental results we conclude that this method gives high detection accuracy without compromising the real-time operation. We also showed that it is very necessary to reduce the number of false positive in order for the proposed method to be practical.

So far we have achieved a total detection accuracy of 97.78%, Sensitivity of 93.84% and Specificity of 99.73%.

We are planning to further improve the firearm detection method presented in this work by using a faster secondary classifier algorithm in a cascade with the CNN to increase specificity and reduce the number of false positives. We are also considering training the CNN on a larger dataset.

References

1. Zivkovic, Z.: Improved adaptive Gaussian mixture model for background subtraction. Paper presented. In: Proceedings of the IEEE 17th International Conference on Pattern Recognition, Cambridge, UK, vol. 2, pp. 28–31 (2004)
2. Olivier, B., Droogenbroeck, M.V.: Vibe: a universal background subtraction algorithm for video sequences. In: IEEE Transactions on Image processing, vol. 20, no. 6, pp. 1709–1724 (2011)
3. Sieradzki, R., Grega, M., Lach, S.: Automated recognition of firearms in surveillance video. In: IEEE International Multi-Disciplinary Conference on Cognitive Methods in Situation Awareness and Decision Support (CogSIMA), San Diego, CA, USA, pp. 45–50 (2013)
4. Megherbi, N., Flitton, G.T., Breckon, T.P.: A classifier based approach for the detection of potential threats in ct based baggage screening. In: 17th IEEE International Conference on Image Processing (ICIP), Hong Kong, China, pp. 1833–1836 (2010)
5. Akcay, C., Kundergorski, M.E., Devereux, M., Breckon, T.P.: Transfer learning using convolutional neural networks for object classification within X-ray baggage security imagery. In: IEEE International Conference on Image Processing (ICIP), Phoenix, AZ, USA, pp. 1057–1061 (2016)
6. Merry, D., Mondragon, G., Riffo, V., Zuccar, I.: Detection of regular objects in baggage using multiple X-ray views. Insight-Non-Destr. Test. Cond. Monit. **55**(1), 16–20 (2013)
7. Roomi, M., Rajashankarii, R.: Detection of concealed weapons in X-ray images using fuzzy k-nn. Int. J. Comput. Sci. Eng. Inf. Technol. **2**(2), 65–70 (2012)
8. Lai, J., Maples, S.: Developing a real-time gun detection classifier. Accessed on August 2017 (2017). http://cs231n.stanford.edu/reports/2017/pdfs/716.pdf
9. O'Reilly, D., Bowring, N., Harmer, S.: Signal processing techniques for concealed weapon detection by use of neural networks. In: IEEE 27th Convention of Electrical & Electronics Engineers in Israel (IEEEI), Eilat, Israel, pp. 1–4 (2012)
10. Olsmos, R., Tabik, S., Herrera, F.: Automatic handgun detection alarm in videos using deep learning. Neurocomputing **275**, 66–72 (2018)

11. Canny, J.: A computational approach to edge detection. IEEE Trans. Pattern Anal. Mach. Intell. **6**, 679–698 (1986)
12. Brutzer, S., Benjamin, H., Gunther, H.: Evaluation of background subtraction techniques for video surveillance. In: IEEE Conference on Computer Vision and Pattern Recognition (CVPR), Colorado Springs, CP, USA, pp. 1937–1944, 20th–25th June 2011
13. Gun Video Database. Accessed Sept 2017. http://kt.agh.edu.pl/grega/guns/

Segmentation of Optic Disc by Localized Active Contour Model in Retinal Fundus Image

Shreenidhi H. Bhat and Preetham Kumar

1 Introduction

Nowadays, digital retinal fundus images are rapidly used as one of the primary diagnostic tool for detection of retinal related diseases such as glaucoma and diabetic retinopathy. These retinal images provide the objective analysis which help the associated pathologist for early detection of the diseases. Among the major eye-related diseases in the world population, glaucoma and diabetic retinopathy are the main retinal diseases [1]. Both of the diseases have to be early detected to slow down the progression by giving suitable treatment and medication or else leads to complete vision loss. The changes in the anatomical structures of retina that consists of optic disc, optic cup, optic nerve and blood vessels are key indicators for the early diagnosis of the diseases. Thus, the computer-aided diagnostic analysis of retinal fundus images helps in the detection and localization of these structures in retina which is the first step in screening of any retinal related diseases [2]. The process of screening includes detection and segmenting the optic disc and optic cup, followed by calculating the segmented area of optic disc and optic cup respectively and then estimating the cup-to-disc ratio.

The optic disc (OD) of retina is most important feature among all the part in retina which often looks bright and oval like shape. This yellowish region appears in various size and is converged by the blood vessels at its centre. Above this physical characteristic of OD, make an account while detecting and localization of OD automatically. There are various challenges faced for detecting and segmenting the OD [3]. First, poor contrast and variations in the intensity and quality of images during

S. H. Bhat (✉) · P. Kumar
Department of Information and Communication Technology, Manipal Institute
of Technology, Manipal Academy of Higher Education, Manipal, Karnataka, India
e-mail: shrinidhi.h.bhat@gmail.com

P. Kumar
e-mail: preetham.kumar@manipal.edu

© Springer Nature Singapore Pte Ltd. 2019   35
S. Tiwari et al. (eds.), *Smart Innovations in Communication and Computational
Sciences*, Advances in Intelligent Systems and Computing 851,
https://doi.org/10.1007/978-981-13-2414-7_4

the acquisition of an image. Second, the presence of the blood vessel in the centre region of the OD which leads to false detection of the boundary of the OD. Third, the presence of the papillary atrophy located outside in the OD region appears brighter sometimes causes boundary deformation while segmenting. Taking account of all the above challenges, this paper has overcome by presenting a new approach for detection and segmentation of OD from retinal fundus images automatically.

This paper is organized as follows. Literature Survey deals with the different past approaches for the detection and segmentation which are available. Next, the main aim of this paper is proposed in 'Objective' and followed by the detailed methodology which is explained in methodology. In the 'Result Analysis' performance analysis is made and even a comparison between the past approaches. Lastly, the paper is concluded by 'Conclusion'.

2 Literature Survey

Various approaches have been proposed for retinal detection and segmentation of OD over the last few years. Among all these approaches, it can be classified mainly into two groups, i.e. location and segmentation methods respectively based on localizing the OD region from the fundus image and another on estimating the boundary of OD for segmenting it.

2.1 Detection of Optic Disc (OD)

Adaptive mathematical morphological operations which include closing and opening operations were used for automated detection of OD in the retinal fundus image. These morphological operations are basically designed for finding the structures. These approaches was defined in Angulo et al. [4] where Principal Component Analysis (PCA) and morphological operations was performed along with different watershed transformation technique for detecting the OD. In Marn et al. [5], a method was proposed on processed morphological fundus image by applying edge detection technique with Circular Hough Transform (CHT) and feature extraction technique such as template matching method for identifying the boundary of OD. Further, in another approach, automatic thresholding technique along with CHT was used to find the centre of OD and region of interest for detection of OD with help of Prewitt edge detector [6].

2.2 Segmentation of OD

Several methods are proposed for segmenting OD from the previous step of automatic detection of boundary of OD. A technique called sliding band filter was proposed in Campilho et al. [7] which was applied on preprocessed fundus image and smoothing algorithm was exerted for segmenting OD. The novel method for automatic segmentation of OD using modified multi-thresholding method was presented in Kankamala and kubakuddi [8]. In Salazar Gonzalez et al. [9], segmentation of OD was suggested by using graph cut technique for delineating the OD and compensation factor method for segmenting the localized OD. In Khande and Mittapalli [2], OD segmentation was advanced by using active contour model using the extended local binary fitting method. An approach called grow-cut algorithm was discussed in Fraz et al. (2016) for segmentation of OD [3]. Bharkad [10] proposed the method for automatic segmentation of OD using equiripple low-pass finite impulse response (FIR) filter for removing the blood vessels present and threshold calculation method along with morphological operations and median filtering for the segmentation of OD from DRIVE, DRIONS, DIRATEDB0 databases. Using edge-based and active contour fitting model technique, OD segmentation approach was defined in Singh et al. [11]. This work even focused on applying the image processing technique such as filters for deletion of blood vessels and morphological operations for detection of OD boundary rejecting the false positives.

Although sufficient amount of research has been done in this area by positing different approaches on extraction of OD from fundus image. The main aim of this work is to propose the strategic method of mathematical morphological operation for inpainting the blood vessel to overcome its influence of presence and Circular Hough transformation (CHT) for detection of OD. Another contribution of this paper is applying localized region-based active contour model method for segmenting the OD from the fundus image.

3 Methodology

In the medical field, fundus images are base of anything as they are used for primary diagnosis of any diseases. The fundus images of retina known as retinal fundus images are studied for identification of retinal related diseases with help of image processing techniques. This paper proposes a novel method which is mainly categorized into three major steps. They are,

  i. Preprocessing
 ii. Localization of OD
iii. Segmentation of OD.

The work of this paper is explained in the block diagram as shown in the Fig. 1. In the preprocessing steps, the retinal fundus images will be normalized and the contrast

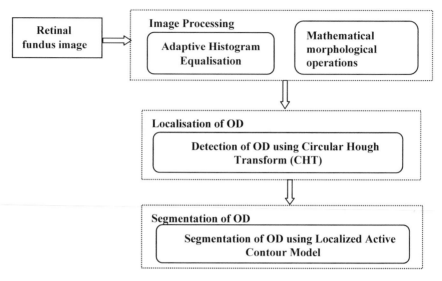

Fig. 1 Block diagram of the proposed method

of the image is enhanced. The iterative mathematical morphology operations are used for removal of blood vessels. Then Circular Hough Transform (CHT) is applied for the detection and localization of OD. The detected OD is then segmented by using localized region-based active contour model.

3.1 Preprocessing

The input fundus image (as shown in Fig. 2a) is preprocessed by applying image processing techniques. Since there are variations on image contrast and illumination as well as the blood vessels present in the region of the OD. The red channel of RGB is extracted from the fundus images for normalization (as shown in Fig. 2b). Then the normalized images are converted to grey scale for further steps [12].

An adaptive histogram equalization is applied on normalized grey scale image so that the appearance of the image is improved and pdf of a fundus image is uniformed [7]. The blood vessels present in the region of optic disc usually split into various region with common gradient values, therefore smoothing of the image is made by employing that generates an equitably constant region (as shown in Fig. 2c).

The mathematical morphological operations in image processing are generally designed for finding the structures for studying the formation of the objects. It consists of two fundamental processes, i.e. erosion and dilation. The erosion operations are used to shrink the objects whereas the dilation operations is for increasing the size of the objects. These two operations are combined together to form two mechanism, i.e. opening and closing. In opening operations, erosion is followed by the dilation

(a)   **(b)**   **(c)**

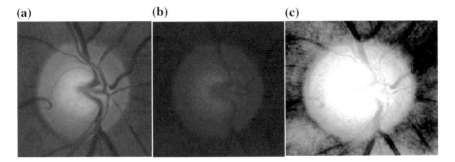

Fig. 2 **a** Input image, **b** normalized image by extracting red channel, **c** enhanced image after histogram equalization and median filter

that removes the noises. In closing operation, dilation is followed by erosion where it removes the holes present in the region and infused by small gaps. Thus, combining these algorithms is used for background removal and finding the specific shapes in images.

Morphological operations are defined for grey-scaled image $f(m, n)$ and the binary structuring element S [4].

Erosion

$$f \otimes S = \min f(m - j, n - k) - S(j, k) \tag{1}$$

Dilation

$$f \oplus S = \max f(m - j, n - k) - S(j, k) \tag{2}$$

Opening

$$foS = (f \otimes S) \oplus S \tag{3}$$

Closing

$$f \cdot S = (f \oplus S) \otimes S \tag{4}$$

In this paper, the closing operation is applied on the enhanced fundus image for elimination of holes and to fill the gaps. Then the opening operation is employed to the image to remove the noise. For both, the operation disc shaped structuring element was used of radius ranging from 5 to 8. The reconstruction operation is called for combining both closing and opening operation (as shown in Fig. 3a). The output image obtained will be free of blood vessels and optic disc will be clearly visible.

(a)   (b)   (c)

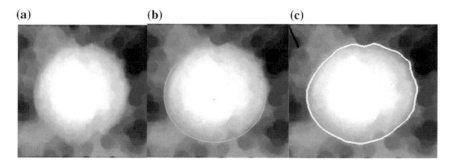

Fig. 3 **a** Preprocessed image after iterative morphological operation, **b** detection of OD by CHT, **c** segmentation of OD by localized ACM

3.2 Localization of Optic Disc

Circular Hough Transform (CHT) is a technique used for feature extraction to detect the circular objects in images. In two-dimensional space [3], circle is defined as

$$(x - p)^2 + (y - q)^2 = r^2 \tag{5}$$

where (p, q) is the centre of the circle and r is the radius. The processes of CHT is followed by fixing the radius and finding the centre of the circles in two-dimensional and then locating the radius in one-dimensional space. If radius is already fixed then parameter space is reduced to two-dimensional space. Then it starts to define a circle centred at (x, y) with initial radius $r = 1$ for each point (x, y) on the original circle and the point of interest (POI) is denoted by the accumulator matrix. The voting is then followed on the parameter space and every circle passing through POI, the voting number is increased by one. The maximum voted circle of accumulator results in Hough space and the maximum voted circle in this space will provides the circle [13]. In this proposed method, the CHT is applied by initializing the radius (min to max). The OD centre is selected through the obtained accumulator matrix and edge detected are used to calculate the radius of the circle. Once the circle is identified (as shown in Fig. 3b), then the circle points are localized using the calculated radius and used to initialize the active contour model.

3.3 Segmentation of OD

After the localization of the position of the optic disc, the segmentation process is achieved by acquiring the boundary of the optic disc using the framework called as active contour model (ACM) [14]. The basic idea of active contour model is to match a deformable model to image by minimizing the energy and the process takes by introducing the initial contour into image and starts to evolve by fitting the boundary of the image. The active contour model are of two types, they are edge based and region based, respectively. The region-based active contour model is more efficient than the edge based as it accurately segments the objects by finding the optimum energy where the model best fit the image [12].

Consider a function which is vector (image), $I: \Omega \to S$ n where CS^n is the domain of image and $n \geq 1$ is the vector dimension. S^n represents the parameter of the contour C curvature. For grey-scaled image, the active contour model is defined according to Chan-Vese model [15] for energy function as:

$$E\left(c^+, c^-, C\right) = \chi^+ \int_{in(c)} \left|I(x) - c^+\right|^2 dx + \chi^- \int_{out(c)} \left|I(x) - c^-\right|^2 dx$$
$$+ \mu \, length(C) \tag{6}$$

where $in(c)$ and $out(c)$ are the region inside and outside the contour C. $c-$ and $c+$ are constants used for approximate the intensity of an image present inside and outside of the contour. $\lambda+$, $\lambda-$ and $\mu > 0$ are weights for proper fitting and regularizing the contour. Similarly, the region-based active contour model uses local information of an image and the local function is defined as f for two points (x, y) on image I as [12].

$$f(x, y) = \begin{cases} 1, & if \, \|x - y\| \leq r \\ 0, & otherwise \end{cases} \tag{7}$$

where f will the local image domain within the radius r around the point x. The mean intensities which are depended on foreground and background regions of a contour C, known as *mean separation energy* (as defined in Eq. 8) is introduced on region active contour model for better optimum energy [16].

$$E_m s = \int_{\Omega y} \left(c^+ - c^-\right)^2 \tag{8}$$

The above equation Eq. 6 is redefined for point x by combining Eqs. 7 and 8 as

$$E\left(c^+, c^-, C\right) = \chi^+ \int_{in(c)} \left(\frac{|I(x) - c^+|^2}{A_{c^+}}\right) dy + \chi^- \int_{out(c)} \left(\frac{|I(x) - c^-|^2}{A_{c^-}}\right) dy \tag{9}$$

where $A+c$ and $A-c$ are area of inside and outside region of contour C [16]. The above-optimized energy influences only the local image information present within the local domain and causes to evolve where there is possible difference of the interior and exterior region. This energy also helps in capturing local boundaries efficiently which were missed by C-V model without getting any disturbance from the inside and outside region even when these are not uniformed in contour C.

In this proposed method, the localized active contour is applied to the preprocessed image by localizing the mask region which is used to initialize the active contour method. The coordinate values obtained from the previous step (by circular Hough transform) are used to as input for initializing the mask region. Then, the number of iteration for evolving the active contour method is also initialized and active contour model is employed and the segments the OD (as shown in Fig. 3c).

4   Result Analysis

4.1   Database

This proposed method uses RIM-ONE online database [17] which consists high-resolution images of both healthy eyes and eye which are effected by different levels of glaucoma (as shown in Fig. 2), along with gold standards of each images. It consists of 169 optic disc cropped images from full retinal fundus images that are captured from three different hospital located in Spanish region. All the steps in methodology were implemented using MatLab R2017b.

4.2   Segmentation of OD

The results of OD segmentation are differentiated into OD pixel region and non-OD pixel region. According to this, it can be further classified into *true-positive (tp)*, *true-negative (tn)*, *false-positive (fp)* and *false-negative (fn)*. *tp* is when the system detects the actual OD pixel region which will be the same as OD region of ground truth image whereas *tn* both the system and ground truth image detects as non-OD region. In *fp*, the system detects OD pixel region but it will be the non-OD pixel region while detecting in ground truth image whereas in *fn* ground truth image identified as OD region but the system fails to identify [3]. Thus, all these above standards are used to calculate accuracy as given below [3]:

$$accuracy = \frac{(\text{tp} + \text{tn})}{(\text{tp} + \text{tn} + \text{f p} + \text{fn})} \tag{10}$$

Furthermore, the overlap score is evaluated as the ratio between the segmented OD and the ground truth image and it is given below [3]:

$$overlap\ score = \frac{\text{area}(\text{segmentedOD} \cap \text{ground truth})}{\text{area}(\text{segmentedOD} \cup \text{ground truth})} \tag{11}$$

The proposed method attained 0.98 for average accuracy in RIM-ONE database and the average overlap score for 169 images present in the database is 0.905. The

Table 1 Comparison of performance analysis

Database	Methodology	Overlap score (in percent)
Gold standard	Adaptive morphological method (AMM) [4]	82
	Feature extraction technique (FET) [5]	86
	Sliding band filter technique (SBFT) [7]	89
	Grow-cut algorithm (GCA) [3]	87
	Active contour model (ACM) [11]	90
	Bharkad et al. (SB) [10]	62.9
	Localized active contour method (LACM)	90.5

Fig. 4 Bar graph representing the comparison of performance analysis

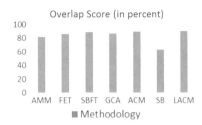

comparison of overlap score has been made between the localized active contour model method and the approaches in the literature survey by taking in account of gold standard which includes standard public database. This is illustrated in the below table (as shown in Table 1) and represented in the bar graph form as shown in Fig. 4.

5 Conclusion

The localization and segmentation of OD region from the retinal fundus image are initial step for the early detection of the retinal based diseases. This paper introduces the procedure for automatic detection and segmentation of OD region from retinal fundus image by using morphological operations, Circular Hough Transform and local region-based active contour model. Here, the proposed active contour model collects the maximum local information over multidimensional feature space using point of interest (POI) of an image and optimize the energy so that model can best fit to an image. Analysis were carried for calculating the segmentation of OD have achieved a better average accuracy and overlap score which were calculated on RIM-ONE database and gave better output compared with the methods described in the literature. The main aim of this paper is to segment out the OD to calculate cup-to-disc (CDR) ratio as a future work.

References

1. Sonka, M., Abramoff, M.D., Garvin, M.K.: Retinal imaging and image analysis. IEEE Rev. Biomed. Eng. 169–208 (2010)
2. Kande, G.B., Mittapalli, P.S.: Segmentation of optic disk and optic cup from digital fundus images for the assessment of glaucoma. Elsevier J. Biomed. Signal Process. Control **24**, 34–46 (2016)
3. Fraz, M.M., Abdullah, M.: Application of grow cut algorithm for localization and extraction of optic disc in retinal images. In: 12th International Conference on High-Capacity Optical Networks and Enabling/Emerging Technologies (HONET), pp. 1–5 (2015)
4. Angulo, J., Alcaiz, M., Morales, S., Naranjo, V.: Automatic detection of optic disc based on PCA and mathematical morphology. IEEE Trans. Med. Imaging 786–796 (2013)
5. Marn, D., Aquino, A., Gegndez-Arias, M.E.: Detecting the optic disc boundary in digital fundus images using morphological, edge detection, and feature extraction techniques. IEEE Trans. Med. Imaging 1860–1869 (2014)
6. Suero, A., Bravo, J.M., Marin, D., Gegundez-Arias, M.E.: Obtaining optic disc center and pixel region by automatic thresholding methods on morphologically processed fundus images. Comput. Methods Programs Biomed. 173–185 (2015)
7. Campilho, A., Dashtbozorg, B., Mendona, A.M.: Optic disc segmentation using the sliding band filter. Comput. Biol. Med. 1–12 (2014)
8. Kanakamala, M., Kubakaddi, S.: Automatic segmentation of optic disc using modified multi-level thresholding. In: 2014 IEEE International Symposium on Signal Processing and Information Technology (ISSPIT) (2014)
9. Salazar-Gonzalez, A., et al.: Segmentation of blood vessels and optic disc in retinal images. IEEE J. Biomed. Health Inform. **99**, 1–14 (2014)
10. Bharkad, S.: Automatic segmentation of optic disk in retinal images using dwt. In: 2016 IEEE 6th International Conference on Advanced Computing (IACC) (2016)
11. Singh, A., Agarwal, A., Issac, A.: Automatic imaging method for optic disc segmentation using morphological techniques and active contour fitting. In: 2016 Ninth International Conference Contemporary Computing (IC3), pp. 1–5 (2016)
12. Karan, K., Prashanth, R., Joshi, G.D., Sivaswamy, J., Krishnadas, R.: Vessel bend-based cup segmentation in retinal images. In: Proceedings of International Conference on Pattern Recognition (ICPR), pp. 2536–2539 (2010)
13. Yazid, H., Saad, P., Shakaff, A.Y.Md., Saad, A.R., Sugisaka, M., Yaacob, S., Mamat, M.R,. Karthigayan, M., Rizon, M.: Object detection using circular Hough transform. Am. J. Appl. Sci. (2005)
14. Dhanapackiam, C., Geetharamani, R.: Automatic localization and segmentation of optic disc in retinal fundus images through image processing techniques. In: 2014 International Conference on Recent Trends in Information Technology (2014)
15. Chan, T., Vese, L.: Active contours without edges. IEEE Trans. Image Process. (2001)
16. Lankton, S., Tannerbaum, A.: Localizing region based active contours. IEEE Trans. Image Process (2008)
17. Fumero, F., Alayn, S., Sanchez, J.L., Sigut, J., Gonzalez-Hernandez, M.: RIM-ONE: an open retinal image database for optic nerve evaluation, vol. 7, pp. 1–6 (2011)

A Novel Genetic Algorithm Based Scheduling for Multi-core Systems

Aditi Bose, Tarun Biswas and Pratyay Kuila

1 Background and Motivations

The MCS may consist of several processors which are heterogeneous in nature in terms of resources capacity, working platform, topology, etc. Let, an application is submitted to the system and one core or processor is not sufficient to execute the application. Then, the computationally challenged application is distributed among the multiple cores so the system as a single coherent unit to execute the assigned application as early as possible. Based on the characteristic of an application, the execution is performed in two different ways like static and dynamic scheduling [1–6]. The main aim of task scheduling is to how well task assignment is done to the processors? The assignment of task should be done in such a way that it should lead to faster possible completion time. If the tasks are assigned to the processors in an uneven manner, then one or two processors became heavily loaded. Moreover, it may also happen that some of the resources are extensively used and some of the costly resources remain unused. As a result, the performance of the system may be degraded in terms of overall completion time, response time, resource utilization, load balancing, speedup factor, etc. [1–3, 7–9]. Therefore, we are motivated to design a scheduling algorithm for a multi-core system by addressing makespan, utilization of resources, and speedup ratio. It should be noted that the scheduling on a multi-core systems (MCS) is a NP-complete problem. In this paper, we proposed a genetic algorithm based scheduling technique with the following objectives: minimization of makespan, maximization of resource utilization, and maximization of speedup ratio. Our main contributions in this paper are summarized as follows.

A. Bose · T. Biswas (✉) · P. Kuila
Department of Computer Science and Engineering, National Institute of Technology, Ravangla 737139, Sikkim, India
e-mail: tarun.nitskm@gmail.com

A. Bose
e-mail: aditibasusarkar@yahoo.com

P. Kuila
e-mail: pratyay_kuila@yahoo.com

© Springer Nature Singapore Pte Ltd. 2019
S. Tiwari et al. (eds.), *Smart Innovations in Communication and Computational Sciences*, Advances in Intelligent Systems and Computing 851,
https://doi.org/10.1007/978-981-13-2414-7_5

- A genetic algorithm based scheduling problem is proposed by considering three conflicting objectives.
- An efficient representation of chromosome that gives complete solution to the problem. The validity of the chromosomes can be ensured even after crossover and mutation operation.
- Derivation of multi-objective fitness function by considering all the conflicting objectives.
- The proposed algorithm has been validated through simulated results by considering synthetic as well as benchmark data sets [10] as presented in Sects. 5.1 and 5.2 respectively.

The remainder of this paper is organized as follows. Section 2. describes the state of the art of current literature on scheduling algorithms. In Sect. 3, we present ILP formulation of the problem and terminologies used to describe the work. The proposed GA-based task scheduling algorithm is presented in Sect. 4. In Sect. 5, we have analyzed and validated the proposed work improvements comparing with some existing algorithms with synthetic as well as benchmark data sets. We conclude our work with future possibilities in Sect. 6.

2   Related Works

There are several relevant task scheduling algorithms as follows.

A genetic algorithm based task scheduling is presented for heterogeneous computing system [11]. The authors have done the simulation by considering processing power which improves makespan and average processor utilization simultaneously. In [12], the authors presented a hybrid heuristic genetic algorithm with adaptive parameter (HGAP) which operates based on heuristic earliest finish time and genetic algorithm. The parameters of crossover probability are adaptive according to the current evolution status to promote evolution and find better solution. They have considered only one objective, i.e., minimization of makespan. A genetic algorithm for task scheduling on heterogeneous computing systems using multiple priority queues is presented in [13]. The assignment of task to processor is done by a heuristic based earliest finish time (HEFT) approach where the computation of priority or rank of a task is highly time consuming. It deals with a single objective, i.e., minimization of makespan. An efficient non-dominated sorting genetic algorithm II (NSGA-II) encoding for large-scale multi-objective resource allocation scheme is addressed in [14]. Here, the information of tasks is known in advance. The worked environment is highly heterogeneous in the machines power consumption, task execution time, etc. The authors simultaneously optimized makespan and energy consumption. The authors proposed a template-based genetic algorithm (TBGA) task scheduling by considering users QoS constraints [15]. The algorithm first calculates the template size for each processor. Templates are nothing but the maximal size of tasks that can be allocated to a processor. Then, the algorithm combines tasks to each template according to the size and then assign to the processors by using genetic algorithm. It considered one objective, i.e., minimization of makespan.

The above algorithms considered minimization of makespan or maximization of resource utilization. Our proposed work, genetic algorithm based scheduling for independent task over multi-core system is done by considering three objectives, minimization of makespan, maximization of utilization, and maximization of speedup ratio.

3 Problem Statement and Terminologies

In the proposed work, a set of n tasks, $\theta = \{\theta_1, \theta_2, \theta_3, \ldots \theta_n\}$ is to be scheduled on different m processors, i.e., $\lambda = \{\lambda_1, \lambda_2, \lambda_3, \ldots \lambda_m\}$. The communication channels are assumed to be contention free. The task is dispatched to the processor and scheduled by the scheduling algorithm. We formulate the addressed problem in terms of linear programming problem (LPP). Let α_{ir} be a Boolean variable defined as follows:

$$\alpha_{ir} = \begin{cases} 1, & \text{if } \theta_i \text{ is assigned to } \lambda_r \\ \\ 0, & \text{Otherwise.} \end{cases} \tag{1}$$

Then, the linear programming problem (LPP) can be formulated as follows:

$$\text{Minimize } F_1 = Max\{FT(\theta_i)|\forall \theta_i \in \theta\}$$
$$\text{Maximize } F_2 = UT$$
$$\text{Maximize } F_3 = SP$$
$$\text{Subject to}$$
$$\sum_{r=1}^{m} \alpha_{ir} = 1, \forall i, 1 \leq i \leq n \tag{2}$$

The following terminologies are used to describe the proposed algorithm:

- The execution time $\xi T(\theta_i, \lambda_r)$ of task θ_i on the processor λ_r is calculated by the following equation.
$$\xi T(\theta_i, \lambda_r) = \frac{\omega_i}{f_r^{op}} \tag{3}$$

where ω_i is the instruction size of the task and θ_i and f_r^{op} are the operating frequency of the processor λ_r.
- The actual start time, $AST(\theta_i, \lambda_r)$ of θ_i on processor λ_r means the task is started its execution on the processor at that time.

Fig. 1 Representation of chromosome

- The release time of the processor λ_r initially is 0. After execution of task θ_i, it is calculated by the following equation:

$$RT(\lambda_r) = RT(\lambda_r) + \xi T(\theta_i, \lambda_r) \tag{4}$$

- The finish time $FT(\theta_i)$ means the sum of the execution time and the actual start time of the task θ_i on processor λ_r which is given as follows:

$$FT(\theta_i) = \xi T(\theta_i, \lambda_r) + AST(\theta_i, \lambda_r) \tag{5}$$

4 Proposed Work

The proposed work genetic algorithm based scheduling in multi-core system consists of the following sections as discussed below.

4.1 Representation of Chromosome

The chromosome should always give a complete solution to the problem. Here, a chromosome is a string of randomly generated processor number. Initial population is created by randomly generated P_{SIZE} of chromosomes, where P_{SIZE} is the population size. Let us assume a system with a set of 10 tasks $\theta = \{\theta_1, \theta_2, \theta_3, \ldots, \theta_{10}\}$ and 3 processors $\lambda = \{\lambda_1, \lambda_2, \lambda_3\}$. It can be observed from Fig. 1 that the 2nd gene position is one, which indicates that the θ_2 is assigned to λ_1. Accordingly, θ_1, θ_3, and θ_5 is assigned to λ_2, λ_3 and λ_2 respectively.

4.2 Derivation of Fitness Function

To derive the fitness value, we have considered three objectives like minimization of makespan, maximization of processor utilization, and maximization of speedup ratio. Here, each chromosome is evaluated by the fitness value and finds out the best solution for our problem. The detailed derivation of the fitness function is given below.

- **Minimization of Makespan**: Each task $\theta_i \in \theta$ should be assigned to the processors so that the execution can be completed within the shortest possible time. Therefore, the first objective can be described as follows:

$$\text{Objective 1 } (O_1) : \text{ Minimize } (Mks) \tag{6}$$

where, $Mks = Max \ (FT(\theta_1), FT(\theta_2), FT(\theta_3) \dots FT(\theta_n))$.

- **Maximization of Processor Utilization**: It can be defined as the ratio between average amount of time in which the processors are busy by the overall system schedule time. It is calculated as follows:

$$UT = \frac{1}{Mks} \times \frac{\sum_{r=1}^{m} \xi T(\lambda_r)}{m} \tag{7}$$

Now, the second objective can be stated as:

$$\text{Objective 2} (O_2) : \text{ Maximize } (UT) \tag{8}$$

- **Maximization of Speedup**: The speedup factor can be defined as the ratio between sequential execution time by makespan (parallel schedule length) as given below.

$$SP = \frac{\sum_{i=1}^{n} \xi T(\theta_i, \lambda_r)}{Mks} \tag{9}$$

$$\text{Objective 3} (O_3) : \text{ Maximize } (SP) \tag{10}$$

Note that the above objectives are conflicting with each other. Here, we have used weighted sum approach (WSA) [13] to compute the fitness function. WSA is a classical approach for optimizing the multi-objective task scheduling problem, where each objective is assigned a weight. Our goal is to choose the solution with minimum fitness value. Let the weight value of the parameters be $\beta_1, \beta_2, \beta_3$, respectively, where $\sum_{j=1}^{3} \beta_j = 1$. The fitness function of the proposed work is as follows:

$$Fitness = \beta_1 \times Mks + \beta_2 \times (1 - UT) + \beta_3 \times (1 - SP) \tag{11}$$

Our final objective is to minimize *Fitness*.

4.3 Crossover, Mutation, and Selection

In crossover, two parent chromosomes mate to produce two child chromosomes. There are many crossover operations like, one-point, two-point, hybrid, etc. Here, we have used one-point crossover as shown in Fig. 2a. The crossover operation is followed by mutation, where the child chromosomes are mutated for possible betterment. In mutation, a gene position is randomly selected and the value is replaced by another valid value as shown in Fig. 2b. In the selection operation, better chromosomes are selected based on the fitness value.

Fig. 2 **a** Crossover and **b** mutation operation

The pseudo code of proposed scheduling is given in the Algorithm 1.

5   Result and Discussions

We have done the simulations to demonstrate our proposed work on a system which is running on Intel i7 3.4 GHz CPU with 4 GB of RAM. The proposed work is implemented by using C programming on Ubuntu 14.04 operating system. We have used a robust selection mechanism, Roulette Wheel selection procedure, that selects parents based on their fitness values. Here, we have used the weighted sum approach (WSM) for calculating the fitness of a chromosome by considering the mentioned conflicting objectives. Our algorithm is analyzed and validated by the synthetic as well as benchmark data set [10] as follows.

Algorithm 1 : Genetic algorithm based task scheduling

Input: (1) A List of n tasks θ_i where, ω_i denotes ith task instruction size and $\forall \theta_i \in \theta$
    (2) A List of m processor λ_r corresponding frequencies.
Output: An efficient mapping of tasks to the processors, $\theta_i \rightarrow \lambda_r$.

1: Create Initial population, say P_{INI}.
2: Apply Selection operation to find new population P_{NEW} with P_{SIZE} number of chromosomes.
3: Initialize $g = 0$;
4: **while** $(g \neq Terminate)$ **do**
5:   Randomly select two parent chromosome (say p_1 and p_2) from P_{NEW}.
6:   Apply crossover on p_1 and p_2 to generate offsprings (Ch_1 and Ch_2).
7:   **if** (fitness(Ch_1) < fitness(p_1)) **then**
8:    Ch_1 replaces p_1 in P_{NEW}.
9:   **end if**
10:   **if** (fitness(Ch_2) < fitness(p_2)) **then**
11:    Ch_2 replaces p_2 in P_{NEW}.
12:   **end if**
13:   $g + +$;
14: **end while**
15: Select the minimum fitness value chromosome as final solution.

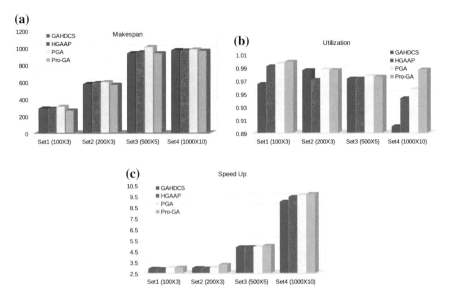

Fig. 3 Simulation results on synthetic data sets **a** makespan, **b** utilization, and **c** speedup ratio

5.1 Simulation on Random Data Set

The proposed work is simulated by considering four randomly generated data sets like, (1) 100 tasks with 3 processors (100×3), (2) 200 tasks with 3 processors (200×3), (3) 500 tasks with 5 processors (500×5), and (4) 1000 tasks with 10 processors (1000×10). Here, the size of the tasks is also randomly generated and the used weight values $0.5, 0.3, 0.2$ for the calculation of fitness value. The pictorial representation of the simulation results is shown in Fig. 3a–c. To show the effectiveness of the proposed work, we have also validated our work by benchmark data set [10] as discussed below.

5.2 Simulation on Benchmark Data Set

Here, two data sets 256 tasks on 8 processors (256×8) and 512 tasks on 16 processors (512×16) have been considered to validate the work. The data sets contain 12 instances which helps in scheduling [10]. The instances are expressed as u_x_yyzz where u denotes instances generated by the uniform distribution, x denotes the consistency as follows, consistent (c), inconsistent (i), or semi-consistent (s)]. The yy shows heterogeneity [i.e., high (hi) or low (lo)] of the task and zz shows the heterogeneity [i.e., high (hi) or low (lo)] of the processor. The u_c_hihi, u_c_hilo, u_c_lohi, u_c_lolo, u_i_hihi, u_i_hilo, u_i_lohi, u_i_lolo, u_s_hihi, u_s_hilo, u_s_lohi, and

(a)

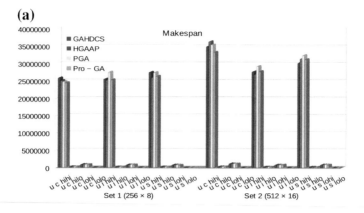

(b)

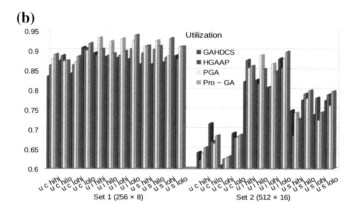

(c)

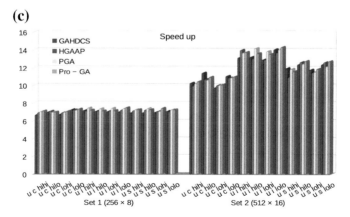

Fig. 4 Simulation results on benchmark data sets **a** makespan, **b** processor utilization, and **c** speedup ratio

u_s_lolo are the generated 12 different instances of the set. We have evaluated our work by considering the sets based on makespan, utilization, and speedup ration. It has been observed that the proposed algorithms $Pro - GA$ perform better than GAHDCS [11], HGAAP [12], and PGA [16] in terms of considered objectives.

It observed that the proposed algorithm produces lesser makespan due to adjustment of tasks to the available processors based on the earliest finish time. Therefore, before assigning a task to the processor, the actual start time is calculated which helps in finding the EFT. So, the overall execution is minimized by utilization of the cores and improve the speedup ratio. We have shown the pictorial representation of performance study of the proposed work as in Fig. 4a–c.

6 Conclusions

In this paper, we have designed a genetic algorithm based independent task scheduling for multi-core system. The experimental results show that the proposed algorithm has considerable improvements over the *GAHDCS*, *HGAAP*, and *PGA* in terms of the considered objectives. To show the effectiveness, the proposed work is validated by synthetic and benchmark data set. However, the proposed algorithm does not consider dependency of the tasks and energy optimization of the processor. In future, our attempt will be to develop a new energy-efficient scheduling algorithm with dependency constraints.

References

1. Jiang, J., Lin, Y., Xie, G., Fu, L., Yang, J.: Time and energy optimization algorithms for the static scheduling of multiple workflows in heterogeneous computing system. J. Grid Comput. 1–22 (2017)
2. Gogos, C., Valouxis, C., Alefragis, P., Goulas, G., Voros, N., Housos, E.: Scheduling independent tasks on heterogeneous processors using heuristics and column pricing. Future Gener. Comput. Syst. **60**, 48–66 (2016)
3. AlEbrahim, S., Ahmad, I.: Task scheduling for heterogeneous computing systems. J. Supercomput. **73**(6), 2313–2338 (2017)
4. Biswas, T., Kuila, P., Kumar Ray, A.: Multi-level queue for task scheduling in heterogeneous distributed computing system. In: 2017 4th International Conference on Advanced Computing and Communication Systems (ICACCS), pp. 1–6. IEEE (2017)
5. Amalarethinam, D.G., Kavitha, S.: Priority based performance improved algorithm for meta-task scheduling in cloud environment. In: 2017 2nd International Conference on Computing and Communications Technologies (ICCCT), pp. 69–73. IEEE (2017)
6. Alkayal, E.S., Jennings, N.R., Abulkhair, M.F.: Efficient task scheduling multi-objective particle swarm optimization in cloud computing. In: 2016 IEEE 41st Conference on Local Computer Networks Workshops (LCN Workshops), pp. 17–24. IEEE (2016)
7. Liu, Y., Zhang, C., Li, B., Niu, J.: Dems: a hybrid scheme of task scheduling and load balancing in computing clusters. J. Netw. Comput. Appl. **83**, 213–220 (2017)

 8. Vasile, M.-A., Pop, F., Tutueanu, R.-I., Cristea, V., Kołodziej, J.: Resource-aware hybrid scheduling algorithm in heterogeneous distributed computing. Future Gener. Comput. Syst. **51**, 61–71 (2015)
 9. Biswas, T., Kumar Ray, A., Kuila, P., Ray, S.: Resource factor-based leader election for ring networks. In: Advances in Computer and Computational Sciences, pp. 251–257. Springer (2017)
10. Braun, T.D., Siegel, H.J., Beck, N., Bölöni, L.L., Maheswaran, M., Reuther, A.I., Robertson, J.P., Theys, M.D., Yao, B., Hensgen, D.: A comparison of eleven static heuristics for mapping a class of independent tasks onto heterogeneous distributed computing systems. J. Parallel Distrib. Comput. **61**(6), 810–837 (2001)
11. Jooyayeshendi, A., Akkasi, A.: Genetic algorithm for task scheduling in heterogeneous distributed computing system. Int. J. Sci. Eng. Res. **6**(7), 1338 (2015)
12. Ding, S., Wu, J., Xie, G., Zeng, G.: A hybrid heuristic-genetic algorithm with adaptive parameters for static task scheduling in heterogeneous computing system. In: Trustcom/BigDataSE/ICESS, 2017 IEEE, pp. 761–766. IEEE (2017)
13. Yuming, X., Li, K., Jingtong, H., Li, K.: A genetic algorithm for task scheduling on heterogeneous computing systems using multiple priority queues. Inf. Sci. **270**, 255–287 (2014)
14. Friese, R.D.: Efficient genetic algorithm encoding for large-scale multi-objective resource allocation. In: 2016 IEEE International Parallel and Distributed Processing Symposium Workshops, pp. 1360–1369. IEEE (2016)
15. Sheng, X., Li, Q.: Template-based genetic algorithm for qos-aware task scheduling in cloud computing. In: 2016 International Conference on Advanced Cloud and Big Data (CBD), pp. 25–30. IEEE (2016)
16. Kwok, Y.-K., Ahmad, I.: Efficient scheduling of arbitrary task graphs to multiprocessors using a parallel genetic algorithm. J. Parallel Distrib. Comput. **47**(1), 58–77 (1997)

Various Preprocessing Methods for Neural Network Based Heart Disease Prediction

Kavita Burse, Vishnu Pratap Singh Kirar, Abhishek Burse and Rashmi Burse

1 Introduction

A disease is a pathological condition in which the organ or body part or immune system of the body does not work properly. This is due to some infection, genetic heredity, or stress. According to WHO the most dangerous chronic diseases, which causes death, are breast cancer, heart disease, and diabetes. Most of the patients with these diseases belong to developed countries. Heart disease is also referring as coronary artery disease. The coronary arteries and veins become narrow which affects the circulation of blood which may cause a heart attack. The common symptoms of heart disease like chest pain, high or low blood pressure varies person to person. High cholesterol level is the most common cause of heart disease apart from smoking, eating habits, and lifestyle.

Perelson and Oster proposed artificial Immune System (AIS) in 1979 which is based on the shape-space model that can explain the mechanism and interconnection

K. Burse
Department of Electronics & Communication, Technocrats Institute
of Technology, Bhopal, India
e-mail: kavitaburse14@gmail.com

V. P. S. Kirar
Department of Computer Science, University of Bedfordshire, Luton, UK
e-mail: vishnupskirar@live.com

A. Burse
Department of Computer Science, Oriental Institute of Science
and Technology, Bhopal, India
e-mail: abhishekburse25@gmail.com

R. Burse (✉)
Department of Computer Science, Maulana Azad National Institute
of Technology, Bhopal, India
e-mail: rashmeeburse@gmail.com

© Springer Nature Singapore Pte Ltd. 2019
S. Tiwari et al. (eds.), *Smart Innovations in Communication and Computational
Sciences*, Advances in Intelligent Systems and Computing 851,
https://doi.org/10.1007/978-981-13-2414-7_6

between two neurons in the human immune system [1]. Shape-space can provide different identity to attributes of the same class. Attribute with the function of weight than AIS is known as Attributed Weighted AIS (AWAIS). Şahan et al. proposed a new neural network in which the weight attributes are used to calculate the Euclidean distance [2]. Das et al. proposed multiple predecessor model in which, class and interval target is combined together. In this method, two different models are trained with partition of the dataset. The outcome of both models is merged together and ensembled for a new model [3]. Karayılan and sKılıç proposed a ANN predictive model using back propagation algorithm for Cleveland dataset and achieve accuracy nearly 95% [4].

Jabbar et al. proposed a feature subset selection method. It only selects the data which can help in the prediction process. PCA extracts the information from the dataset and compresses the useful attributes. Covariance matrix is calculated Eigen value and Eigen vector, which is derived from the matrix of a feature vector. The classification accuracy of proposed work is 99.38% [5]. Kavitha et al. proposed a new model of feedforward ANN which emerge with a genetic algorithm. Instead of BP algorithm, hybridization is used to train the ANN. The reason behind this change is the trapping tendency of local minima. The classification accuracy of the proposed model is 93.45% [6]. Polat et al. proposed AIS and fuzzy AIS k-NN with accuracy of 87% [7]. Uyar and İlhan used genetic algorithm (GA) with recurrent fuzzy neural network for diagnosis of heart disease and achieve accuracy of 97.78% [8]. Shah et al. proposed Probabilistic PCA approach for prediction of heart disease. PPCA helps in missing value of attributes. Projection vector of PPCA calculate the covariance and also reduce the dimensionality. Its accuracy for Cleveland dataset is 82.18% [9]. Nilashi et al. proposed a data mining technique of classification and regration tree for medical diagnosis. This knowledge-based system performs the clustering, noise reduction, and prediction techniques. For Cleveland dataset, its accuracy is 92.8% [10]. Kelwade and Salanker proposed a redial basis function neural network (RBFN) which works on both linear and nonlinear heart disease dataset. The accuracy of network is 95.9% [11].

From this survey, we found that authors proposed a dedicated network for diagnosis of heart disease. In this research proposal, we try to develop a network which can achieve and perform at a high accuracy than previously proposed models. We also demonstrate the effect of various preprocessing methods and validation methods on proposed model and its results, which is the main objective of this research study.

2 Proposed Neural Network Architecture for Diagnosis of Heart Disease

The basic building block of the multiplicative neuron model is single neuron. All neurons in this network are fully connected to each other and form a layered structure. In proposed model, three-layer model is developed. These layers are input layer,

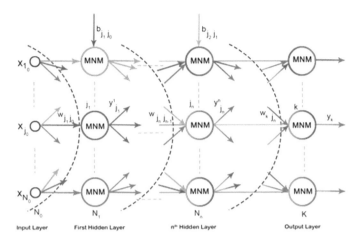

Fig. 1 Architecture of multi-layer Pi-Sigma neuron network

hidden layer, and output layer. All the medical data is received by input layer and processed by the network. Output layer provides the decision for medical condition. Multi-Layer Pi-Sigma Neuron Network is a feedforward network. The mapping and prediction ability of feedforward network are greater than other neural networks. It contains higher order information and takes less time in learning and training. In MLPPNM the weights of neurons are automatically updated to minimize the error [12]. In the learning process, we have used the BP learning algorithm. In BP algorithm the actual output and target output is compared, and error is minimized by learning computation. All the medical data is classified by the learning process and when the network is fully trained then its knowledge is stored, and it is used for the future decision-making process. The architecture of MLPPNN is shown in the Fig. 1.

During the training, phase we use the BP algorithm for weight update. As we know, the BP algorithm suffers from local minima problem. To avoid this situation, a proportional factor and momentum term is used at the time of weight and bias update. The momentum term is derived from previous weight change. Due to change of slope of error function the oscillation of the gradient function is suppressed. The momentum term opposes the change in gradient function and this prevents the problem of local minima. On the other hand, due to gradient descent the activation function goes slow down and learning rate also become saturated. To avoid these both problems and improve BP algorithm, a proportion factor and momentum is used.

2.1 Principal Component Analysis

Principal Component Analysis (PCA) is a technique of data preprocessing. When the data is collected from different sources then there are chances of inconsistency

Fig. 2 The flow of principal component analysis

in the stored data. Different sources use the different techniques for data handling. Sometimes the data is incomplete, inconsistent, and noisy. The data should be preprocessed before the use. The task that involves in data preprocessing is data cleaning, data integration, data transformation, data dimensionality reduction and data decentralization. The flow of data preprocessing is shown in Fig. 2.

In data cleaning method, unformatted data is removed, and missing value is replaced by logical values. All data is arranged in the same format as binary or with same units. In data dimensionality reduction process the dimension of data is reduced for meaningful calculations.

2.2 Performance Evaluation

For this research, we select the Mean Square Error to evaluate the performance. The Accuracy proposed network could be derived from Eq. (1) and the standard deviation is derived from Eq. (2). The sensitivity (ϱ) and specificity (ς) are calculated from Eqs. (3) and (4)

$$\% \, Accuracy = \frac{correctly \; identified \; events \; during \; testing}{total \; number \; of \; events \; during \; testing} \tag{1}$$

$$Standard \; deviation \, (\sigma) = \sqrt{\frac{sum \; of \; squared \; deviation \, from \; the \; sample \; mean}{number \; of \; obervations}} \tag{2}$$

$$Sensitivity \, (\varrho) = \frac{TP}{TP + FN} \tag{3}$$

$$Specificity \, (\sigma) = \frac{TN}{TN + FP}, \tag{4}$$

where,

TP True Positive
FP False Positive
TN True Negative
FN False Negative

Table 1 Description of UCI heart disease dataset

Attribute ID	Attribute name	Attribute description
1	age	Age of patient in years
2	sex	Gander of patient
3	cp	Chest pain type
4	trestbps	Resting blood pressure
5	chol	Serum cholesterol
6	fbs	Fasting blood sugar level
7	restecg	Resting electrocardiographic results
8	thalach	Maximum heart rate achieved
9	exang	Exercise include angina
10	oldpeak	ST depression induced by exercise relative to rest
11	slope	The slop of the peak exercise ST segment
12	ca	Number of major vessels (0–3) colored by flourosopy
13	thal	3 = normal; 6 = fixed defect; 7 = reversible defect
14	num	Diagnosis of heart disease

3   Simulation and Results

In this research for diagnosis of heart disease, we are using UCI machine learning repository for heart disease dataset [13].

The total number of attributes in original data is 76 and the total number of instances is 303. In the original dataset, many attributes are not used and have missing values. Thus, the original data set is processed, and 14 attributes are selected for formation of a new dataset. Description of the dataset is presented in Table 1.

3.1   MLPSNM for Diagnosis of Heart Disease

MLPSNM is tested on Cleveland heart disease dataset. Network architecture of MLPSNM is shown in Fig. 3.

The number of neurons in the input layer is 14, hidden layer 1 has t neurons, hidden layer 2 has two neurons and output layer has one neuron. To achieve a maximum performance and accuracy, different preprocessing methods are applied on MLPSNM. Block diagram of diagnosis of heart disease is shown in Fig. 4.

Heart disease dataset is divided into six parts. Training and testing are performed on these different datasets. The results of diagnosis of heart disease are shown in Table 2.

Different graphs of MSE of convergence and performance is described below in Figs. 5 and 6.

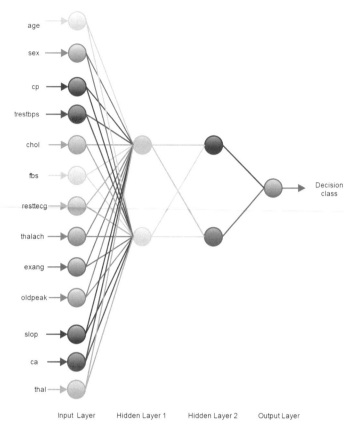

Fig. 3 Neural network architecture of MLPSNM for diagnosis of heart disease

Fig. 4 Diagnosis of heart disease using data preprocessing

3.2 K-Fold Cross Validation for Diagnosis of Heart Disease

For diagnosis of heart disease, we use threefold cross validation method. Out of 303 instances, we tool 200 instances for training and rest 100 for testing purpose. Table 3 shows the performance of validation process. MSE convergence of threefold cross validation by using normalization and the scatter graph between first principle component and second component for PCA for different values of k is shown below in Figs. 7 and 8.

Table 2 Training and testing performance for diagnosis of heart disease

Training			Testing	
Epochs	MSE (Normalization)	MSE (PCA)	MSE (Normalization)	MSE (PCA)
50	0.0101	5.9597e−07	0.0435	2.5863e−07
100	0.0057	3.9146e−07	0.0021	1.7562e−08
150	0.0034	2.7224e−07	0.0013	3.7793e−07
200	0.0027	2.5647e−07	9.7652e−05	3.1113e−08
250	0.0020	2.4190e−07	5.2836e−04	2.3878e−07
303	0.0017	1.9822e−07	1.5862e−05	7.2298e−08

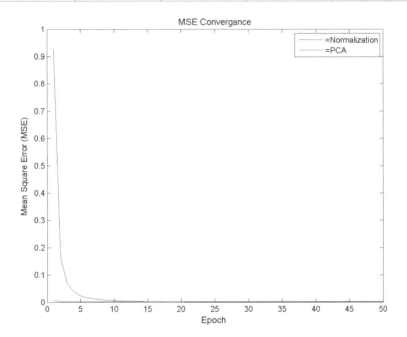

Fig. 5 Training performance for PCA and normalization (303 dataset)

3.3 *Diagnosis of Heart Disease by Using Support Vector Machnie-LDA*

In this section, we are using the linear SVM to separate two different classes that are situated on both sides of the hyperplane. This hyperplane is divided by training datasets. The main aim of this technique is to minimize the distance between two known hyperplanes. The validation process used here is holding out validation. For the validation process, data is divided into two parts. Generally, two-third part is used for training and one-third part is used for testing. During the training phase,

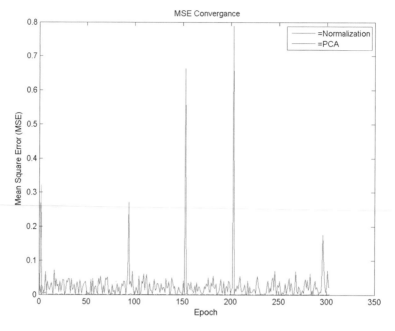

Fig. 6 Testing performance for PCA and normalization (303 dataset)

Table 3 Training and testing performance for diagnosis of heart disease using 3-fold cross validation

Training			Testing	
Value of K	MSE (Normalization)	MSE (PCA)	MSE (Normalization)	MSE (PCA)
1	0.0025	3.2415e−07	0.0020	4.3486e−08
2	0.0025	2.7010e−07	4.5272e−04	3.9828e−09
3	0.0022	2.5534e−07	1.2474e−05	2.3899e−09

the network is trained and results for variables are stored. In testing phase, network predicted the output value for the dataset. These values are compared with target output for computation of accuracy of the network. In this research, we divided 202 datasets for training and 101 datasets for testing.

The total attributes without LDA are fourteen namely age, sex, Fbs, Ca, Trestbps, Chol, Oldpeak, Slop, Exang. Cp, Num with LDA the features are reduced to four only Cp, Trestbps, Chol, and Thalach. The selected attributes are major cause of heart disease. On the other hand, rest of the attributes are not as harmful. The prediction of these features can easily predict the chances of this heart. The classification accuracy for Linear SVM is shown in Table 4. A comparison Table 5 shows the different proposed model for diagnosis of Heart Disease.

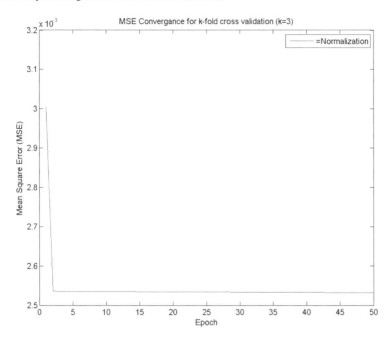

Fig. 7 MSE convergence of 3-fold cross validation (Normalization)

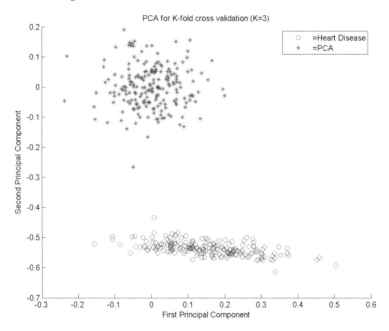

Fig. 8 First principle component and second principal component for PCA

Table 4 Classification with SVM-LDA for heart disease dataset

Total events	Training set	Testing set	Total attributes	Reduced attributes	Accuracy (Training)	Accuracy (Testing)
303	202	101	14	4	85.64%	88.32%

Table 5 Proposed models and classification accuracies for diagnosis of heart disease

Author	Proposed method	Accuracy (%)
Polat et al.	Artificial Immune System (AIS)	84.50
Şahan et al.	AWAIS	82.59
Polat et al.	Fuzzy AIS k-NN	87.00
Das et al.	Neural Network Ensembles	89.01
Kavitha et al.	ANN-GA	93.45
Jabbar et al.	Naïve Bayes	72.50
This study	MLPSNM + Normalization	91.34
This study	MLPSNM + PCA	94.53
This study	MLPSNM + k-fold	90.44
This study	SVM-LDA	88.32

4 Conclusion

We have proposed a Multi-Layer Pi-Sigma Neuron Model (MLPSNM) for heart disease diagnosis. The proposed model is based on PI-Sigma model in which, the architecture and calculation are less complex as compared to other previously proposed models. For the learning of network, BP algorithm is used with bipolar sigmoid function activation function. PCA and LDA preprocessing methods are used to reduce the dimensionality of the dataset. In SVM-LDA method, the attributes that are closer to hyperplane are selected. For validation of the network, k-fold validation method is used. The network converges after 50 iterations. The proposed model achieves 94.53% classification accuracy for diagnosis of heart disease by using PCA.

References

1. Perelson A.S., Oster G.F.: Theoretical studies of clonal selection: minimal antibody repertoire size and reliability of self-nonself discrimination. J. Theor. Biol. **81**, 645–670 (1979)
2. Şahan, S., Polat, K., Kodaz, H., Gunes, S.: The medical application of attribute weighted artificial immune system (AWAIS): diagnosis of heart and diabetes disease. In: ICARIS' vol. 2005, pp. 456–468 (2005)
3. Das, R., Turkoglu, I., Sengur, A.: Diagnosis of valvular heart disease through neural networks ensembles. Comput. Methods Programs Biomed. **93**(2), 185–191 (2009)

4. Karayilan, T., Kiliç, Ö.: Prediction of heart disease using neural network. In: 2017 International Conference on Computer Science and Engineering (UBMK), pp. 719–723 (2017)
5. Jabbar, M.A., Deekshatutu, B.L., Chandra, P.: Classification of heart disease using artificial neural network and feature subset selection. Glob. J. Comput. Sci. technol. Neural Artif. Intell. **13**(3), 5–14 (2013)
6. Kavitha, K.S., Ramakrishnan, K.V., Singh, M.K.: Modeling and design of evolutionary neural network for heart disease detection. Int. J. Sci. Issues **7**(5), 272–283 (2010)
7. Polat, K., Güneş, S., Tosun, S.: Diagnosis of heart disease using artificial immune recognition system and fuzzy weighted pre-processing. Pattern Recogn. **39**(11), 2186–2193 (2006)
8. Uyar, K., İlhan, A.: Diagnosis of heart disease using genetic algorithm based trained recurrent fuzzy neural networks. Procedia Comput. Sci. **120**, 588–593 (2017)
9. Shah, S.M.S., Batool, S., Khan, I., Ashraf, M.S., Abbas, S.H., Hussain, S.A.: Feature extraction through parallel probabilistic principal component analysis for heart disease diagnosis. Phys. Stat. Mech. Appl. **482**, 796–807 (2017)
10. Nilashi, M., Ibrahim, O.B., Ahmadi, H., Shahmoradi, L.: An analytical method for diseases prediction using machine learning techniques. Comput. Chem. Eng. **106**, 212–223 (2017)
11. Kelwade J.P., Salankar S.S.: Comparative study of neural networks for prediction of cardiac arrhythmias. In: International Conference on Automatic Control and Dynamic Optimization Techniques (ICACDOT), pp. 1062–1066 (2016)
12. Burse K., Yadav, R.N., Srivastava, S.C.: Fully complex multiplicative neural network model and its application to channel equalization, in advances of neural network research and applications. In: Zeng Z., Wang, J. (eds.) Springer Lecture Notes in Electrical Engineering, vol. 67, pp. 493–501 (2010)
13. UCI Machine Learning Repository. Heart disease data set. https://archive.ics.uci.edu/ml/datasets/Heart+Disease (2015)

Intelligent Passenger Information System Using IoT for Smart Cities

Polamarasetty Anudeep and N. Krishna Prakash

1 Introduction

Nowadays transportation plays a key role for humans in their day-to-day life. For work and needs, people use different modes of transportation. Public Transport Systems (PTS) are the systems which are mostly available to all the people [1]. People usually prefer PTS for its low cost and comfortable travel. So there will be an increasing load on PTS [2] and efficiency has to be increased necessarily on these systems. But due to the increasing population there are several factors that come into consideration, such as traffic, less number of public vehicles, crowd density, etc. One of the most important factors is that the public buses lack punctuality and reliability. Invulnerable, fast, and decisive transportation can provide a considerable contribution towards the development of smart cities. Especially for commuters, mode of transport should be very fast and convenient. Commuters can plan their work accordingly as per the transport or plan their transport as per the work. This can save a lot of time and cost for them.

In order to help the people from the difficulties, Internet of Things (IoT) plays a very significant role. IoT technology [3, 4] can easily help the people to track the location of their vehicles. Using IoT people can access this system at anywhere and at any place. IoT can also improve the passenger experience for better public transportation. Mobile devices [2, 5] like smart phones and tablets can remove barriers in creating, accessing and sharing the data among the passengers.

Intelligent Public Transport Systems (IPTS) [6, 7] are the foremost applications which dispense innovation services relating to different modes of transportation and

P. Anudeep (✉) · N. Krishna Prakash
Department of Electrical and Electronics Department,
Amrita School of Engineering, Amrita Vishwa Vidyapeetham, Coimbatore 641112, India
e-mail: anudeep.polamarasetty@gmail.com

N. Krishna Prakash
e-mail: n_krishnaprakash@cb.amrita.edu

© Springer Nature Singapore Pte Ltd. 2019
S. Tiwari et al. (eds.), *Smart Innovations in Communication and Computational Sciences*, Advances in Intelligent Systems and Computing 851,
https://doi.org/10.1007/978-981-13-2414-7_7

also provide transport information. These systems provide the real-time tracking of the vehicle and personalized information of the passengers. They can send the improved responses about the unexpected events such as road closures, breakdowns, etc. These systems also enable various users to be better reliable. The main objective of this paper is to design an application user interface which shows the status of the bus and estimated arrival time.

2   Literature Review

In this present-day scenario, passengers spend more time in waiting for the transport due to lack of passenger information system. Many authors have come with innovative solutions to make the mode of transport intelligible. If the real-time information is known, it will be very useful for the people. As technology is evolving, there are different technologies that can be used for calculating the information about the bus.

Transport information system developed in [4] use RF (Radio Frequency) for communication and when the bus arrives in the range of the bus stop, information is displayed in the bus stop. This updates the current location of the bus on the website of Bus Management System (BMS) through Ethernet. Simultaneously the bus stop sends a message to the next successive bus stop informing about the arrival of bus at the stop. A smart tracking system was proposed [5] where Wi-Fi routers are used at bus terminals and replacing the Ethernet modules in the bus stop [4]. When the bus comes to the bus terminal, the Wi-Fi module [8] gets connected to the router and sends the location, i.e., latitude and longitude of the bus terminals to the cloud displaying to the user in the mobile application. Later a system was developed to use public transport buses as a Mobile Enterprise Sensor Bus (M-ESB) [3], consisting of onboard Sensor Networks (SNs) [9] for environment and road condition monitoring along with a data exchange interface to feed a data cloud computing system. Then Smart Bus System (SBS) developed in [10] can model a city-wise bus transport system. A framework has been designed to integrate the heterogeneous sensor data with modules for the software facilities provided at the bus stops, to the driver's cabin and to the central facility operators at the bus depots. Also interfacing biometric authentication devices using RFID (Radio frequency identification), NFC (Near-field communication) technologies to the central server and Aadhar database had also been considered.

Developments in IoT [7, 11] has made an evolution in the transport. IoT [12–14] and Real-Time Public Vehicle Mobile Tracking System have been implemented using GPS (Global Positioning System) Technology [2]. The user can track the bus in an android application using the Bus Simulator Tracking Algorithm. The travelers get prior information about the current location, the next location of bus and crowd level [13] inside the bus. Computation of Estimated Time of Arrival (ETA) [2, 12, 15] in micro controller is done based on the GPS data. Similarly a Tracking System [16] was developed for installation in the bus, to enable the Administrator/User to track the bus location with an installed Android App on any smart phone. Buses carry GPS devices

[2, 17, 18] to track their positions. Client application displays a map showing the position of the bus. Later a solution [1] was found with the existing Internet-enabled devices on the bus (like the e-ticketing system) to track the real-time location and update in the servers by harnessing IoT technology [19, 20] stack. Accessing this location data from servers is facilitated by Representational State Transfer (REST) APIs where users will access through web-portals. This system is based on MQTT (Message Queuing Telemetry Transport) [9] protocol instead of the traditional HTTP (Hypertext Transfer Protocol) protocol. Hardware solution was proposed which can compute the dynamic shortest path to reach the destination in real time and sends the information to the bus driver. Artificial Neural Networks (ANN) is used to give an accurate ETA to the commuter by means of an application. ETA to the next stop is communicated to the commuter using the MQTT protocol [1, 12] by the hardware mounted on the bus.

There are different methods to calculate the ETA which was briefly explained in [21]. A survey has been done on analyzing historical data with statistical methods and Kalman Filters. ANN has been applied to GPS data collected from transit vehicles with an emphasis on their model and architecture. Then the results have obtained to show the complexity of estimating the speed and travel time of public transport vehicles for traffic modeling [22]. Also, parameters such as average bus stop service time, number of bus stops are taken into consideration by using statistical distribution.

3 System Overview

3.1 System Architecture

The intelligent public transport system will have a bus module for each bus, cloud module and an Android application. Figure 1 Shows the overview of the system which is implemented. Here N1 is considered as the node in the Bus1 and N2 is considered as the node in the Bus2. It can be extended up to the K number of buses. From each node parameters such as Sensor data, GPS data, Speed, Time will be sent to the cloud server for every second. Whenever the user sends the request from the Android app, the cloud server will analyze the data received from each bus and send back the response to the Android app of the user. Here each and every node is connected to the cloud server through a Wi-Fi module. Since a transport system is implemented for smart cities, it was assumed that the Wi-Fi was available everywhere.

3.2 Vehicle Node

Vehicle node shown in Fig. 2 consists of Node MCU interfaced with GPS module and four Infrared (IR) Proximity sensors placed at the front door and back door of

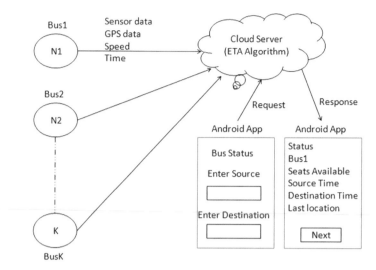

Fig. 1 System architecture for an intelligent public transport system

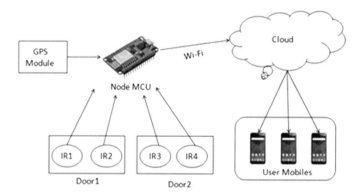

Fig. 2 Vehicle node

the bus. Each vehicle will send the sensor data in specific formats to the cloud. The message is split into different fields after it was sent into IoT cloud. The splitting of data frame was done using the JavaScript language. The coordinates of all the bus stops are stored in the IoT cloud database. Using Google API the distance between the two coordinates, i.e., the live bus location coordinates and the bus stop coordinates is calculated and stored in the database. In Average Speed Algorithm, the average speed is calculated by summing all the historical speeds of the bus received from the GPS module and dividing by the number of samples taken as given in Eq. (1). Using the average speed and distance, estimated time arrival of the bus to each and every bus stop is calculated and stored in the database.

$$v = \frac{\sum_{i=1}^{n-1} vai + vr}{n},$$ (1)

where

vai	average speed of route segment i ($i = 1, 2, …, n − 1$)
vr	current speed of the bus obtained from GPS data, and
$n − 1$	number of route segments to the bus stop.

3.3 Android Application

The users can also make use of an Android application. Entering the source and destination, the mobile app displays details such as the upcoming buses, estimated time taken by the upcoming buses to reach the arrival stop, estimated time taken by the upcoming buses to reach the destination, total seats available in the bus at each stop and location of the bus after the exit of each stop.

4 Layered Architecture

4.1 Perception Layer

Perception layer which is also known as sensing layer is used to perceive the physical properties of things surrounding us that are integral part of the IoT. In addition, this layer is also responsible for converting the information to digital signals, which is more suitable for network transmission. Perception layer will identify the object and collect information through the sensors.

4.2 Network Layer

Network layer consists of communication networks making accurate processing and transmission of information obtained from the perception layer as shown in Fig. 3. This layer is used for translating the logical addresses into physical addresses and also responsible for addressing messages and data to the appropriate source and destination. This layer is responsible for sending the data to the cloud in specific frame format using IoT lightweight data protocol.

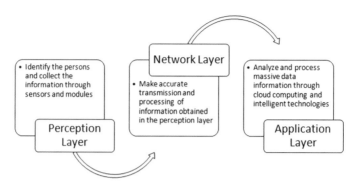

Fig. 3 Layered architecture for the proposed system

4.3 Application Layer

Application layer mainly deals with the management of application such as smart city and interacts directly with all the end users. It analyzes and processes the massive data information through cloud computing and intelligent technologies. This layer is responsible for delivering application specific services to the user and can also provide applications with access to network services.

5 System Implementation

5.1 Bus Module

In this module two IR Sensors are deployed at front and back door of the bus to count the number of passengers entering and leaving the bus. After departure from bus stop, the node will send sensor data to the cloud for every one minute. The node will send the sensor data, speed and GPS location of the bus to cloud server using appropriate IoT lightweight protocol.

For identifying the message from each node, a frame format as shown in Fig. 4 has been created for each message. The total size of this frame consists of 19 Bytes for three fields, i.e., SoF, Node_ID, Payload. In this the Payload field consists of six data fields. First field in the frame format is the Start of Frame which is of 8 bits. For both the nodes N1 and N2 the start of frame will be 0x11. The second field is Node_ID which is 0x01 for Node1 and 0x02 for Node2. The maximum value can be 0xFF, so a total of 255 nodes can be connected to the cloud. The first field in Payload is RTC where Hours, Minutes and Seconds are stored in each byte. The second field

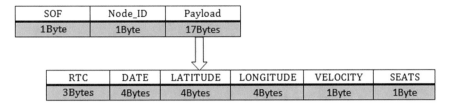

SOF	Node_ID	Payload
1Byte	1Byte	17Bytes

RTC	DATE	LATITUDE	LONGITUDE	VELOCITY	SEATS
3Bytes	4Bytes	4Bytes	4Bytes	1Byte	1Byte

Fig. 4 Sensor node frame format

is DATE where date and month will take one byte each and year will take two bytes to store. Latitude and Longitude fields will take 4 bytes each where there can be a precision of up to 6 decimal points. Velocity and seat takes 1 byte to store the value.

5.2 Cloud Module

The average speed of the bus will be calculated for every one minute and will be updated in the server. Estimated timings are calculated and stored in the cloud server. The details like number of seat availability, estimated arrival at source and destination, location are updated to the server. MQTT and CoAP are the two prominent IoT data protocols which are commonly used for smaller devices. These protocols are open standards and are also befitted to constrained environments than HTTP protocol. MQTT is a many to many communication protocol for transferring messages among multiple different clients through a central broker which gives lots of flexibility in communication patterns. CoAP is a one-to-one protocol for transferring state information between client and server, which is purely based on states but not on events. MQTT protocol is a lightweight publish/subscribe messaging protocol, broker based protocol and can run on low power and on bandwidth. So considering these characteristics, MQTT protocol is ideal for use even in constrained environments.

6 Results and Discussion

The proposed system is implemented by considering two bus modules, a cloud module and user module (mobile app). For cloud module, IBM Watson IoT platform has been considered. Both bus modules will send the data to the cloud. The message format obtained from the node is sent to the cloud as shown in Fig. 5.

The message format obtained is converted into hexadecimal format. 11 represent the starting frame, 01 represents first node, then 1E17E2 represents the Indian time. 1E17E2 represents the date. A665BE and 4956380 represents the latitude and longitude of the node. Then speed of the bus is 0 kmph and 32 represent the number of seats in the bus.

```
TIME: 14364 HEX TIME: E244
Date: 3012018 HEX Date: 1E17E2
Lat: 10905022///A665BE
Long: 76899200///4956380
Speed in kmph: 0///0
No of seats: 50///32
Message Format:
11010E24041E0107E200A665BE049563800032
```

Fig. 5 Message format obtained from the hardware

	_id	▼	deviceId	▼	payload	▼	deviceType	▼	format
☐ 📋	383c36810d4d44986b...		IoTNode1		[{ "sof": 17, "nodeid": ...		NodeMCU		json
☐ 📋	383c36810d4d44986b...		IoTNode2		[{ "sof": 17, "nod				json
☐ 📋	383c36810d4d44986b...		IoTNode2		[{ "sof": 17, "nod				json
☐ 📋	4253bb9ab1a89e077a...		IoTNode2		[{ "sof": 17, "nod				json
☐ 📋	4253bb9ab1a89e077a...		IoTNode2		[{ "sof": 17, "nod				json
☐ 📋	4253bb9ab1a89e077a...		IoTNode2		[{ "sof": 17, "nod				json
☐ 📋	4253bb9ab1a89e077a...		IoTNode2		[{ "sof": 17, "nod				json
☐ 📋	64604c5144f1a2a578b...		IoTNode1		[{ "sof": 17, "nod				json
☐ 📋	6ef89142a1fc8b3f7ed...		IoTNode1		[{ "sof": 17, "nod				json
☐ 📋	6ef89142a1fc8b3f7ed...		IoTNode2		[{ "sof": 17, "nodeid": ...		NodeMCU		json

[
{
"sof": 17,
"nodeid": 1,
"gpstime": {
"hr": 19,
"min": 38,
"sec": 10
},
"gpsdate": {
"day": 5,
"month": 2,
"year": 2018
},
"lat": 10.905267,
"long": 76.899215,
"speed": 0,
"seats": 50
} ...
]

Fig. 6 IoT cloud database

The message format has been sent in Jason format. A flow-based programming tool named Node-RED has been used for wiring together the hardware devices, API's and various online services. This is an open-source graphical development tool mainly used for event-driven applications such as IoT. The Jason string obtained is converted into the Jason object and using the split function, the Jason object is split into different fields and stored in the database. The messages are stored in the form of documents in the IBM Cloudant NoSQL DB database. Figure 6 shows the IoT cloud database for two nodes. The sample mobile application is shown in Fig. 7. Whenever the passengers enter the details, respective buses with respect to the location are shown in the mobile application. In this paper, an intelligent public transport system is implemented using IoT.

Fig. 7 Android app with passenger information and bus details

7 Conclusion

A prototype has been developed using an Internet of Things (IoT) based intelligent real-time passenger information system for Smart City applications. NodeMCU with GPS module is used as vehicle node and data is sent to cloud using MQTT protocol. The prototype is tested for its operation in a campus environment. The system is validated for displaying the current location of bus, estimated arrival time and vacant seats in the Mobile App developed.

References

1. Jay Lohokare, F., Reshul Dani, S., Sumedh Sontakke, T.: Scalable tracking system for public buses using IoT technologies. In: International Conference on Emerging Trends and Innovation in ICT, vol. 01, pp. 104–109 (2017)
2. Ogbonna, J.C., Nwokorie, C.E., Odii, J.N.: Real-time public vehicle mobile tracking system using global positioning system technology. Int. J. Comput. Trends Technol. **38**(2), 71–80 (2016)
3. Lin Kang, F., Stefan Poslad, S., Weidong Wang, T.: A public transport bus as a flexible mobile smart environment sensing platform for IoT. In: 12th IEEE International Conference on Intelligent Environments, vol. 01, pp. 1–8 (2016)
4. Priti Shende, F., Pratik Bhosale, S., Shahnawaz Khan, T.: Bus tracking and transportation safety using internet of things. Int. Res J. Eng. Technol. **03**, 1–4 (2016)
5. Durga Bhavani, D., Ravi Kiran, S.C.V.S.L.S.: Implementation of smart bus tracking system using Wi-Fi. Int. J. Innov. Res. Sci. Eng. Technol. **6**, 1–7 (2017)

6. Thiyagarajan Manihatty, F., Bojan Umamaheswaran, S., Raman Kumar, T.: An internet of things based intelligent transportation system. In: International Conference on Vehicular Electronics and Safety, vol. 01, pp. 174–179 (2014)
7. Datta, S.K., Da Costa, R.P.F., Harri, J.: Integrating connected vehicles in internet of things ecosystems: challenges and solutions. In: IEEE 17th International Symposium on a World of Wireless, Mobile and Multimedia Networks, vol. 01, pp. 1–6 (2016)
8. Yogavani, D., Krishna Prakash, N.: Implementation of wireless sensor network based multi-core embedded system for smart city. Int. J. Control Theory Appl. **10**(2), 119–123 (2017)
9. Katsikeas, S.: Lightweight and secure industrial IoT communications via the MQ teleme-try transport protocol. In: IEEE Symposium on Computers and Communications, vol. 01, pp. 1193–1200 (2017)
10. Giridhar Maji, F., Sharmistha Mandal S., Soumya Sen, T.: A conceptual model to implement smart bus system using Internet of Things (IoT). In: 32nd International Conference on Computers and Their Applications, vol. 1, pp. 1–7 (2017)
11. Munoz, R.: Integration of IoT, transport SDN, and Edge/Cloud computing for dynamic distribution of IoT analytics and efficient use of network resources. J. Light Wave Technol. **36**(7), 1420–1428 (2018)
12. Sharad S., Bagavathi Sivakumar P., Anantha Narayanan V.: The smart bus for a smart city—a real-time implementation. In: IEEE International Conference on Advanced Networks and Telecommunications Systems (ANTS), vol. 1, pp. 1–6 (2016)
13. Ramachandra, M.: IoT solution for scheduling in transport network. In: International Conference on Internet of Things and Applications (IOTA), vol. 01, pp. 327–331 (2016)
14. Mondal, A., Bhattacharjee, S.: A reliable, multi-path, connection oriented and independent transport protocol for IoT networks. In: 9th International Conference on Communication Systems and Networks (COMSNETS), vol. 01, pp. 590–591 (2017)
15. Lavanya R., Sheela Sobana Rani S., Gayathri R.: A smart information system for public transportation using IoT. Int. J. Recent Trends Eng. Res. **03**, 222–230 (2017)
16. Sridevi K., Jeevitha A., Kavitha K.: Smart bus tracking and management system using IoT. Int. J. Res. Appl. Sci. Eng. Technol. **1**, 372–374 (2017)
17. Lakshmi Sadanandan, F., Nithin S.: A smart transportation system facilitating on-demand bus and route allocation. In: International Conference on Advances in Computing, Communications and Informatics, vol. 1, pp. 1000–1003 (2017)
18. Sun D., Luo, H., Fu, L.: Predicting bus arrival time on the basis of global positioning system data. Transp. Res. Rec. J. Transp. Res. Board **2034**, 62–72 (2007)
19. Zambada, J., Quintero, R.R., Isijara, R.: An IoT based scholar bus monitoring system. In: IEEE First International Smart Cities Conference (ISC2), vol. 01, pp. 1–6 (2015)
20. Dlodlo, N.: The internet of things in transport management in South Africa. In: International Conference on Emerging Trends in Networks and Computer Communications, vol. 1, pp. 19–26 (2015)
21. Mehmet Altinkaya, F., Metin Zontul, S.: Urban bus arrival time prediction: a review of computational models. Int. J. Recent Technol. Eng. **2**(4), 1–6 (2013)
22. Krystian Birr, F., Kazimierz Jamroz, S., Wojciech Kustra, T.: Travel time of public transport vehicles estimation. In: 17th Meeting of the EURO Working Group on Transportation, vol. 3, pp. 359–365 (2014)

Continuous Monitoring and Detection of Epileptic Seizures Using Wearable Device

Darshan Mehta, Tanay Deshmukh, Yokesh Babu Sundaresan and P. Kumaresan

1 Introduction

Epilepsy is a serious, potentially fatal neurological disorder which often requires the immediate attention of a caregiver. It affects nearly 50 million people worldwide. It is the fourth most common neurological disorder. It was responsible for 125,000 deaths in 2015 [1]. People who have experienced a seizure before have a 40–50% chance of experiencing another [2]. Common symptoms of epileptic seizures include sudden fluctuations in heart rate and involuntary muscular contractions (seizures). They are visible as brief to extended periods of prolonged shaking. Seizures tend to occur periodically and have no immediate underlying cause. The actual causes of seizures may range from brain tumors, stroke to birth defects, and genetic anomalies.

There are six different types of seizures: tonic-clonic, tonic, clonic, myoclonic, absence, and atonic seizures. All of them involve loss of consciousness and most of them involve contraction of the limbs followed by the arching of the back. This phase is called the tonic phase and may last up to 30 s. After the shaking has stopped the person enters a state called postictal state which may last up to 30 min. It is only after this phase that normal level of consciousness can return. The person may still have symptoms like tiredness, headaches, and nausea.

D. Mehta (✉) · T. Deshmukh · Y. B. Sundaresan · P. Kumaresan
School of Computing Science and Engineering, VIT University,, Vellore 632014,
Tamil Nadu, India
e-mail: mehtadarshan@icloud.com

T. Deshmukh
e-mail: tanay.deshmukh2013@vitalum.ac.in

Y. B. Sundaresan
e-mail: yokeshbabu.s@vit.ac.in

P. Kumaresan
e-mail: pkumaresan@vit.ac.in

© Springer Nature Singapore Pte Ltd. 2019
S. Tiwari et al. (eds.), *Smart Innovations in Communication and Computational Sciences*, Advances in Intelligent Systems and Computing 851,
https://doi.org/10.1007/978-981-13-2414-7_8

Since the onset of seizures is sudden and its consequences can be severe, it is risky to leave the patient alone. For instance, the sudden onset of a seizure during driving may lead to accidents and its occurrence during sleeping hours could prove fatal if no immediate, proper attention is provided by a care-giver or a doctor. Epilepsy causes social stigma and put limits on the person's daily activities. It can have an effect on their educational development. The stigma also affects the patient's family members and caregivers. Thus it is important to relieve some of the anxiety and tension surrounds the patient's safety. This will ensure that the patient regains confidence in them and is able to go about their daily lives performing activities unhindered.

We propose a wireless electronic monitoring system that can accurately detect the onset of seizures at an early stage. We will measure two critical health indicators–brain and cardiac bio-potential signal in the form of EEG and ECG using body sensor networks. We will also utilize accelerometer sensors as an additional measure to detect if the person has fallen down and is incapacitated.

2   Literature Survey

In the S.M.A.R.T Belt [3] proposed by the students of Rice University, the idea revolved around a wearable device embedded with heart rate and EDA sensor which used a simple threshold to determine whether or not a seizure has occurred and signaled a Bluetooth message in the case of a seizure. The simple thresholding algorithm was quite basic and failed to understand the pattern in the data which a machine learning algorithm could have. Moreover, the Bluetooth signal imposed a restriction on the location of the caregiver to a few hundred meters. Nigam et al. in his paper on neural network based approach to epilepsy [4] noted that parametric methods such as Fast Fourier transform (FFT) and Autoregressive (AR) are not suitable for analysis of the non-stationary EEG signal and proposed that FFT be performed on smaller chunks of time. Shoeb et al. in his paper [5] presented patient specific classifier which used SVM coupled with a Gaussian kernel which showed promising accuracy but had a high latency (>5 s) especially when using both ECG and EEG sensors for detection. Patel et al. [6] leveraged accelerometer and electromyography (EMG) which probed over extended periods of time to detect epileptic seizures. They used PCA and C4.5 Decision tree to isolate daily-life activities from seizures and were able to identify seizures ~98% cases. But the issue here was that epileptic seizures do not always reflect as outward muscle activity and hence EMG was not an effective choice. Also, this model was patient-independent, which was based on the hypothesis that all people would react in the same way physically during the seizure. In a study by Subasi et al. [7], they used a hybrid model for seizure detection. They combined traditional optimization techniques such as Genetic Algorithms (GA) and Particle Swarm Optimization (PSO) with Support Vector Machines (SVMs) for classification. They found that the PSO-based approach was much better at determining the optimal hyper parameters and significantly improved classification accuracy. They were able to achieve an accuracy of 99.38%. They used discrete wavelet transform to preprocess the data before feeding it into the classifier. Their usage of optimization techniques to determine hyper parameters was a novel one. While all of the above methods usually suffered from low performance, they were however, non-invasive in nature. Meanwhile, in-parallel, there have been research on invasive epilepsy detection mechanisms where intracranial EEG is used. This can be seen in the paper by Gardner et al. [8] where one class SVM is trained over one second window of the training data. Unlike non-invasive procedures, this method does not suffer from the noise and interference of signals from facial muscles, and hence is able to achieve extremely high levels of performance. However, intracranial EEG being an invasive procedure, requires an extremely risky surgery, which makes puts it down the preference list of people.

3 Proposed System

3.1 Introduction

There are three basic components to our system—training the classification model, acquiring the real-time data from the sensors, and using the model and real-time data to predict and detect seizures.

We hypothesize that the EEG waveforms before and during an epileptic seizure vary significantly from the baseline, and that this can be used to train a classification model using an existing dataset [9]. This model, once trained, will be used to classify real-time data and predict the onset of a seizure. This phase of the project involves acquiring the dataset, training it, and verifying its accuracy. Once we are satisfied with the results, we will move onto the hardware implementation phase.

One key component of our training methodology was the use of under sampling. Since the data was so heavily skewed towards negative samples, we selectively removed parts of our training data from the dataset which did not have any positive samples. This helped in the learning rate and prevented over-fitting of the model.

Another important decision point was whether to train a patient- specific classifier or a multi-patient general classifier. The EEG and seizure waveforms are different between individuals so it may not be possible to build a general model. Thus we trained on both a per-patient basis and on multiple patients at once. The scores for the multi-patient classifier dropped significantly, which confirmed our hypothesis.

In the next phase, we will port the model to an embedded system such as a microcontroller. The microcontroller will interface with the sensor control unit to obtain real-time information about the patient. This data will be fed into the model which will then classify the status of the patient as baseline, or about to enter a state of seizure. We will then attach the GSM and GPS modules to send the information to a doctor's or caregiver's mobile number via SMS.

3.2 System Model

The system will consist of one microcontroller (Fig. 1) which will handle processing the data from the sensors. A level shifter will convert and step up the voltage from the EEG output in order to amplify the signals. The location of the patient will be obtained from the GPS module which is connected to the microcontroller in TX mode. The microcontroller will then transmit the coded signal to the GSM module so that an alert message can be sent to the caregiver's phone number. An LCD will provide graphical output of the system. In the final product all these components will be packed into a compact package which can be worn on the arm or the waist.

3.3 Algorithm

1. First the EDF data [9] is read from the disk and split into 2-second epochs.
2. The spectral components are extracted by taking the FFT (Fast Fourier Transform) of every epoch (Fig. 3).
3. Only frequencies between 0.5–25 Hz are selected (Fig. 2).
4. The frequencies are then split into $M = 8$ filter banks.
5. The entire EEG signal is thus parsed by taking a sliding window of size 3 epochs (or 6 s).
6. The data is shuffled and split into training and validation sets by performing k-fold cross-validation. We chose the value of k as 10. Same training and validation data is provided to all the classifiers for a justified comparison.
7. This preprocessed data is fed into the SVM classifier and trained.
8. Once the training is complete, the model will be serialized and saved to disk
9. The serialized model will be flashed to the microcontroller.
10. Inputs from the EEG sensor will undergo the same pre- processing outlined in steps 2–4, then is fed into the model which will then classify the input as seizure or non-seizure state.
11. If a seizure is detected, the system will instruct the GSM module to send an SMS containing the location of the patient taken from the GPS module, if available.
12. Replace the classifier to Naïve Bayes and Random Forest and repeat the steps 7 through 11 again.

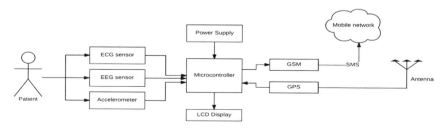

Fig. 1 Proposed system

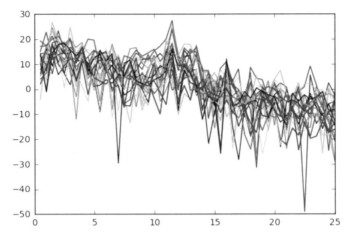

Fig. 2 Visualization of the spectral components of the EEG signal

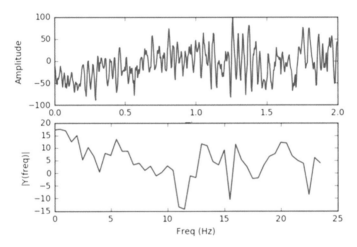

Fig. 3 Sample EEG signal in time (top) and frequency (domain)

3.4 Metrics

We will be using the following performance metrics to evaluate the accuracy of the model.

Accuracy. The accuracy score of a binary classification model is defined as

$$Accuracy = \frac{(TP + TN)}{n},\tag{1}$$

where n is the number of samples, TP refers to number of true positives, TN refers to number of false positives. For our purposes, accuracy score is not a very good measure of the models performance. This is because the negative samples (without seizure) far outnumber positive samples (with seizure). The proportion of positive samples present in the dataset is well below 0.01%. Thus a high accuracy reading is expected.

Precision. Precision refers to the number of positively detected samples to the total number of detected samples. It measures how many of the selected items are relevant, i.e., out of the total number of seizures detected, what proportion were actually seizures.

$$Precision = \frac{TP}{TP + FP} \tag{2}$$

Recall. Recall is the measure of the number of relevant items selected from the total dataset. It is the number of positively detected samples divided by the total number of positive samples. This measures how many seizures in total were identified. This is an important metric because a low recall score means that more seizures were unidentified.

$$Recall = \frac{TP}{TP + FN} \tag{3}$$

F1 Score. F1-Score is a measure of the classifier's accuracy based on the precision and recall of the model. It is a better indicator of the performance than just the raw accuracy score.

$$F_1 = \frac{precision \times recall}{precision + recall} \tag{4}$$

4   Experimental Results and Analysis

Single Patient Classifier. As we can see in Fig. 4 Naïve Bayes gave good accuracy and recall, but precision scores were very low. This was due to the fact that the dataset had a very large number of false positives, meaning that many non-seizure states were identified as seizures. SVM's performed the best, with 99.95% accuracy and recall of 91.23%. Random Forests outperformed SVMs slightly in precision scores, but as explained in the previous section, recall and F1 scores are more important metrics.

Multiple Patient Classifier. As seen in the multi-patient case (Fig. 4), the classifiers performed poorly. Even the SVM had only a 45% recall rate, which meant that over half of the seizures would be unidentified which was not acceptable. Interestingly, Naïve Bayes had the highest recall of all. However, this is statistically significant,

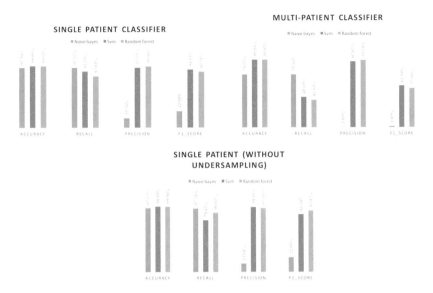

Fig. 4 Comparison graph for single patient classifier (top-left), multi-patient classifier (top-right) and single patient classifier without under sampling (bottom)

since as explained in the previous section, Naïve Bayes tends to misclassify and create lots of false positives, as evidenced by its abysmal 1% precision rate.

Single Patient Classifier (without under sampling). For verification of our results we ran the algorithm without under sampling on a single patient. The recall rate and the F1 scores (as seen in Fig. 4) were lower than with under sampling, thus affirming our decision to utilize under sampling while training the model.

5   Conclusion

Our experiments led us to the conclusion that Support Vector Machines are the optimal algorithm to use while building the classification model. Furthermore, it would be necessary to train the model uniquely on each patient's readings first beforehand. A single generalized model has shown poor results. This would involve monitoring the patient and obtaining at least 24 h of EEG data so that the training model could be built on a powerful external device and then deployed to the microcontroller since the onboard processor is power-efficient and hence not capable of performing compute intensive training. The heart-rate sensor and the accelerometer sensor when used as a part of the prediction system serve as an added benefit to monitor the patient and send an alert in case of a seizure. Also, Long Short Term Memory networks (LSTMs)—a variant of Recurrent Neural Networks (RNNs) excel at remembering values for long or short periods of time, whereas RNNs only remember up to the last few inputs [10].

Thus they lend themselves well to signal processing and have been used with great success at audio processing tasks such as speech recognition [11] and might hence prove to be a better model for prediction.

References

1. Wang, H., Naghavi, M., Allen, C., Barber, R.M., Carter, A., Casey, D.C., Charlson, F.J., Chen, A.Z., Coates, M.M., Coggeshall, M., Dandona, L.: Global, regional, and national life expectancy, all-cause mortality, and cause-specific mortality for 249 causes of death, 1980–2015: a systematic analysis for the Global Burden of Disease Study 2015. The Lancet **388**(10053), 1459–1544 (2016)
2. Berg, A.T.: Risk of recurrence after a first unprovoked seizure. Epilepsia **49**(s1), 13–18 (2008)
3. Leng, E., Mongia, M., Park, C., Varughese, T., Wu, A.: SMART Belt: A Low-cost Seizure Detection Device. Rice University. http://aac-rerc.psu.edu/wordpressmu/RESNA-SDC/2013/06/13/smart-belt-a-low-cost-seizure-detection-device-rice-university/
4. Nigam, V.P., Graupe, D.: A neural-network-based detection of epilepsy. Neurol. Res. **26**(1), 55–60 (2004)
5. Shoeb, A., Edwards, H., Connolly, J., Bourgeois, B., Treves, S.T., Guttag, J.: Patient-specific seizure onset detection. Epilepsy Behav. **5**(4), 483–498 (2004)
6. Patel, S., Mancinelli, C., Dalton, A., Patritti, B., Pang, T., Schachter, S., Bonato, P.: Detecting epileptic seizures using wearable sensors. In: 2009 IEEE 35th Annual Northeast Bioengineering Conference, pp. 1–2. IEEE (2009)
7. Subasi, A., Kevric, J., Canbaz, M.A.: Epileptic seizure detection using hybrid machine learning methods. Neural Comput. Appl. 1–9 (2017)
8. Gardner, A.B., Krieger, A.M., Vachtsevanos, G., Litt, B.: One-class novelty detection for seizure analysis from intracranial EEG. J. Mach. Learn. Res. 7, 1025–1044 (2006)
9. Shoeb, A.: CHB-MIT Scalp EEG Database. MIT. https://www.physionet.org/pn6/chbmit/
10. Gers, F.A., Schmidhuber, J., Cummins, F.: Learning to Forget: Continual Prediction with LSTM (1999)
11. Graves, A., Mohamed, A.R., Hinton, G.: Speech recognition with deep recurrent neural networks. In: 2013 IEEE International Conference on Acoustics, Speech and Signal Processing (ICASSP), pp. 6645–6649. IEEE (2013)
12. Shoeb AH, Guttag JV (2010) Application of machine learning to epileptic seizure detection. In: Proceedings of the 27th International Conference on Machine Learning (ICML-10), pp. 975–982
13. Song, Y., Crowcroft, J., Zhang, J.: Automatic epileptic seizure detection in EEGs based on optimized sample entropy and extreme learning machine. J. Neurosci. Methods **210**(2), 132–146 (2012)
14. Goldberger, A.L., Amaral, L.A., Glass, L., Hausdorff, J.M., Ivanov, P.C., Mark, R.G., Mietus, J.E., Moody, G.B., Peng, C.K., Stanley, H.E.: Physiobank, physiotoolkit, and physionet. Circulation **101**(23), e215–e220. http://circ.ahajournals.org/cgi/content/full/101/23/e215 (2000)
15. Singh, M., Kaur, S.: Epilepsy detection using EEG: an overview. Int. J. Inf. Technol. Knowl. Manag. **6**(1), 3–5 (2012)

Brain Tumor Detection Using Cuckoo Search Algorithm and Histogram Thresholding for MR Images

Sudeshna Bhakat and Sivagami Periannan

1 Introduction

Brain tumor, which is a standout among the most widely recognized brain sicknesses, has influenced and crushed many lives. Measurements still shows low survival rate of brain tumor patients. To battle this, as of late, specialists are utilizing multidisciplinary approach including learning in drug, arithmetic, and software engineering to better comprehend the illness and discover more powerful treatment strategies. In medical picture preparing brain tumor discovery is one of the testing errands, since brain pictures are entangled and tumors can be dissected just by expert doctors [1]. Brain tumor shows up when one sort of cell changes from its typical qualities and develops and increases in an unusual way. The not-so-ordinary development of cells inside the brain or skull, which can be malignant or non-carcinogenic. The tumor is one type of growth. Disease begins from cells which comprises of different tissues. Organs of the body are made up of these tissues [2]. MRI can give itemized data about the illness and can distinguish numerous pathological conditions, giving a precise determination. Segmentation of brain from tumor process can be more precise if best optimization calculation is used. Segmentation can be done using different kinds of algorithms.

Nature-Inspired Optimization Algorithm—In the area of medical science an intelligent model can help in cost reduction and efficiency. Different algorithms which are inspired from nature are applied in medical treatment to build such a model and these algorithms have proved to be of large efficiency and accuracy. The different types of algorithms inspired from nature are cuckoo search technique, ant

S. Bhakat (✉) · S. Periannan
VIT University, Chennai, Chennai, India
e-mail: sudeshna.bhakat2015@vitalum.ac.in

S. Periannan
e-mail: msivagami@vit.ac.in

© Springer Nature Singapore Pte Ltd. 2019
S. Tiwari et al. (eds.), *Smart Innovations in Communication and Computational Sciences*, Advances in Intelligent Systems and Computing 851,
https://doi.org/10.1007/978-981-13-2414-7_9

colony optimization, particle swarm optimization, firefly algorithm, etc. Here Cuckoo Search technique is applied for brain tumor detection

Cuckoo Search Optimization—Cuckoo Search (CS) is one of the most recent nature propelled metaheuristic calculations, created by Yang and Deb in 2009. CS depends on parasitism behavior of few cuckoo species. In addition to this, the calculation is upgraded by the supposed Levy flights, as opposed to by basic isotropic random strolls. Late reviews demonstrate that CS is possibly much more productive than PSO and hereditary calculations. Cuckoos are intriguing winged animals, not just as a result of the lovely sounds they can make, but also as a result of their forceful proliferation methodology. A few animal types, for example, eggs of Guira and ani cuckoos are laid in settles of other birds, in spite of the fact that they might evacuate eggs of others to expand the incubating likelihood of their eggs possessed. Certain types of animals are there who connect with this type of parasitism and use the settles of other host winged animals to lay their eggs (frequently different species) [3]. Cuckoo search follows three ideal rules: (1) The settles that have superior characteristics of eggs will persist to the upcoming generation. (2) The host nests accessible is settled and the likelihood in which the host finds the egg laid by the cuckoo is $P_a(0, 1)$. (3) Each Cuckoo winged creature lays one egg at any given moment and relinquishes its egg in an arbitrarily chosen nest. Finding of the eggs is worked on some arrangement of most exceedingly unwanted nests and are dumped from further computations [2–8]. Eggs to be found is worked on arrangement of most exceedingly unwanted nests and are dumped from further computations [2–8]. Cuckoo's each egg shows a solution which is new. The new arrangement might be a superior arrangement and then it supplants the arrangement which is old. As per run of CS, one home can have just a single egg, eggs' position is chosen by brood's position. In this manner the arrangement of CS is shown by vector's position of the settle [8]. By applying these rules updation of cuckoos position is done by the following equation.

$$x(t+1)i = x(t\ i) + \alpha \oplus \text{Levy}(\lambda) \quad i = 1, 2, \ldots, n, \tag{1}$$

where $t+1$ for ith cuckoo is place for stage $xi(t+1)$ and $xi(t)$ denotes place in stage t, $\oplus$ denotes multiplication factor and α is the step length parameter. The first equation is a condition of random walk whose area next just depends on the present area and the probability of transition. Differently Levy Distribution is defined for Levy Flight (Fig. 1).

$$\text{Levy} \sim u = t\lambda, \quad (1 < \lambda \leq 3) \tag{2}$$

Histogram Thresholding—The histogram is a graphical portrayal that demonstrates the heaviness of the picture pixel values. As it were, the histogram is a chart appearing the quantity of shading qualities in the picture. The picture turns out to be more particular in the picture when the histogram values are analyzed [9]. The histogram of a bimodal picture comprises of two peak focuses; the bottom most point (valley) between these two peaks indicates the threshold point [10]. In some data, for

Fig. 1 Cuckoo search
algorithm

Cuckoo Search via Lévy Flights

---

begin
  Objective function $f(\mathbf{x})$, $\mathbf{x} = (x_1, ..., x_d)^T$
  Generate initial population of
    n host nests $\mathbf{x}_i$ $(i = 1, 2, ..., n)$
  while *(t <MaxGeneration) or (stop criterion)*
    Get a cuckoo randomly by Lévy flights
      evaluate its quality/fitness F_i
    Choose a nest among n (say, j) randomly
    if $(F_i > F_j)$.
      replace j by the new solution;
  end
  A fraction (p_a) *of worse nests*
    are abandoned and new ones are built;
  Keep the best solutions
    (or nests with quality solutions);
  Rank the solutions and find the current best
  end while
  Postprocess results and visualization
end

---

example, tops and valleys from the histogram of the smoothed picture are separated and examined on the basis of the threshold point.

2 Background Details and Related Work

Ben George et al. [2] described about the cuckoo search optimization for brain tumor detection using Markov Random Field technique with CSA to perform segmentation on brain tumor image.

Rajabioun [4] described about Cuckoo Search Optimization using k-means clustering for finding the best habitat of cuckoos.

A seed choice calculation utilizing cuckoo search improvement is proposed by Preetha et al. [5]. Entropy thresholding technique is used to generate the seeds to perform best segmentation of the image.

Adnan and Razzaque [6] shown a comparative study between CSA and PSO. How CSA performs better than PSO. CSA converges to global optimum in faster and efficient way than PSO.

Zhao et al. [8]. described a picture segmentation innovation on 2D greatest entropy and cuckoo search. The proposed technique was contrasted with alternate strategies like Genetic Algorithm and Particle Swarm Optimization.

Preetha and Suresh [11] described about brain tumor detection using fuzzy c-means clustering and then performed classification with the help of SVM.

Ulku and Camurcu [9] shown histogram equalization can be utilized to bring out the portion that can be tumor using K-closest neighbor and support vector machine as classification algorithm.

Suresh and Lal [10] proposed a technique of comparison between CS, ABC, DPSO, PSO and a variation of CS calculation known as CS Mantegna for discovering ideal multilevel edge values for fragmenting satellite pictures.

Bouaziz et al. [12] described utilization of a CS calculation for differentiation of gray scale unique finger impression pictures.

3   Proposed Approach

A novel approach to extract the tumor from brain image with the help of Cuckoo Search Algorithm and histogram thresholding is demonstrated. Here, histogram of the input image is calculated and the peak values are randomly taken as input in the cuckoo search algorithm then histogram thresholding is done on the optimized image to extract the tumor (Figs. 2 and 3).

1. 2D Histogram peak values as parameters: Plot the histogram to know and investigate the intensity circulation of the pixels. The intensity values at the peak will be passed as variables in the Cuckoo Search technique.
2. Cuckoo Search Optimization Algorithm: Optimization manages discovering best estimations of factors so that the estimations of a fitness function winds up noticeably ideal. This search begins with an irregular arrangement, and after that neighborhood arrangement is moved from present position and acknowledges the new arrangement only if there is a chance that it enhances the fitness function (Otsus method 2016) [10]. Here $\mu 0$, $\mu 1$, ..., μm is denoted as mean intensity value of the pixels at position 0, 1, ..., m, pi is the probability of intensity of pixel values where i ranges from 0 to 255, N represents total of distinct intensity levels and μT represents global mean. t is threshold value. This value is the first random cuckoo chosen from the host nest and then its fitness, f(i) is calculated. Now choose another random cuckoo from the nest and calculate its fitness value, fj. If fi < fj then, i is replaced by j. Hence a certain amount of worst nests are abandoned and settlement of new ones are done. From the best outcome obtain recent best. Thus this threshold is optimized by cuckoo search technique for initial pixel selection. For utilization of thresholding based division procedure, it is required to apply the right edge values with a specific end goal to accomplish legitimate divisions, if not then obtained results are poor. The basic steps of CSA are as follows.

Step 1   MR image of brain, say I is taken as input.
Step 2   Now introduce n initial solutions where each solution or nest contains 1 egg.
Step 3   Now generate a random cuckoo or solution by Levy flight.

Fig. 2 System architecture

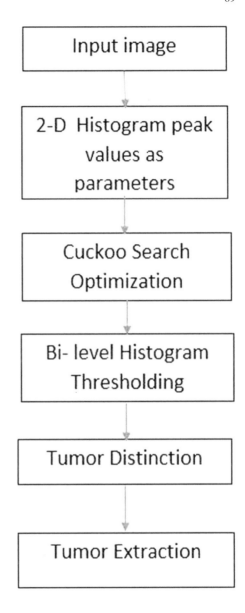

Step 4   Perform segmentation on the input image with the generated cuckoo and calculate its fitness F_i.

Step 5   Now randomly choose a nest or solution from the host (j).

Step 6   Now perform segmentation on the input image and evaluate its fitness with the new chosen nest F_j.

Step 7   Next rank the segmented output and based on its value of fitness find the current best.

Fig. 3 Objective function

$$f(t) = \sum_{i=0}^{m} \sigma_i^2$$

where

$$\sigma_0^2 = w_0(\mu_0 - \mu_T)^2$$

$$\sigma_1^2 = w_1(\mu_1 - \mu_T)^2$$

$$\cdots$$

$$\sigma_m^2 = w_m(\mu_m - \mu_T)^2$$

where

$$\mu_0 = \frac{\left(\sum_{i=t_0}^{t_1-1} ip_i\right)}{w_0}$$

$$\mu_1 = \frac{\left(\sum_{i=t_1}^{t_1-1} ip_i\right)}{w_1}$$

$$\cdots$$

$$\mu_m = \frac{\left(\sum_{i=t_m}^{N-1} ip_i\right)}{w_m}$$

Step 8   Now perform comparison between the current nest position and best position which was obtained last and retain best nest as current which has better value of fitness.

Step 9   Now discard a fraction of worst nest if less than pa.

Step 10   Sustain the nest having solution of high quality.

Step 11   Now, again perform ranking to obtain the current best.

Step 12   Determine if end is met and if it has then go to next step otherwise go to step 4.

Step 13   Get the segmented image as output.

3. Bi-level Histogram Thresholding: Here bi-level histogram thresholding is applied where the background and the foreground is separated by a certain threshold value. Histogram thresholding algorithm can be described as—The MRI picture of the mind is separated into two equivalent parts around its focal axis and the histogram of each part drawn. Then infectious side of the brain is detected. The threshold of the histograms is computed in light of an examination strategy made among the two histograms. Segmentation is done utilizing the limit point for both the parts.

4. Tumor Distinction: Distinct the tumor by focusing on the black and white portions of the binary image obtained from histogram thresholding. All associated segments that have less than P pixels from the binary image obtained is districted and a new binary image is generated. Structures that are lighter than their envi-

Table 1 Confusion matrix

Total	Predicted No	Predicted Yes
Actual No	TN	FP
Actual Yes	FN	TP

ronment and that are associated with the picture boundary are suppressed. The closed areas with certain color is filled to distinct the tumor and other areas.

5. Tumor Extraction: For extracting the tumor, boundary of regions of image is calculated. The picture with the boundary overlaid is shown. The region number alongside each boundary is added. Centroids for associated parts in the picture is figured. The centroids on the image are plotted. Centers and diameters of the regions with centroid obtained in the last step are calculated. All the other properties like perimeter, area and roundness of the region are calculated. The color filled region is shown. The edge of the region using sobel method is computed. Sobel filtered image mask is shown. Borders are removed and final segmented image is printed. Erosion is performed on the image with structuring element. The final binary mask is obtained. The mask with the original gray image is superimposed. This superimposed image is the tumor.

4 Result Analysis

The calculation utilizing CS and histogram thresholding is tested on 30 standard pictures using MATLAB R2015b and different types of tumor images are downloaded from the website of Harvard Medical School where they have provided MR images for different slices of brain from top view (http://www.med.harvard.edu/AANLIB/), from the website of brainweb which provides custom MR Simulator to generate ground truth image (http://brainweb.bic.mni.mcgill.ca/brainweb/) and from the website of radiopaedia (https://radiopaedia.org/). After division the outcomes are tried with the ground truth images and certain parameters are examined. The parameters are accuracy, sensitivity, specificity [2],

$$\text{Accuracy} = (TP + TN)/(TP + TN + FP + FN)$$
$$\text{Sensitivity} = TP/(TP + FN)$$
$$\text{Specificity} = TN/(TN + FN)$$

Here, TP, FN, TN, and FP are True Positive, False Negative, True Negative, and False Positive, respectively. These values are obtained from confusion matrix. It is a matrix that is used to depict the execution of a model on an arrangement of test data for which the genuine esteems and false esteems are known. It is generalized as (Table 1).

Table 2 Comparison with other techniques

Comparison of methods	Accuracy (%)
BTD-MRF-CSA [2]	98
BTD-FCM-SVM [11]	94.68
Proposed technique	98.03

These parameters processed produces the accuracy of 98.03%, sensitivity of percent 99.06, which implies this calculation can section the whitish tumor pictures skillfully. 72.29% of specificity calculation has come, which demonstrates execution of the technique to show the nonappearance of the whitish tumor region. The technique proposed by Ben George et al. [2] had shown accuracy of 98%, sensitivity of 98.8% and specificity of 93.3%. The technique proposed in this paper is compared with the current strategy exhibited by Ben George et al. [2] and Preetha and Suresh [11]. The first technique describes about the cuckoo search optimization for brain tumor detection (BTD) using Markov Random Field technique with CSA to perform segmentation on brain tumor image in two steps. The image is labeled using Markov Random Field and then it is segmented using CS Algorithm (BTD-MRF-CSA) [2]. The second technique describes about brain tumor detection using Fuzzy C-means clustering. The precision of tumor extraction is distinguished by Support Vector Machine classifier. The classification is done on three types glioma, glioblastoma, and meningioma (BTD-FCM-SVM) [11]. It is shown in Table 2.

The properties like radius, roundness and executsion time are also calculated based on the size of the tumor. Some properties of different size of tumors are presented. Figure 4 (I, II, III, IV) are downloaded from weblinks [13, 14]. Different stages of the algorithm is demonstrated below.

Figure 5 (downloaded from [13]) is the input of original image which is fed into the algorithm and the image of ground truth which is used for comparing the technique's output. Figure 6 (generated through MR Simulator from [15]) is the region showing tumor and other white pixel areas and the boundary region showing the tumor. Figure 7 is the boundary region superimposed on the original image.

5 Conclusions

The system proposed in this paper was effective in extracting the tumor divide, it has given an exact division of the limit of the tumor, alongside right visual area of the tumor with the assistance of a boundary. First the histogram peak values are passed as parameters to the CS algorithm then bi-level histogram thresholding is performed by partitioning the brain into left and right brain. Then some morphological operations are done for enhancing the optimized image. Here the tumor is actually there or not is also analyzed by the tumor's size. The proposed calculation can be connected with certain adjustment for location of lungs tumor.

Image	Radius of Tumor	Roundness of Tumor	Execution Time	Present or Absent
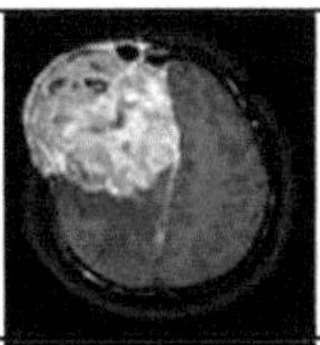	53.06	0.22	10.58	Present
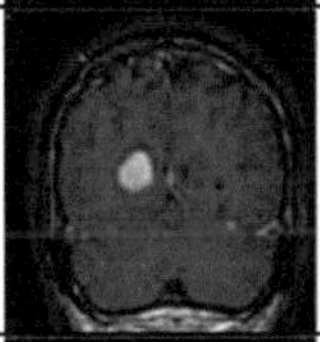	12.79	0.87	10.73	Present
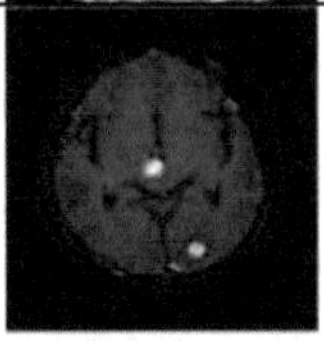	Upper-7.38 Lower-5.53	Upper-0.88 Lower-0.99	13.22	Present
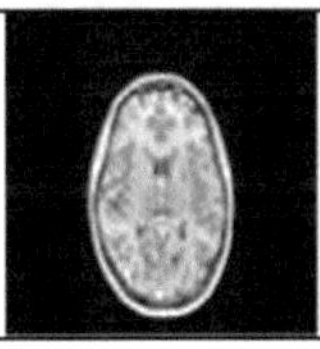	—	—	8.056	Absent

Fig. 4 Properties of different types of tumor

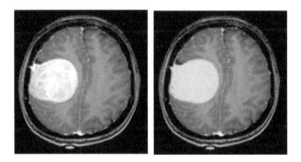

Fig. 5 Original image and ground truth image

Fig. 6 First showing tumor area and second showing tumor boundary

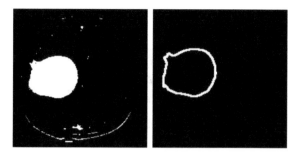

Fig. 7 Tumor boundary superimposed on original image

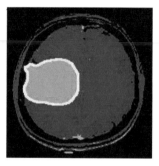

Acknowledgements The calculation is tested on 30 standard images using MATLAB R2015b and different types of tumor images are downloaded from the website of Harvard Medical School where they have provided MR images for different slices of brain from top view [13], from the website of radiopaedia [14] and from the website of brainweb which provides custom MR Simulator to generate ground truth image [15]. The images provided are free to use.

References

1. Noureen, E., Kamrul Hassan, Md.: Brain tumor detection using histogram thresholding to get the threshold point. IOSR J. Electr. Electron. Eng. **9**(5) (2014)
2. Ben George, E., Jeba Rosline, G., Gnana Rajesh, D.: Brain tumor segmentation using cuckoo search optimization for magnetic resonance images. In: Proceedings of the 8th IEEE GCC Conference and Exhibition, Muscat, Oman (2015)
3. Yang, X.S., Dev, S.: Cuckoo search: recent advances and applications. Neural Comput. Appl. **21**(1), 169–174 (2014)
4. Rajabioun, R.: Cuckoo Optimization Algorithm. Elsevier (2011)
5. Preetha, M.M.S.J., Padma Suresh, L., John Bosco, M.: Cuckoo search based threshold optimization for initial seed selection in seeded region growing. Int. J. Comput. Eng. Res. (8) (2014)
6. Adnan, MdA., Razzaque, M.A.: A comparative study of particle swarm optimization and cuckoo search techniques through problem-specific distance function. In: International Conference of Information and Communication Technology (ICoICT). IEEE (2013)
7. Tirpude, N., Welekar, R.: Automated detection and extraction of brain tumor from MRI images. Int. J. Comput. Appl. **77**(4) (2013)

8. Zhao, W., Ye, Z., Wang, M., Ma, L., Liu, W.: An image threholding approach based on cuckoo search algorithm and 2D maximum entropy. In: 8th IEEE International Conference (2015)
9. Ulku, E.E., Camurcu, A.Y.: Computer aided brain tumor detection with histogram equalization and morphological image processing techniques. In: 2013 International Conference on Electronics, Computer and Computation (ICECCO). IEEE (2014)
10. Suresh, S., Lal, S.: An Efficient Cuckoo Search Algorithm Based Multilevel Thresholdng for Segmentation of Satellite Images Using Different Objective Functions. Elsevier (2016)
11. Preetha, R, Suresh, G.R.: Performance analysis of fuzzy C means algorithm in automated detection of brain tumor. In: 2014 World Congress on Computing and Communication Technologies (WCCCT). IEEE (2014)
12. Bouaziz, A., Draa, A., Chikhi, S.: A Cuckoo Search Algorithm for Fingerprint Image Contrast Enhancement. IEEE (2014)
13. http://www.med.harvard.edu/AANLIB
14. https://radiopaedia.org/
15. http://brainweb.bic.mni.mcgill.ca/brainweb/

eMDPM: Efficient Multidimensional Pattern Matching Algorithm for GPU

Supragya Raj, Siddha Prabhu Chodnekar, T. Harish and Harini Sriraman

1 Introduction

With advent of GPUs, the execution time of image processing applications has greatly reduced. Pattern searching is an integral part of many applications. Any improvement in the performance of pattern matching of multidimensional image will greatly improve the performance of many GPU based applications. Pattern matching can be broadly classified into exact bit pattern matching and approximate bit pattern matching algorithms. The main bottleneck with respect to improving the performance pattern searching is reduction operation. In recent times, Shift-Or operation is used to improve the efficiency of reduction operation based pattern matching algorithms.

Parallel pattern matching involves tasks identification, communication identification, task agglomeration and mapping of tasks to processors. GPU is a SIMT architecture that can be utilized effectively for pattern matching algorithms. The matching of individual patterns can happen in the stream processors available inside the stream multiprocessors. The communication involves the reduction process. Agglomeration and mapping depends on the number of processors available. If we consider n to be the number of primitive tasks and p to be the number of processors, then $\frac{n}{p}$ tasks can be assigned per processor.

S. Raj (✉) · S. P. Chodnekar · T. Harish · H. Sriraman
School of Computing Science and Engineering, Vellore Institute of Technology,
Chennai Campus, Chennai 600127, India
e-mail: supragyaraj@gmail.com

S. P. Chodnekar
e-mail: sprabhu.chodnekar@vit.ac.in

T. Harish
e-mail: harini.s@vit.ac.in

© Springer Nature Singapore Pte Ltd. 2019
S. Tiwari et al. (eds.), *Smart Innovations in Communication and Computational Sciences*, Advances in Intelligent Systems and Computing 851,
https://doi.org/10.1007/978-981-13-2414-7_10

2 Related Work

Pattern matching involves the process of identifying all patterns in a given text. There are many practical applications in which pattern matching algorithms are applied. The pattern matching algorithms can be classified with respect to the dimensions of the data. Further classification of these algorithms includes the type of pattern matching namely, exact pattern matching and approximate pattern matching.

2.1 Exact and Approximate Pattern Matching

Exact pattern matching provides a solution by viewing each text position to be a possible pattern start. For exact pattern matching algorithm, the complexity is O(mn) where m * n is the dimension of the text matrix. Many improvements exist in the literature for exact pattern matching algorithms. Usage of linear automata [1] for the purpose is the most popular technique. This reduces the complexity of the pattern matching algorithm to be O(m + n). Good-suffix heuristic algorithm proposed in [2] provides an improved performance of O(n) with constant space requirement. In [3], bad character rule strategy is generalized and improves the performance of the pattern matching algorithm with time complexity of O(n/m). Simple pattern searching algorithm proposed in [4] improves the average time complexity to be O(m + n).

Approximate parallel pattern matching algorithms works on the principle that a error can be tolerated by the pattern match. Say for example, if a particular pattern match algorithm with pattern size of m can tolerate n error bits, approximate pattern matching algorithm will find out all the matches with m to m − n matching bits. This matching will be particularly useful for bigger images. Parallelizing of the above categories of algorithms can be performed in different ways. This is highlighted in Sect. 2.2.

2.2 Parallel Bit Pattern Matching Algorithms

Approximate and exact pattern matching algorithms falls into the category of finite and nondeterministic automata. Parallelizing these algorithms for better performance has evolved and there are many perspectives to this in the literature. Parallel algorithms that consider bit pattern mapping for parallelization is described in this section.

Many traditional algorithms like Shift-Or [5] and Shift-And [6] algorithms form the basis of almost all Deterministic Finite State automata based pattern matching algorithms. These finite state automata are stored in form of bits. Hence, for pattern text of size n, there are n bits to store all the prefixes of match in the text. Algorithms such as KMP are a good example of this. Recent work in the domain of pattern matching like Bit Parallel Wide window [7] and Backward Nondeterministic Matching [8] reduces the n bits requirements in favour of computation cost. These research directions are based on the idea that matches in strings of large alphabet size is often sparse and not all prefixes are needed to be stored for the matching.

Reverse factor [9], linear DAWG matching [10], BSDM using Q-Grams and shift-xor [11] and Two way shift-or [10] are manifestations of the popular Shift-Or algorithm itself and apply the domain-specific constraints to efficiently parallelize the Shift-Or algorithm.

3 eMDPM Algorithm

The eMDMP algorithm rests heavily on the developments done in [1] and takes over much of these concepts to build upon in this paper. The nomenclature and a few of the terms that are used in the algorithm are as follows:

- M are used for denoting masks. Masks are state storing variables and can take any of 2^n combination of bits. They are 0 wherever there is a match of certain character and 1 otherwise. This mask is then reversed. For instance in string "AABAB", the mask M_A is 10100 (swapped from 00101, the positions of A).
- R is the residue vector, bit of which denote the matching of all prefixes of pattern P on text T. (See p. 3 [1] for further reading).
- $\langle a, b \rangle$ denotes the binary operator [dot] used in SO exact matching. (see p. 5 [1] for further reading).
- $a|b$ denotes the Shift-Or operations done in cuShiftOr. This takes advantage of multiple threads and computing S matrix. The S matrix denotes the residual values extracted from R. X is the final result vector.

3.1 Proposed Algorithm

Input: A text T of $n_1 \times n_2$ and pattern P of $m_1 \times m_2$
Output: A matrix X of size $(n_1 - m_1 + 1) \times (n_2 - m_2 + 1)$ indicating matching positions

begin
　　Initialise..Masks.M_1, M_2, ... M_n.*for.the.test.image.T.*
　　for $i := 1$ **to** n_1-m_1-1 **do**
　　　　for $j := 0$ **to** m_1-1 **do**
　　　　　　$A_0 := \langle 0, 1^m \rangle$
　　　　　　$A_k := \langle 1, M_k \rangle for.all.k.in.0 \leq k \leq n1$
　　　　　　for $k := 0$ **to** n_2 **do** *inparallel*
　　　　　　　　$R_k := sum(A_0, A_1, ... A_k)$ Inclusive Scan
　　　　　　end
　　　　　　$\langle u_k, S_{j,k} \rangle := R_k..for.all..0 \leq k \leq n2$
　　　　end
　　　　for $j := m2$ **to** $n2$ **do**
　　　　　　$y_j := S_{i,j} | S_{i+1,j} | ... | S_{i+m1-1,j}$ Shift-Or
　　　　　　$X[i][j$-m_2+$1] := y_j[m_2]$
　　　　end
　　end
end
end

An example run of the above-given algorithm can be summarized using the following Fig. 1.

3.2   *Tile-Based Data Decomposition*

Tile-based data decomposition is used for improved performance in the GPU. Tiles of size $w * w$ are shared among multiple SPs(Stream Processors) for processing.

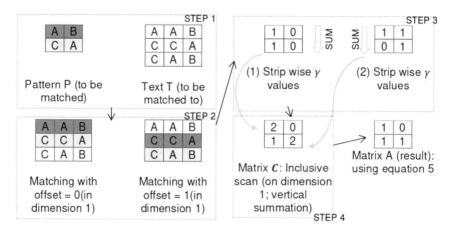

Fig. 1 Matching operation in two dimensional systems. For higher dimensions, multiple summations occur to gather matrix C; first on dimension 1 to give intermediate set C1; on them, summation on dimension 2 is done etc. The process continues till all dimensions are summed upon to give final C, hence A

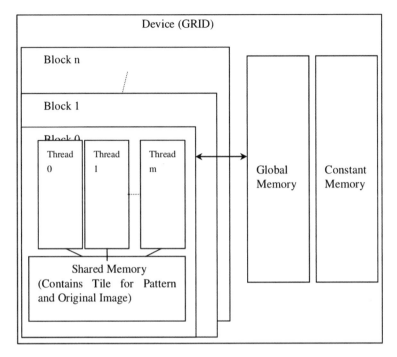

Fig. 2 Storing of image and pattern as tile in shared memory of GPU

This improves the performance by reducing the redundant reads otherwise needed in the system. The parallelism is performed by considering the data are stored as a single linear array of values. The single linear array values are then loaded to the shared memory of the SP to be used by multiple threads of the same block. To synchronize the operation, after each tile read, a $\_syncthreads()$ function is called for performing a barrier like wait till all the threads in the same block have completed loading input to the tile. In the considered algorithm, this tile-based data load has greatly improved the performance of the system. The data update is done in the shared memory is shown in the following Fig. 2.

4 Result and Analysis

For multidimensional images, pattern matching algorithms have plenty of possibilities to parallelize. This helps in utilizing the GPU to the maximum and improves the performance of the algorithm. The main overhead with respect to parallelism is the increased communication overhead. But in the proposed implementation, the latency of this access is reduced by utilizing the shared memory. For the implemented system, the Compute to Global Memory Access (CGMA ratio is 3:1. The compute for global memory access includes the following list of tasks.

- Shift-Or operation to perform pattern matching
- Or operation to update row-wise results of the tile
- Reduction with multiple rows.

In the above case, row-wise operations are performed within thread of the tile. This can also be performed with respect to columns.

Let m be the number of dimensions of the image considered, n represents the total image size, k represents the size of the pattern and w represents the dimension of the tile. The time complexity of the proposed algorithm is given by:

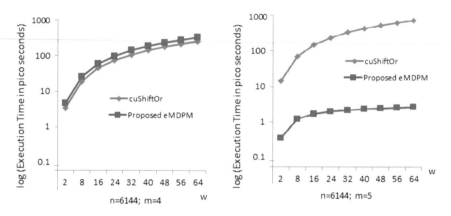

Fig. 3 Comparison between cuShiftOr and eMDPM on dimensions 4 and 5

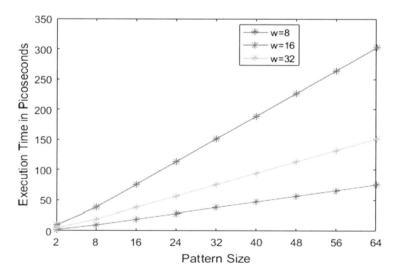

Fig. 4 eMDPM: Execution time versus tile size

$$Execution time = \left(m \times \frac{\log_w(k)}{w \log(n)} \right)$$

Comparison between the performances of the proposed algorithm with cuShiftOr is shown in Fig. 3. The performance of the proposed algorithm is better when the pattern size is less than 64 and the tile size matches the pattern size.

In the Fig. 3, the performance of eMDPM algorithm for different values of w is shown. For this analysis, n is considered to be 6144 and m is set to be 3. For other values also, the algorithm can be analysed. Here for a sample m and n, the analysis is shown (Fig. 4).

5   Conclusion and Future Work

In this paper, an efficient multidimensional parallel pattern matching algorithm is proposed and analysed. The performance of the proposed algorithm with multidimensional images is much improved compared to existing algorithms. The implementation of the proposed algorithm is done using GPU. The proposed algorithm works well with GPU architecture by increasing CGMA ratio. The proposed algorithm makes use of tiles for data caching which improves the performance of pattern matching in GPU based architectures. The proposed algorithm is an exact bit pattern matching one. This can be extended for approximate pattern matching in the future. The role of warp size on the efficiency of the pattern matching algorithm for bigger sized patterns can also be explored in the future work.

References

1. Mitani, Y., Ino, F., Hagihara, K.: Parallelizing exact and approximate string matching via inclusive scan on a GPU. IEEE Trans. Parallel Distrib. Syst. **28**(7), 1989–2002 (2017)
2. Agrawal, J., Diao, Y., Gyllstrom, D., Immerman, N.: Efficient pattern matching over event streams. In: Proceedings of the 2008 ACM SIGMOD International Conference on Management of Data, pp. 147–160. ACM (2008)
3. Cantone, D., Cristofaro, S., Faro, S.: A space-efficient implementation of the good-suffix heuristic. In: Stringology, pp. 63–75 (2010)
4. Sahli, M., Shibuya, T.: Max-shift BM and max-shift horspool: practical fast exact string matching algorithms. J. Inf. Process. **20**(2), 419–425 (2012)
5. Oladunjoye, J.A., Afolabi, A.O., Olabiyisi, S.O., Moses, T.: A comparative analysis of pattern matching algorithm using bit-parallelism technique. IUP J. Inf. Technol. **13**(4), 20–36 (2017)
6. Fredriksson, K.: Shift-or string matching with super-alphabets. Inf. Process. Lett. **87**(4), 201–204 (2003)
7. Goyal, R., Billa, S.L., Cavium Inc.: Generating a non-deterministic finite automata (NFA) graph for regular expression patterns with advanced features. U.S. Patent 9,563,399 (2017)
8. Kida, T., Takeda, M., Shinohara, A., Arikawa, S.: Shift-And approach to pattern matching in LZW compressed text. In: Annual Symposium on Combinatorial Pattern Matching, pp. 1–13. Springer, Berlin, Heidelberg (1999)

9. Park, B., Won, C.S.: Fast binary matching for edge histogram descriptor. In: The 18th IEEE International Symposium Consumer Electronics (ISCE 2014), pp. 1–2 (2014)
10. Hirvola, T., Tarhio, J.: Bit-parallel approximate matching of circular strings with k mismatches. J. Exp. Algorithmics (JEA) **22**, 1–5 (2017)
11. Pfaffe, P., Tillmann, M., Lutteropp, S., Scheirle, B., Zerr, K.: Parallel string matching. In: European Conference on Parallel Processing, pp. 187–198. Springer, Cham (2016)
12. Faro, S.: Evaluation and improvement of fast algorithms for exact matching on genome sequences. In: International Conference on Algorithms for Computational Biology, pp. 145–157. Springer, Cham (2016)
13. Alfred, V.: Algorithms for finding patterns in strings. In: Algorithms and Complexity, p. 255, pp. 1–5 (2014)

Computer Vision-Based Fruit Disease Detection and Classification

Abhay Agarwal, Adrija Sarkar and Ashwani Kumar Dubey

1 Introduction

Since many years, fruit diseases have been detected and identified by eye observation by a human. The practice is still prevalent in developing nations. But it is a costly and very time-consuming process. Therefore, an automatic fruit disease detection and classification method have been proposed as a visual inspection of apples is not reliable.

Partial automation is done to identify the diseases as is important to improve yield and quality. It is also crucial for research papers. The three diseases identified are apple scab, black rot canker, and core rot.

2 Literature Survey

Apples (*Malus pumila*) have been used for research in this paper as it is commercially the most important fruit in the temperate regions [1]. After banana, orange and grape, it is the fourth most widely produced fruit worldwide. It is the ideal fruit with high nutrition components as compared to other fleshy fruits like bananas and mangoes [2] which have a high fat content. From the farmers who grow apples in their orchards

A. Agarwal · A. Sarkar (✉) · A. K. Dubey
Department of Electronics and Communication Engineering, Amity School
of Engineering and Technology, Amity University Uttar Pradesh, Noida, Uttar Pradesh, India
e-mail: 19adrijasarkar@gmail.com

A. Agarwal
e-mail: abhay.agarwal4001@gmail.com

A. K. Dubey
e-mail: dubey1ak@gmail.com

© Springer Nature Singapore Pte Ltd. 2019   105
S. Tiwari et al. (eds.), *Smart Innovations in Communication and Computational Sciences*, Advances in Intelligent Systems and Computing 851,
https://doi.org/10.1007/978-981-13-2414-7_11

to the factory workers producing apple juice, cedar, jellies, etc., apple quality is a major concern for them. In India, almost 60% of the apples produced are of inferior quality and they are only grown in the Himalayan regions [3]. Hence, demand always surpasses supply. Therefore, apples produced must be free from diseases and of good quality and have a longer shelf life.

There are many types of diseases that affect apples, but this paper deals with the three major diseases that are described as follows:

Apple Scab—It is caused by a fungus called *Venturia inaequalis* that affects both leaves and fruits. In young apples, the scab infections usually start out as olive green to brown spots. As the infected regions enlarge, they harden and become dull black or gray-brown as well as corky [4]. Scabby apples are mostly unfit for consumption as they become inedible. Apples infected with scab are deformed and may crack open.

Black Rot Canker—This is also a fungal disease caused by the fungus *Botryosphaeria obtusa* that affects leaves, fruits, and branches. It results in purple, black, or dark brown spots on leaves, fruit rot and branch cankers [5].

Core Rot—There are two types of Core Rots [5]: dry and wet. The former develops in the orchard while the latter is a postharvest disease. In dry core rot, this is also called, moldy core, brown lesions form around the core which is dry and spongy. In wet core rot, the core of the apple appears to be light brown in color.

3   Methodology

The proposed methodology is based on nondestructive approach. In which, apple images are first preprocessed so that unwanted noise would be removed [6]. For segmentation, various techniques are there like Otsu, wavelet-based multivariate statistical [7] but k-means clustering [8] is used to detect the defective portion of the apple fruit because it is best compatible with the database and the classifier. The cluster selection was done after the segmentation. The feature extraction was done on the selected cluster by using the GLCM algorithm [9, 10]. The affected region of the apple was also calculated. Then, the type of disease which was present in the apple was detected by using SVM Classifier [11]. The accuracy of the classifier was also determined [12]. The accuracy of the classifier achieved is 98.387% as compared to the previous method used. Finally, the proposed method was implemented on a GUI, which is very user friendly (Fig. 1).

3.1   Preprocessing

There are several types of noises which are present in the image which results in unwanted pixels. The contrast enhancement is applied to remove the noises and

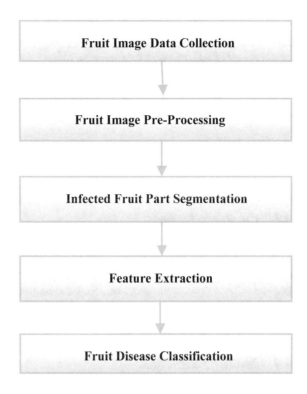

sharpen the image. Images were reconstructed and shadows were removed. PSNR value was also calculated to check filtering accuracy [13].

3.2 Segmentation

The purpose of segmentation is to make an image into something that has more meaning and is easier to analyze. K-means clustering is used to segment the images. The input RGB image of the fruit is converted to L*a*b color space [14]. Here, L is the luminosity layer and a, b are chromaticity layer. The squared error function [15] is the objective function of the K-means clustering algorithm, which is shown in (1):

$$J = \sum_{\substack{i=1 \\ j=1}}^{\substack{n \\ k}} \left(\left\| \alpha_i - \beta_j \right\| \right)^2 = 1 \tag{1}$$

where $\left(\left\| \boldsymbol{\alpha}_i - \boldsymbol{\beta}_j \right\| \right)$ is the Euclidean distance measured between a point $\boldsymbol{\alpha}_i$ and a centroid $\boldsymbol{\beta}_j$ iterated over all k points in the ith cluster, for all n clusters [8]. Recalculation of the center of new cluster [8] is given by (2):

$$V_i = (1/C_i) \sum_{j=1}^{C_i} x_i \tag{2}$$

where 'C_i' is the number of data points in the cluster. Finally, recalculation of the distance was done between new cluster centers and each of the data points obtained.

3.3 Cluster Selection

After K-mean clustering, features would be extracted but there was a need to select one cluster out of 'n' clusters obtained, on which the feature extraction and classification will be processed. One can adjust the number of clusters according to his need and in this proposed system we have selected 3 clusters.

3.4 Feature Extraction

Feature extraction plays a crucial role in the whole proposed system, as on the basis of these features, the classifier detects the diseases and hence the robustness of the system is achieved. For feature extraction, GLCM algorithm is used. It is the most classical second-order statistical method for texture analysis [16]. The calculations of texture feature use the contents of GLCM for measurement of the variations in intensity value at the pixel of interest.

3.5 Classification

For the classification, SVM has been used because it is among the widely known ways for pattern recognition and image classification. It is a supervised machine learning based algorithm [17]. As we know in supervised learning, the data is labeled hence clustering of data becomes the necessity therefore k-means was used. SVM builds the optimum separating hyperplanes. In SVM classification, kernel trick is additionally enforced. The kernel ways are primarily instance-based learners [18]. Rather than learning some fixed set of parameters with respect to the features of their inputs, they remember the ith training sample (x_i, v_j) and learn a corresponding weight w_i for it. For instance, a kernelized binary classifier calculates a weighted sum of similarities as given in (3):

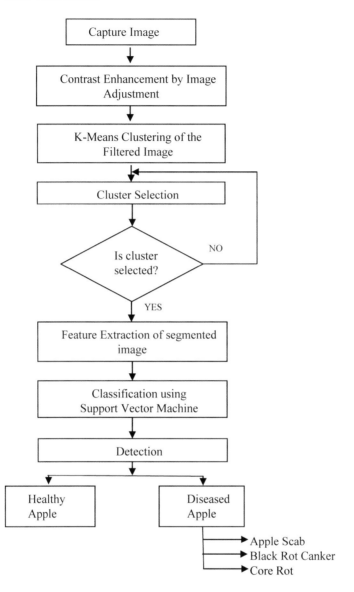

Fig. 2 Flowchart of the proposed method

$$y = sgn \sum_{i=1}^{n} w_i y_i k(x_i, x') \qquad (3)$$

where $y \in \{-1, +1\}$ is the kernelized binary classifier's foretold label for the unlabeled input x' (Fig. 2).

4   Simulation and Results

4.1   K-Means Clustering

The data set that we have collected contain healthy apple and unhealthy apple. There were three different categories of unhealthy apple: apple scab, black rot canker, core rot. Table 1 represents the four samples, one of every class that we have collected, i.e., three classes of diseased apple and one class of healthy apple. It is also representing the contrast-enhanced image of the input image and the three clusters that are obtained by K-means clustering. Now selection of the cluster, for feature extraction, should be done very carefully as in Img 2, the segmented defective region is shown in cluster 2 whereas in Img 3 it is in cluster 3.

4.2   Extracted Features

Table 2 shows the GLCM features that are calculated on the segmented image of the four different categories that we have selected. GLCM algorithms are more efficient

Table 1   Segmentation results

S. no.	Input image	Contrast enhancement	Cluster 1	Cluster 2	Cluster 3
Img 1					
Img 2					
Img 3					
Img 4					

Table 2 Extracted features

S. No.	Feature extracted	Img 1	Img 2	Img 3	Img 4
1	Mean	27.67	42.8668	104.14	58.4
2	Standard deviation	53.551	80.501	113.798	91.3512
3	Entropy	3.03156	3.06184	5.01455	4.42163
4	Root mean square	5.87778	8.05669	11.7112	9.83271
5	Variance	1356.2	5383.15	9304.28	4412.06
6	Smoothness	1	1	1	1
7	Kurtosis	6.58175	3.93649	1.16644	2.87302
8	Skewness	2.06552	1.60718	0.306686	1.25958
9	Inverse difference moment	255	255	255	255
10	Contrast	0.176348	0.163358	0.206771	0.30516
11	Correlation	0.950932	0.984677	0.990708	0.97797
12	Energy	0.512879	0.565796	0.401274	0.43862
13	Homogeneity	0.97061	0.981689	0.974948	0.96908

and more reliable which thus help the classifier to produce better results. These features are extracted on one of the three clusters obtained. Hence the selection of cluster plays the vital role here.

4.3 Classification

Table 3 shows the classifier results of the four segmented images which we got after K-means clustering. One image out of all four classes is selected, for which the classifier is trained. The affected region was also calculated and the highest accuracy of the classifier was achieved 98.3871% for healthy apple and black rot canker. The accuracy of the classifier was determined after 500 successful iterations

Table 3 Classification of the diseases

Input image	Disease detected (if any)	Affected region	Accuracy of the classifier
Img 1	Healthy apple	None	98.3871
Img 2	Apple scab	23.4937	95.1613
Img 3	Black rot canker	43.8557	98.3871
Img 4	Core rot	26.2115	96.7742

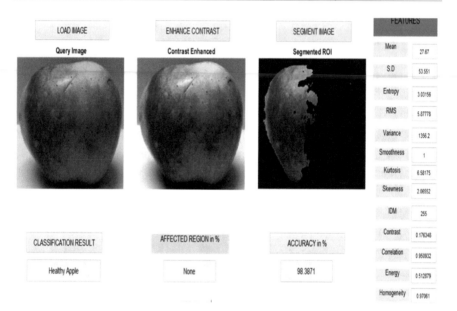

Fig. 3 GUI with results (Img 1)

4.4 GUI Implementation

The complete proposed method was implemented on a GUI as shown in Fig. 3. The GUI was made very user friendly, anyone can operate it without any in-depth knowledge of MATLAB. In Fig. 4, black rot canker diseased apple is shown which is implemented in GUI and Fig. 3 shows healthy apple (Figs. 5 and 6).

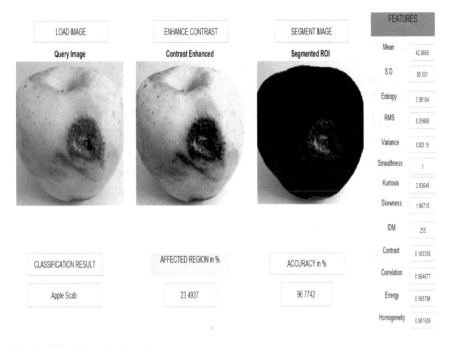

Fig. 4 GUI with results (Img 2)

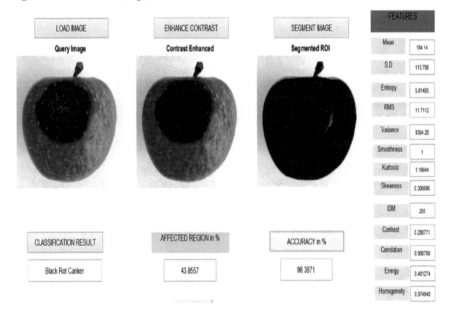

Fig. 5 GUI with results (Img 3)

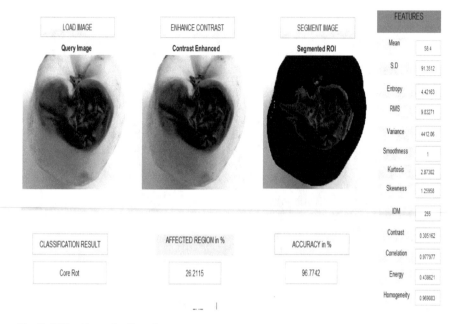

Fig. 6 GUI with results (Img 4)

5 Conclusions

As there are many techniques by which we can detect the disease of an apple. But the challenge is to make a system which is cheaper, more efficient, and easily accessible by anyone. Hence, in this paper, an automatic apple fruit disease detection method is proposed and validated through image processing techniques. There are mainly three steps—Image Segmentation, Feature Extraction, and Classification. The accuracy of the classifier is up to 98.387%. Thus, making it more accurate as compared with the existing classifiers , the complete algorithm runs pretty faster and does not hang even when tested for thousands of samples simultaneously. The SVM classifier correctly detected the diseases namely Apple scab, black rot canker, and core rot as well as healthy apples. The GUI makes it user friendly and anyone can operate it without any significant errors. In the future, it will be extended to the several other fruits like mangoes, pears, etc., and also on the internal deficiencies of apple and mango fruit.

References

1. Samajpati, B.J., Degadwala, S.D.: Hybrid approach for apple fruit diseases detection and classification using random forest classifier. In: 2016 International Conference on Communication and Signal Processing (ICCSP), Melmaruvathur, pp. 1015–1019 (2016)

2. Ashok A., Vinod, D.S.: A comparative study of feature extraction methods in defect classification of mangoes using neural network. In: 2016 Second International Conference on Cognitive Computing and Information Processing (CCIP), Mysore, pp. 1–6 (2016)

3. India Rank in Agriculture. http://www.apeda.gov.in/apedawebsite/six_head_product/FFV.htm

4. Bhargava, K., Kashyap A., Gonsalves, T.A.: Wireless sensor network based advisory system for Apple Scab prevention. In: 2014 Twentieth National Conference on Communications (NCC), Kanpur, pp. 1–6 (2014)

5. Awate, A., Deshmankar, D., Amrutkar, G., Bagul, U., Sonavane, S.: Fruit disease detection using color, texture analysis and ANN. In: 2015 International Conference on Green Computing and Internet of Things (ICGCIoT), Noida, pp. 970–975 (2015)

6. Dubey, A.K., Jaffery, Z.A.: Maximally stable extremal region marking-based railway track surface defect sensing. IEEE Sens. J. **16**(24), 9047–9052 (2016)

7. Jaffery, Z.A., Dubey, A.K.: Scope and prospects of non-invasive visual inspection systems for industrial applications. Indian J. Sci. Technol. **9**(4) (2016)

8. Masazhar, A.N.I., Kamal, M.M.: Digital image processing technique for palm oil leaf disease detection using multiclass SVM classifier. In: 2017 IEEE 4th International Conference on Smart Instrumentation, Measurement and Application (ICSIMA), Putrajaya, Malaysia, pp. 1–6 (2017)

9. Jaffery, Z.A., Dubey, A.K.: Architecture of noninvasive real time visual monitoring system for dial type measuring instrument. IEEE Sens. J. **13**(4), 1236–1244 (2013)

10. Mukherjee, G., Chatterjee, A., Tudu, B.: Study on the potential of combined GLCM features towards medicinal plant classification. In: 2016 2nd International Conference on Control, Instrumentation, Energy and Communication (CIEC), Kolkata, pp. 98–102 (2016)

11. Mizushima, A., Lu, R.: An image segmentation method for apple sorting and grading using support vector machine and Otsu's method. Comput. Electron. Agric. **94**, 29–37 (2013)

12. Li, D., Shen, M., Li, D., Yu, X.: Green apple recognition method based on the combination of texture and shape features. In: 2017 IEEE International Conference on Mechatronics and Automation (ICMA), Takamatsu, pp. 264–269 (2017)

13. Öziç, M.Ü., Özşen, S.: A new model to determine asymmetry coefficients on MR images using PSNR and SSIM. In: 2017 International Artificial Intelligence and Data Processing Symposium (IDAP), Malatya, pp. 1–6 (2017)

14. Madina, E., El Maliani, A.D., El Hassouni, M., Alaoui, S.O.: Study of magnitude and extended relative phase information for color texture retrieval in L*a*b* color space. In: 2016 International Conference on Wireless Networks and Mobile Communications (WINCOM), Fez, pp. 127–132 (2016)

15. Shanker, R., Singh, R., Bhattacharya, M.: Segmentation of tumor and edema based on K-mean clustering and hierarchical centroid shape descriptor. In: 2017 IEEE International Conference on Bioinformatics and Biomedicine (BIBM), Kansas City, MO, pp. 1105–1109 (2017)

16. Saroja, G.A.S., Sulochana, C.H.: Texture analysis of non-uniform images using GLCM. In: 2013 IEEE Conference on Information and Communication Technologies, JeJu Island, pp. 1319–1322 (2013)

17. Venkatesan, C., Karthigaikumar, P., Paul, A., Satheeskumaran, S., Kumar, R.: ECG signal preprocessing and SVM classifier-based abnormality detection in remote healthcare applications. IEEE Access **6**, 9767–9773 (2018)

18. Deng, Y., Xu, T., Li, Y., Feng, P., Jiang, Y.: Icing thickness prediction of overhead transmission lines base on combined kernel function SVM. In: 2017 IEEE Conference on Energy Internet and Energy System Integration (EI2), Beijing, pp. 1–4 (2017)

Automated Brain Tumor Detection Using Discriminative Clustering Based MRI Segmentation

Abhilash Panda, Tusar Kanti Mishra
and Vishnu Ganesh Phaniharam

1 Introduction

The human body is made up of numerous kinds of cells and every cell is having a unique set of functionalities. Eminently, all the cells in the body performs cell division to sustain their normal growth, this keeps human body in good condition [1]. During this process of this growth, rarely cells drain their capability leading to abnormal growth of the cells. The mass of tissue caused due to abnormal growth of extra cells is called a "Tumor". Tumor can be classified as three types, i.e., Benign tumor, Pre-malignant, and Malignant tumors. Benign tumor is a tumor where it does not enlarge in a brusque way and it does not cause any health issues. Moles are the best examples for the benign tumor [2]. Pre-malignant tumor can be considered as pre-cancer stage; if it is not medicated can lead to cancer. Malignant tumor is a tumor with abnormal growth of cells and can be considered as cancer stage, often leading to the death of a person.

In recent years, one of the main reasons for rising level of morality i.e., reduction in the lifespan of the adolescents is suffering from the brain tumor disease. It has been observed from contemporary studies that the enumeration of the people vanishing due to the brain tumor has risen to 300% [2]. So, brain tumor detection is an urgent need for today's smart world as radiation growing into a dangerous case of causing sudden deaths of birds. Brain tumor detection has lot of applications such as clinical

A. Panda
Gandhi Engineering College, Bhubaneswar, Odisha, India
e-mail: inc.abhilash@gmail.com

T. K. Mishra · V. G. Phaniharam (✉)
Anil Neerukonda Institute of Technology and Sciences, Vizag, India
e-mail: vishnuganesh93@gmail.com

T. K. Mishra
e-mail: tusar.k.mishra@gmail.com

© Springer Nature Singapore Pte Ltd. 2019
S. Tiwari et al. (eds.), *Smart Innovations in Communication and Computational Sciences*, Advances in Intelligent Systems and Computing 851,
https://doi.org/10.1007/978-981-13-2414-7_12

outlining and medication devising [3]. Brain tumor detection faces a lot challenges as tumors in the brain are size variant, shape variant, shape variant, location variant and image intensity variant.

1.1 Scopes and Challenges

Automated Brain Tumor Detection has evolved into an eminent research oriented field in the area of image processing and medicinal pathological research. Image processing is a sub discipline of signal processing, where useful information is being extracted by applying computing algorithms. Brain MRI segmentation plays a prominent role in brain tumor detection, where segmentation can be explained as dividing an image in different parts based on feature homogeneity. MRI technique is used in bio-medical analysis to visualize the cryptic elements of the imbricated regions of brain. When compared to the computed tomography, MRI is a far better technique to uncover the heterogeneities in the tissues. This quality makes MRI a special choice for segmentation of brain internal tissues and detecting a likely tumor in the brain. MRI aligns nuclear magnetization by using magnetic field whereas CT uses ionizing radiation. The scanner detects the change in the alignment of magnetization caused by radio frequencies and this signal is processed further to get internal structure details of the brain. Analyzing MR images manually is very difficult because of its large amount of data in it. So, automatic segmentation has become mandatory for brain tumor detection and clinical diagnostics. MRI segmentation has its own challenges namely acquisition noise, partial volume effect and bias effect. Acquisition noise arises because ideal conditions are never expected. Bias field, also known as intensity heterogeneity arises due to non-magnetic field thereby increasing the heterogeneity. Partial volume effect arises when different types of tissues occur in single voxel. To overcome these challenges, this method uses superpixel level zoning and discriminative clustering (Fig. 1).

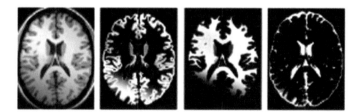

Fig. 1 Challenges faced during brain MRI segmentation

1.2 Paper Organization

The remaining section of this paper is systemized as follows. A literature review of the various previously proposed brain tumor detection algorithms is described in Sect. 2. Our automated brain tumor detection method which is proposed in this paper is explained in Sect. 3. Simulation and experimental results of the proposed method are being discussed in Sect. 4. In the last Sect. 5, conclusion and future work of our paper is being presented.

2 Literature Survey

This section of paper consists of review on previously contemplated automatic brain tumor detection algorithms. Radhakrishna Achanta et al. proposed a new brain MRI segmentation method which uses superpixels generated by clustering in the image plane space and color in five-dimensional space [4]. The simplicity and efficiency of this algorithm [4] are its advantages over advanced methods. Rajeev Ratan et al. proposed a Brain tumor detection method which uses multiple parameters to analyze the image and this method which uses watershed segmentation [1]. Of all these parameters, intensity is taken into consideration to for MR image segmentation. This method can detect the tumor in both 2D and 3D, which can be considered as fringe benefit [1]. Anam Mustaqeem et al. [2] used morphological operators for detecting a tumor from a brain MR image using Watershed and Thresholding Based Segmentation [2]. Prastawa et al. [3] considered T2 MR Image channel as an input to segmentation so as to propose a segmentation of brain using outlier detection. This method is used for diagnosis, planning, and treatment of brain tumor as it reveals the extent of edema. Dahab et al. [5] made use of neural network to propose an automatic detection of brain tumor from an MRI scanned image. The aforementioned works are proposed using several segmentation methods such as k-means and SVM method, neural network, fuzzy methods etc. Although these methods have produced desirable results, they are complex and high computational overhead. So, in this proposed methods we have used discriminative clustering which gives better variance in tissues of brain MR image and AdaBoost is pretty simple for classifying an image into normal or abnormal.

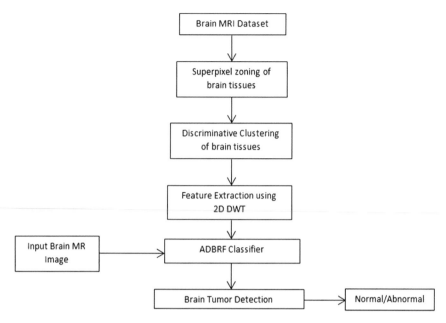

Fig. 2 Proposed system architecture

3 Proposed Model

3.1 System Architecture

Automated brain tumor detection using discriminative clustering based brain MRI segmentation architecture comprises of three stages especially, superpixel zoning, discriminative clustering, feature extraction using discrete wavelet transform (DWT) and classification using AdaBoost with random forests (ADBRF) algorithm. Here, Superpixel zoning is used as initial segmentation and discriminative clustering as secondary segmentation. These first two steps removes tissue heterogeneity in a single superpixel thereby increasing the clarity and correctness of brain MR image which helps in analysis of the image. First, a brain MRI dataset is inputted into the superpixel level feature zoning and the superpixel patches divided by taking cluster center as its lone parameter. These zones are clustered by using discriminative clustering, where these clusters are formed using the homogenous features in the brain tissues. These clustered MR image dataset performs feature extraction using level-3 2-D DWT forming the trained classifier. Using ADBRF as its base classifier, the feature vector can be classified as normal/abnormal, the system classifies it as normal, if it does not contain tumor and abnormal, if it contains tumor (Fig. 2).

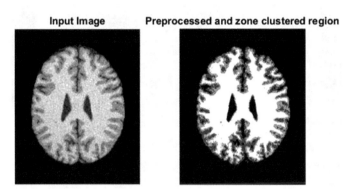

Fig. 3 Sample superpixel level zoning

3.2 Superpixel Level Zoning

Superpixels are becoming highly popular in medical image analysis applications; superpixels provide us ability to capture local image features and capture redundancy there by reducing computational complexity. Although they are popular, they face challenges such as high computational cost, poor quality segmentation, inconsistent size, and shape [4]. Superpixels are used in applications such as depth estimation [6], image segmentation [7, 8], skeletonization [9], body model estimation [10], and object localization [11]. Superpixel can be defined as group of image pixels which has homogenous pixel intensities. Pixels in the superpixels show same properties. If you consider an image grid of same intensity values and also neighborhood values which together grouped to form a superpixel. Let us assume a pixel intensity value of 1 and if we consider all the intensity values of neighboring pixels forming an edge of same values. We group these pixels to form a superpixel. Superpixels extract homogenous features so they account for bias field. In this paper, we have used the simple linear iterative clustering (SLIC) method to generate 2D superpixels for the inputted brain MR Image. This SLIC method uses consider a pixel as its center and initialize with "n" number of centers. It then calculates the Euclidean distance with nearer neighboring pixels and the pixels with same intensity values are grouped as matching pixels. This process is stopped when the distance is greater than threshold value. Figure 3 shows the brain MR image which is segmented using superpixel zoning.

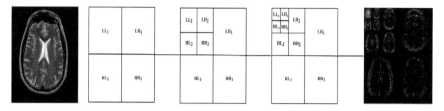

Fig. 4 Decomposition using Haar wavelet

3.3 Feature Extraction Using Haar Wavelet Transform

The Haar Wavelet Transform is an effective method for feature extraction [12]. It has many applications in image processing and signal processing as it considers local features of images. Haar wavelet transform is simple and computationally efficient. It offers decomposition of MR images using time scale representation, which is very useful in classification scenarios [13]. DWT is accepted as new compression standard JPEG2000. At level zero, This wavelet transform has two filters to pass through the image i.e., low pass filter and high pass filter. The input signal is sent into low pass filter where a low resolution signal is extracted whereas when the input signal is sent into high pass filter it extracts a difference signal. At level 1, the output signals of the high pass filters are applied by another pair of filters [13]. This process repeats until the level 3 is completed. It is the best suitable dwt for the classification process as it is very fast, symmetric and orthogonal in nature, it also used to extract structural information from the images and performs well in the presence of noise. DWT is decomposed in this process up to three levels. Firstly, a brain MR image of size M × M is segregated to (M/2) × (M/2) of four sub bands namely, LL (low-low), LH (low-high), HL (high-low) and HH (high-high). The LH, HL and HH are the sub-band images that contain the edges in the vertical, horizontal and diagonal directions respectively. The low-low sub-band image contains maximum information and it can also be treated as output of high filters and sent into next level for decomposition. This low-low sub-band image is also called approximation image (Fig. 4).

Algorithm: Superpixel level Discriminative Segmentation based brain tumor detection.

Input: Brain MR Image dataset (D) with each sample (di), with dimension M x M.

Output: Feature Matrix (f).

Step 1: Initialize f→ φ

Step 2: Repeat for each image (I).

Step 3: Put X superpixel seed points.

Step 4: Label each pixels using

D (x, l) = I (x, l) + C (x, l)

Step 5: Update seed points.

Step 6: Update each pixel label using

D (x, l) = wb×B (x, l)+wi×I (x, l)

Step 7: Repeat step5-6 until pixel label is optimized.

Step 8: Connect the neighboring pixel surroundings the seed   pointers.

Step 9: Apply discriminative clustering to the zones obtained above.

Step 10: Generate final segmented   image Iseg with distinct colors for each segment.

Step 11: Apply Level3 2D-DWT to I_{seg} and the feature to f.

f← f υ fi , Where fi is feature set from I_{seg}

Step 12: final feature set F is given to ADBRF classifier.

3.4 Classification Using ADBRF

ADBRF is an acronym for AdaBoost with random forests algorithm. In this method, AdaBoost is combined with random forests algorithm to build a classifier, which is used to categorize the regions of brain by using the features extracted by haar wavelet transform. It is used to better the results of the accuracy and stability of the any learning algorithm. When we consider several weak classifiers of high error rates and to make them useful, ADB combines them and generate a classifier with a small training error rate [14, 15]. ADB is easy to implement, fast and simple. This algorithm can be integrated with other classifiers as it is non-parametric in nature [16]. RF is a machine learning algorithm which is simple, effectively estimates the missing data, robust to outliers and noise [12]. It can run efficiently on large datasets, estimates important features for classification. Here, we use this algorithm for binary

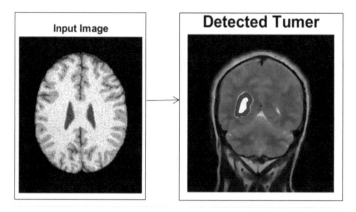

Fig. 5 Sample input and output

classification; it results in two class labels 0 and 1. The class label 0 denotes the normal class whereas the class label 1 denotes the abnormal class (Fig. 5).

4 Simulation Results

Simulation is carried using MATLAB codes and the output figures are attached with each section separately for better understanding. Its performance is being evaluated on the BrainWeb dataset where it acquired 100% accuracy. BrainWeb database is a brain MRI database with several parameters and the images provided with tissue label for each brain tissue voxel, with a size of 181 × 217 pixels which can be retrieved from http://brainweb.bic.mni.mcgill.ca/brainweb/. In this dataset, the echo time and the repetition time have been set to 10 ms and 18 ms respectively. A total of 200 samples are analyzed for the purpose. For computing the accuracy the k-fold (k = 5) cross-validation strategy has been adopted. Overall accuracy rate of 96% has been reported for the proposed scheme on the said dataset. This is quite satisfactory.

A comparative analysis is also performed with other proposed algorithms on same problem statement. The proposed scheme outperforms the rest of the state of the art methods as shown in below figure.

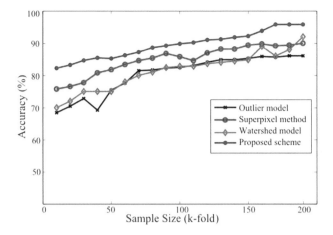

5 Conclusion and Future Work

In this paper, a method for detecting a tumor from a brain MR Image is proposed. This method uses superpixel level zoning of the brain MR Image and performs clustering using the discriminative clustering. These clusters divide the scanned image of brain into different brain tissues into White Matter, Gray Matter, and Cerebro Spinal Fluid. Extraction of features from the structures of brain is carried out by Haar Wavelet Transform. ADBRF is used as a base classifier where it classifies a brain MR Image into normal or abnormal. This method achieves 100% of accuracy on BrainWeb MRI dataset. How to extend our proposed method by integrating with deep learning algorithms and to use in real-world applications constitutes our future work.

References

1. Ratan, R., Sharma, S., Sharma, S.K.: Brain tumor detection based on multi- parameter MRI image analysis. ICGST-GVIP J. **9**(III) (2009). ISSN 1687-398X
2. Mustaqeem, A., Javed, A., Fatima, T.: An efficient brain tumor detection algorithm using watershed & thresholding based segmentation. Int. J. Image Graph. Signal Process. **10**, 34–39 (2012) (Published Online September 2012 in MECS). http://www.mecs-press.org/. https://doi.org/10.5815/ijigsp.2012.10.05
3. Prastawa, M., Bullitt, E., Ho, S., Gerig, G.: A brain tumor segmentation framework based on outlier detection. Med. Image Anal. **8**, 275–283 (2004)
4. Achanta, R., Shaji, A., Smith, K., Lucchi, A., Fua, P., Susstrunk, S.: SLIC Superpixels, EPFL Technical Report 149300, June 2010
5. Dahab, D.A., Ghoniemy, S.S., Selim, G.M.: Automated brain tumor detection and identification using image processing and probabilistic neural network techniques. Int. J. Image Process. Vis. Commun. **1**(2) (2012). ISSN 2319-1724

6. Hoiem, D., Efros, A., Hebert, M.: Automatic photo pop-up. SIGGRAPH (2005)
7. Li, Y., Sun, J., Tang, C.K., Shum, H.Y.: Lazy snapping. SIGGRAPH 303–308 (2004)
8. He, X., Zemel, R., Ray, D.: Learning and incorporating top-down Cues in image segmentation. ECCV 338–351 (2006)
9. Levinshtein, A., Sminchisescu, C., Dickinson, S.: Multiscale symmetric part detection and grouping. ICCV (2009)
10. Mori, G.: Guiding model search using segmentation. In: ICCV 1417–1423 (2005)
11. Fulkerson, B., Vedaldi, A., Soatto, S.: Class segmentation and object localizationwith super-pixel neighborhoods. In: ICCV (2009)
12. Nayak, D.R., Dash, R., Majhi, B.: Brain MR image classification using two-dimensional discrete wavelet transform and AdaBoost with random forests. Neurocomputing **177** 188–197 (2016)
13. Kour, P.: Image processing using discrete wavelet transform. IPASJ Int. J. Electron. Commun. (IIJEC) **3**(1) (2015). ISSN 2321-5984
14. Freund, Y., Schapire, R.E.: Experiments with a new boosting algorithm. In: International Conference on Machine Learning, vol. 96, pp. 148–156 (1996)
15. Duda, R.O., Hart, P.E., Stork, D.G.: Pattern Classification. Wiley, New York (2012)
16. Schapire, R.E.: A brief introduction to boosting. In: International Joint Conference on Artificial Intelligence, vol. 99, pp. 1401–1406 (1999)
17. Achanta, R., Shaji, A., Smith, K., Lucchi, A., Fua, P., Susstrunk, S.: Slic superpixels compared to state-of-the-art superpixel methods. IEEE Trans. Patt. Anal. Mach. Intell. **34**(11), 2274–2282 (2012)
18. Breiman, L.: Random forests. Mach. Learn. **45**(1), 5–32 (2001)
19. Kapoor, L., et al.: A survey on brain tumor detection using image processing techniques. In: 2017 7th International Conference on Cloud Computing, Data Science & Engineering—Confluence. https://doi.org/10.1109/confluence.2017.7943218
20. Rao, H., et al.: Brain tumor detection and segmentation using conditional random field. In: 2017 IEEE 7th International Advance Computing Conference (IACC). https://doi.org/10.1109/iacc.2017.0166
21. Santosh, S., et al.: Implementation of image processing for detection of brain tumors. In: 2017 International Conference on Computing Methodologies and Communication (ICCMC). https://doi.org/10.1109/iccmc.2017.8282559
22. Somwanshi, D., et al.: An efficient brain tumor detection from MRI images using entropy measures. In: 2016 International Conference on Recent Advances and Innovations in Engineering (ICRAIE). https://doi.org/10.1109/icraie.2016.7939554
23. Singh, B., et al.: Detection of brain tumor using modified mean-shift based fuzzy c-mean segmentation from MRI Images. In: 2017 8th IEEE Annual Information Technology, Electronics and Mobile Communication Conference (IEMCON). https://doi.org/10.1109/iemcon.2017.8117123
24. Krishnamurthy, T.D., et al.: Automatic segmentation of brain MRI images and tumor detection using morphological techniques. In: 2016 International Conference on Electrical, Electronics, Communication, Computer and Optimization Techniques (ICEECCOT). https://doi.org/10.1109/iceeccot.2016.7955176
25. Nanda Gopal, N., et al.: Diagnose brain tumor through MRI using image processing clustering algorithms such as Fuzzy C Means along with intelligent optimization techniques. In: 2010 IEEE International Conference on Computational Intelligence and Computing Research (ICCIC). https://doi.org/10.1109/iccic.2010.5705890

Cancer Prediction Based on Fuzzy Inference System

Soumi Dutta, Sujata Ghatak, Abhijit Sarkar, Rechik Pal, Rohit Pal and Rohit Roy

1 Introduction

One of the most common diseases in female worldwide is breast cancer. It has become the primary reason for women death in developed countries and is still increasing per annum. In order to reduce the risk of disease, every woman needs to be attentive and be aware of her chances for developing breast cancer. To avoid the risk of death due to breast cancer, early detection is highly required. Although the causes of this disease are not thoroughly and totally understood, numerous factors have been identified in various studies which can raise or decline breast cancer developing chances. Their risk factors can be analyzed to judge risk level for the development of breast cancer. Though it occurs mostly in women, according to medical statistics in recent times, breast cancer is common in both men and women. To overcome this disease, in early

S. Dutta (✉) · S. Ghatak · A. Sarkar · R. Pal · R. Pal · R. Roy
Institute of Engineering & Management, Kolkata 700091, India
e-mail: soumi.it@gmail.com

S. Ghatak
e-mail: ghatak.sujata07@gmail.com

A. Sarkar
e-mail: abhijit.sarkar@iemcal.com

R. Pal
e-mail: rechik27@gmail.com

R. Pal
e-mail: pal.rohit2009@gmail.com

R. Roy
e-mail: rohit.roy080@gmail.com

S. Dutta
Indian Institute of Engineering Science and Technology Shibpur,
Howrah 711103, India

© Springer Nature Singapore Pte Ltd. 2019
S. Tiwari et al. (eds.), *Smart Innovations in Communication and Computational Sciences*, Advances in Intelligent Systems and Computing 851,
https://doi.org/10.1007/978-981-13-2414-7_13

stage, detection of breast cancer is necessary to increase the life expectancy of the patient.

Breast cancer is caused by uncontrolled growth of breast cells, beginning with formation of small mass or tumor in milk glands or ducts. Breast cancer causes the abnormal growth of cells and uncontrolled behavior that changes the characteristics of the cells. It occurs due to malignant tumor in breast. A mass or tumor with a regular, definite border is noncancerous (benign) whereas, a mass or tumor with speculations or irregular border may be cancerous (malignant). One of the most effective ways to reduce the breast cancer is early detection and diagnosis of the disease. Physicians require a trustworthy procedure for early prognosis to differentiate between malignant breast tumors and benign ones. The patients are assigned to one of the two groups: (i) Benign (noncancerous) and (ii) malignant (cancerous) based on the predictions done. The major scope of this research work is to predict the possibilities of breast cancer based on existing clinical records.

There are several research challenges faced by researchers in detection of breast cancer in early stage. For effective extraction of features or patterns to predict breast cancer, large databases are considered. To find such patterns, various data mining functions are used by several researchers such as association rule, classification, prediction, and clustering. Fuzzy inference based classification system is proposed in this work to predict unseen instances. The tumors genetic behaviors are analyzed for prediction modeling. This research work carries an outline on the enhancement of the breast cancer diagnosis and prediction on the basis of current researches accomplished using a popular experimental breast cancer datasets.

The rest of the paper is organized as follows. A brief literature survey on breast cancer prediction is presented in Sect. 2. Section 4 describes the proposed clustering approach. Section 5 discusses the results of the comparison among the various classification algorithms. In Sect. 6, the paper is concluded with some potential future research directions.

2 Literature Review

In recent times, many researchers are focusing on breast cancer prediction. Various prior works have been done using supervised and unsupervised learning techniques such as decision tree, Artificial Neural Network (ANN), statistical technique, and Unsupervised Artificial Neural Network, a classification model was built by Hota [1]. The efficiency of the ensemble model came into focus because of its accuracy result of 97.73% on experimental testing. The performance of PART, c4.5, and RIPPER algorithm was analyzed and classification rule algorithm was studied by VijayaRani and Divya [2]. Breast cancer data were analyzed by the measures of number of rules generated previously. An improvised ID3 algorithm was intended by Anumith et al. [3] for building a perfect decision tree in a short span of time with the help of Fisher Filtering, Step Disc, Forward Logit, ReliefF, and Backward Logit. Experimental results show that accuracy is enhanced considering five levels decision tree and error rate can be minimized also for using ReliefF feature selection algorithm. With the help of chaste Adaboost, K-means, and ReliefF, Thongkam et al. [4] intended a chaste AdaBoost algorithm for extracting survivability patterns of breast cancer. Specificity, accuracy, and sensitivity were the measures analyzed. For achieving better classification accuracy, the efficiency of chaste AdaBoost algorithm was described and explained by computational results. For evaluating breast cancer data, Paulin and Santhakumaran [5] considered classification technique using feed forward artificial network. The network was trained by various training algorithms such as backpropagation Network. Levenberg–Marquardt algorithm, also a training algorithm, gave a highest accuracy of 99.28%. As a result, the accuracy of classification of breast cancer increased. With the help of neuro fuzzy system and Artificial Neural Network (ANN), Shukla [6] proposed a knowledge-based system for early diagnosis of breast cancer in March 2009. No. of epochs, accuracy of diagnosis, No. of neurons, and training time were the measures analyzed. Simulations show that the survival rates raised effectively due to this knowledge-based system. Narang [7] intended a system in Nov 2012 for detecting the cancer stage as malignant or benign with the help of ART2 (Adaptive Resonance Theory) neural network. For handling breast cancer data, neural network approach was adopted. Accuracy, precision and recall were the measures analyzed and clustering was applied for extracting knowledge as it is a continuous data. Results show that accuracy rate can be improved using ANN model. For a specific medical application, some practical guidelines were provided to select an algorithm by Andreeva et al. [8] and she made a comparison of various learning models in data mining. For breast cancer, various classification algorithms were applied. Highest accuracy was achieved by SMO and Bayesian classification. For classification purpose, a comparative study was conducted on four data mining tools namely, KNIME, weka, Tanagra, and orange by Wahbeh [9]. Nine different datasets were feed in the tool kits. Results demonstrated that in terms of classifiers applicability issue, Weka tool kit proved to be the best. Associating the prediction of multiple classifiers, Pujari [10] developed an ensemble model for improving classification accuracy. With the help of QUEST, CART and CHAID classification algorithm, sensitivity, gain, specificity and accuracy were analyzed to handle ionosphere data. Experimental results demonstrated that accuracy of 93.84% was achieved on test data by the ensemble model with feature selection. Lavanya and Usha Rani [11] found that classification accuracy improved by some data mining task with classification and she intended a hybrid approach. Das et al. [12] used machine learning approach for predicting trends on microblogging data. In terms of accuracy, a comparative study of this method was made with classification with clustering, without feature selection and with feature selection. Experimental results showed that hybrid approach was the best classifier for breast cancer data and it yielded much better result than CART with feature selection.

3 Experimental Dataset

In this work, real-time laboratory dataset has been used which is collected from University Medical Centre, Institute of Oncology, Ljubljana, Yugoslavia. The dataset consists of 10 features/attributes and 286 instances. These features are Age (10–19, 20–29, 30–39, 40–49, 50–59, 60–69, 70–79), Menopause (premeno), Tumor to Size (between 0 and 4, between 5 and 9, between 10 and 14, between 15 and 19, between 20 and 24, between 25 and 29, between 30 and 34, between 35 and 39, between 40 and 44, between 45 and 49, between 50 and 54, between 55 and 59), Inv to Nodes (between 0 and 2, between 3 and 5, between 6 and 8, between 9 and 11, between 12 and 14, between 15 and 17, between 18 and 20, between 21 and 23, between 24 and 26, between 27 and 29, between 30 and 32, between 33 and 35, 36 and 39), Node to Caps (yes, no), Deg to Malig (1, 2, 3), Breast (left, right), Breast to Quad (left and up, left and low, right and up, right and low, central), Irradiat (yes, no), and class (no to recurrence to events, recurrence to events). The "class" feature/attribute signifies the decision as no-recurrence and recurrence events. Based upon the class distribution, no-recurrence-events instances are 70.3% or 281 and recurrence-events instances are 29.7% or 85.

All the values of the attributes in the experimental dataset are crisp in nature, so fuzzification method is applied on the attribute values to represent each attribute in terms of fuzzy values.

4 Proposed Methodology

In this work, a cancer prediction system is proposed based on fuzzy inference system. Due to the fuzzy nature of the dataset, it is one of the effective ways to apply fuzzy Logic in this work. Fuzzy logic is an statistical computational approach which outcome based on "degrees of truth" rather than the usual "true or false". The experimental dataset consists of all scalar values for all the features, so initially for all the feature's value are fuzzified. To design fuzzy inference system, a set of rules are evaluated which will come up with fuzzy output. Then, defuzzification approach is applied for conversion of the fuzzy output into crisp values.

To describe the fuzzy inference process, an example of 2 inputs, 1 output, and 10 rules problem has been taken. Age and tumor size have been taken as inputs, denoted by X and Y which is further divided into three fuzzy groups as low, medium, and high. Output is represented in percentages as chances of occurrence of breast cancer and is denoted by Z. Rules have been used for evaluation of fuzzy output. A pictorial representation of the methodology has been given in Fig. 1.

Initially, using few feature ranker (Chi Square, Infogain) all the attributes are ranked and accordingly attributes are selected for rule generation as conditional attribute. Using an example, the proposed method is demonstrated.

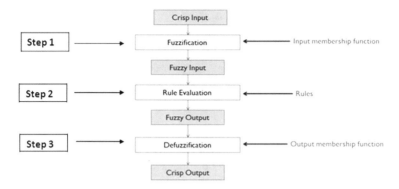

Fig. 1 Example of fuzzy inference system

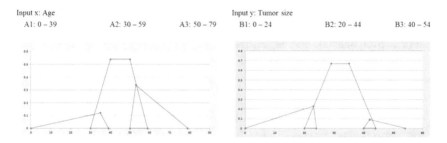

Fig. 2 Attribute fuzzification: **a** Attribute age fuzzy graph and **b** Attribute tumor size fuzzy graph

In this example, two inputs are considered, such as age and tumor size. So, all the crisp values for the attributes are converted into fuzzy value followed by fuzzification method. Age is divided into three fuzzy groups: low (0–39)A1, medium (30–59)A2, and high (50–79)A3 (shown in Fig. 2). Tumor size is also divided into three fuzzy groups: low (0–24)B1, medium (20–44)B2, and high (40–54)B3 (shown in Fig. 2). Output is being determined as percentages of occurrence of breast cancer which is further divided into three fuzzy groups: low (0–30), medium (15–50), and high (35–100). Accordingly, membership functions are also designed for all the attributes. Then to design the fuzzy knowledge-base system fuzzy inference rules are generated which are composed of expert IF <antecedents> THEN <conclusions> rules. The collection of rules extracted, constitutes the knowledge base of the risk model. The ambiguity associated with the explanation of heuristic knowledge is handled by using fuzzy logic in the progress of the heuristic IF-THEN rules.

To predict any the degree of recurrence of cancer for an arbitrary patient, initially all the attribute's values should be converted into fuzzy values. Then, fuzzy inference rules are applied. After the implementation of rules, a combined graph has been derived which is used to calculate output by using a well-known method Center of Gravity (CoG) method also known as Centroid method and Center of Sums (CoS).

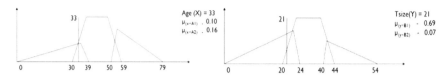

Fig. 3 Mapping of sample data in attribute age fuzzy graph and attribute tumor size fuzzy graph

The Center of Sums (CoS) is a defuzzification method, where the geometric center of area is evaluated for each membership function by the fuzzy logic controller, i.e., the area under the scaled membership functions and within the range of the output variable are evaluated by the fuzzy system using Eq. 1. In this method, the overlapping area is counted twice. Whereas, the Center of Gravity (CoG) can be considered as the weighted average of the membership function. The center of the gravity of the area bounded by the membership function curve is calculated to be the mainly crisp value of the fuzzy quantity shown in Eq. 2. Using the defuzzification method, the final prediction outcome can be generated.

$$Center \ of \ Gravity \ (CoG) = \frac{\int_a^b \mu_A(x) x \, dx}{\int_a^b \mu_A(x) \, dx} \tag{1}$$

$$Center \ of \ Sums \ (CoS) = \frac{\sum_{i=1}^M X_i \cdot \sum_{j=1}^m \mu_{A_j}(x_i)}{\sum_{j=1}^m \mu_{A_j}(x_i)} \tag{2}$$

Here, m is the number of fuzzy sets, M is the number of fuzzy variables, and $\mu_A(X_i)$ represents the membership function for the j-th fuzzy set.

For experimental purpose, a sample data is taken to explain the proposed method, where age is 33 and tumor size is 21. So, initially these data are mapped on the attribute age fuzzy graph, and attribute tumor size fuzzy graph, and corresponding values are evaluated to represent membership functions (shown in Fig. 3). Based on the fuzzy graph shown in Fig. 2, fuzzy inference rules are designed before. For this experiment, a set of rules are evaluated which constitute of nine "AND" rules and one "OR" rule. The values of M and N will be mapped at X and Y axis accordingly.

The sample rules are given as follows.

Rule 1. If the value of M is A1 and the value of N is B1 then P is low
Rule 2. If the value of M is A1 and the value of N is B2 then P is low
Rule 3. If the value of M is A1 and the value of N is B3 then P is low
Rule 4. If the value of M is A2 and the value of N is B1 then P is medium
Rule 5. If the value of M is A2 and the value of N is B1 then P is high
Rule 6. If the value of M is A2 and the value of N is B1 then P is low
Rule 7. If the value of M is A3 and the value of N is B1 then P is medium
Rule 8. If the value of M is A3 and the value of N is B1 then P is low
Rule 9. If the value of M is A3 and the value of N is B1 then P is low
Rule 10. If X is A1 or Y is B3 then the value of P is low

Fig. 4 Mapping of sample data in fuzzy rules

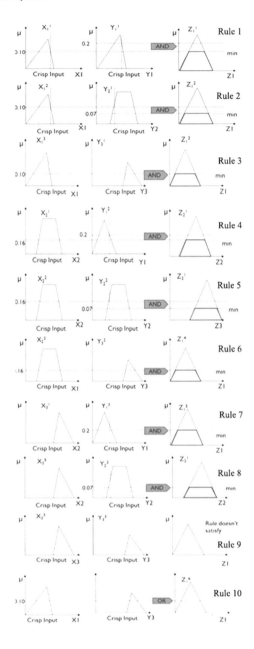

Figure 4 shows the implementation of each rule for the sample input. Then, a combined graph has been generated to evaluate the area delimited by the all membership function curve. Then, Center of Gravity (CoG) and Center of Sums (CoS) are evaluated for defuzzification of the output. The output shows that degree of satisfaction, CoG is 35.08%. It shows the chances of occurrence of breast cancer expressed in percentages.

5 Experiments and Results

5.1 Performance Measuring Metrics

In the no-recurrence-events and recurrence-events classification problem, to determine the performance of different classifiers, the standard methodology has been followed to represent the outcome in the form of a confusion matrix. The standard metrics for performance evaluation are used such as precision, recall, and accuracy.

The *precision* can be represented as the ratio of the number of items that are correctly predicted to be in that class and the total number of items (correct and wrong) predicted in that class with respect to a certain class (no-recurrence-events or recurrence-events). The precision (P) for the correct class is $P = correct/(correct + wrong)$. The *recall* (R) for a certain class is the ratio of the number of items that are correctly predicted to be in that class, to the total number of items (correct and missed) actually belonging to that class. The recall of the positive class is $R = correct/(correct + missed)$. The *accuracy* (A) is the proportion of the total number of items that are correctly classified. The F-measure is represented as the harmonic mean of precision and recall: $F = \frac{2PR}{P+R}$.

5.2 Results

This section describes the experiments using a dataset of cancer features and comparing the proposed methodology with some baseline classification techniques. This section describes the comparison of the proposed method with a set of classical classifiers from the well-known Weka machine learning tool kit [13]. For experiments, information from a dataset provided by University Medical Centre, Institute of Oncology, and Ljubljana, Yugoslavia of 286 patients has been used. From the entire dataset 70% are chosen as training set and rest 30% are considered as test case.

Table 1 reports the precision, recall, and the accuracy metrics for the no-recurrence-events and recurrence-events classification using various classifiers such as Naive Bayes, RBF Network, Logistic, LWL, LogitBoost, DecisionTable, and RandomTree. It can be observed that the proposed fuzzy based system achieves 84.68%

Table 1 Results for the no-recurrence-events and recurrence-events classification, using some state of art classifiers from the Weka machine learning toolkit

Classifiers	Precision	Recall	F-measure	Accuracy (%)
NaiveBayes	0.741	0.755	0.747	75.52
RBFNetwork	0.753	0.776	0.758	77.62
Logistic	0.723	0.759	0.729	75.87
LWL	0.699	0.731	0.71	73.08
LogitBoost	0.72	0.752	0.728	75.17
DecisionTable	0.768	0.79	0.753	79.02
OneR	0.722	0.762	0.723	76.22
RandomTree	0.706	0.724	0.713	72.38
Proposed method	0.79	0.82	0.846	84.68

accuracy, which is better than all other classifiers. The result signifies that to predict the cancer the proposed fuzzy based classification methodology can be adopted with reasonable accuracy.

6 Conclusion

This study proposes a fuzzy model for breast cancer assessment, whose features include the representation of the demonstration of the probabilistic relationships between the factors age and tumor size, using fuzzy logic. This prediction system may provide an effective way for predicting breast cancer and help any patient for earlier diagnosis process for breast cancer. For further research, this method can be extended to include different parameters like inv-nodes, node caps, breast quad, etc. For a more sensible and correct assessment of breast cancer, the research work can be enhanced with optimization algorithms like genetic algorithm for efficient feature selection. Experimental results show the effectiveness of the proposed method. More simulation is needed to enhance the accuracy of satisfaction level. Validated computational simulation tools can be incorporated in this model for better results.

References

1. Hota, H.S.: Diagnosis of breast cancer using intelligent techniques
2. Vijayarani, S., Divya, M.: An efficient algorithm for generating classification rules. Int. J. Comput. Sci. Technol. (2011)
3. Anumitha, S., Diana, S., Suganya, D., Shanthi, S.: Improvisation of ID3 algorithm explored on Wisconsin breast cancer dataset. In: International Conference on Computing and Control Engineering (2012)

4. Thongkam, J., Xu, G., Zhang, Y., Huang, F.: Breast cancer survivability via AdaBoost algorithms. School of Computer Science and Mathematics
5. Paulin, F., Santhakumaran, A.: Classification of breast cancer by comparing back propagation training algorithms. Int. J. Comput. Sci. Eng. (2011)
6. Shukla, A., Tiwari, R., Kapur, P.: Knowledge based approach for diagnosis of breast cancer. In: Advance Computing Conference (2009)
7. Narang, S., Verma, H.K., Sachdev, U.: Breast cancer detection using ART2 model of neural networks. Int. J. Comput. Appl. (2012)
8. Andreeva, P., Dimitrova, M., Radeva, P.: Data mining learning models and algorithms for medical application
9. Wahbeh, A.H., Al Radaideh, Q.A., Al-Kabi, M.N., Al-Shawakfa, E.M.: A comparison study between data mining tools over some classification methods. Int. J. Adv. Comput. Sci. Appl
10. Pujari, P., Gupta, J.B.: Improving classification accuracy by using feature selection and ensemble model. Int. J. Soft Comput. Eng. (2012)
11. Lavanya, D., UshaRani, K.: A hybrid approach to improve classification with cascading of data mining tasks. Int. J. Appl. Innov. Eng. Manag. (2013)
12. Das, A., Roy, M., Dutta, S., Ghosh, S., Das, A.K.: Predicting Trends in the Twitter Social Network: A Machine Learning Approach, pp. 570–581. Springer (2014). http://dblp.uni-trier.de/db/conf/semcco/semcco2014.html#DasRDGD14
13. Hall, M., Frank, E., Holmes, G., Pfahringer, B., Reutemann, P., Witten, I.H.: The weka data mining software: an update. SIGKDD Explor. Newsl. 11(1), 10–18 (2009). https://doi.org/10.1145/1656274.1656278. http://doi.acm.org/10.1145/1656274.1656278

Energy Consumption Data Analysis and Operation Evaluation of Green Buildings

Weiyan Li, Dong-Lin Wang, Chen-Fei Qu, Xuantao Zhang, Wen-Jing Wu and Pengcheng Zhao

1 Introduction

At present, although China has adopted many green energy-saving technologies and has issued various energy-saving and green building codes in the management, control, and advocacy of building energy consumption, there is no clear standard for the total energy consumption of buildings that truly reflects the final energy use of buildings [1]. It was not until December 1, 2016 that the "Civil Building Energy Consumption Standard" that was implemented took a clear guidance on the total energy consumption of buildings. By comparing the energy consumption of the building with the guidance value and the constraint value, it is analyzed whether the building is really energy-efficient, or whether only using some green building

W. Li (✉) · D.-L. Wang · C.-F. Qu · X. Zhang · W.-J. Wu · P. Zhao
Tianjin Architecture Design Institute, Tianjin, China
e-mail: lwytju1@163.com

D.-L. Wang
e-mail: wangdonglin3898@126.com

C.-F. Qu
e-mail: chinaqchf@163.com

X. Zhang
e-mail: 398601931@qq.com

W.-J. Wu
e-mail: 398601931@qq.com

P. Zhao
e-mail: 695533464@qq.com

W. Li · D.-L. Wang · C.-F. Qu · X. Zhang · W.-J. Wu · P. Zhao
Tianjin Intelligent Building Design, Operation, Maintenance and Management Integration
Engineering Technology Center, Tianjin, China

© Springer Nature Singapore Pte Ltd. 2019
S. Tiwari et al. (eds.), *Smart Innovations in Communication and Computational Sciences*, Advances in Intelligent Systems and Computing 851,
https://doi.org/10.1007/978-981-13-2414-7_14

technologies actually does not save energy. Therefore, it is very meaningful to carry out the analysis and evaluation of the energy consumption data of green buildings.

Data analysis processing is a necessary step before the evaluation of energy consumption of green buildings [2]. Due to various reasons such as equipment maintenance and transmission instability, there are various degrees of data loss and abnormality in the energy consumption data variables. Some equipment is in an abnormal state even when they are put into use, or the acquired data is anomalous data in the beginning, which will directly lead to the inability to understand the true operating state of the building, and even result in the wrong evaluation of the construction operation. Therefore, it is necessary to perform data analysis and processing before the evaluation of the green building energy consumption data [3].

2 Preparation

2.1 Standard Specification

This paper is based on the "energy standard for civil buildings" numbered GB/T 51161-2016 [4]. It changes from the previous specification process to the specification result, using the total energy consumption index of the building to evaluate the energy use of the building, and achieve the goal of transformation of building energy efficiency from "process control" to "target effect orientation".

The evaluation of the energy consumption of the building operation takes into account various factors such as the climatic zone to which the building belongs, its function and building classification. The energy consumption index is expressed in terms of annual energy consumption per unit of building area, and the energy consumption index of buildings in different climate zones is different. According to the classification of buildings, there are differences in the values of constrained values and guidance values between A-type public buildings and B-type public buildings. Due to different uses of buildings, such as office buildings, hotels, shopping malls, parking lots, etc., there are slight differences in the energy consumption index constraints and guide values.

According to the climate zone where the building is located, the building classification and the use purpose, the standard gives the constraint value and guide value of the building energy. Here H denotes the constraint value and L denotes the guide value (Table 1).

2.2 Theoretical Algorithm

For those variables whose historical data can reflect the real situation, through the analysis of data characteristics, data processing is performed using outlier monitoring

Table 1 Energy consumption evaluation

No	Measured value	Evaluation
1	>H	Building is not energy saving, there is a waste of energy
2	(L, H]	Building energy efficiency, there is room for energy saving
3	<L	The building is very energy efficient

algorithm and missing value algorithm, which provides good data support for subsequent operation evaluation. The commonly used anomaly monitoring algorithms [5] include cluster-based DBSCAN algorithm, monotone sequence logic monitoring method, threshold method, five-number generalization method, density-based LOF anomaly monitoring method, etc. Missing value processing [6] according to the missing data and data characteristics mainly include deletion method, mean filling method, heat card filling method, regression filling method, K nearest neighbor method, etc. The energy consumption data is dominated by active energy. Considering that the research object is a green building, there are many energy monitoring platforms or intelligent system integration platforms. There are few outliers in the data, and most of them are maintenance outages that result in null data or zero values. Comparing and analyzing the advantages and disadvantages of various algorithms [7, 8], we adopt isolated forest algorithm and mean interpolation method to deal with outliers and missing values.

1. Isolation Forest

Isolated Forest [9] is an Ensemble-based rapid anomaly detection method for continuous data anomaly detection. The algorithm defines anomalies as "outliers that are easily isolated," which can be understood as points that are sparsely distributed and that are farther away from high-density groups. The core idea of iForest's design is to randomly cut the data space. The core idea of iForest design is to randomly cut the data space, use a random hyperplane to cut the data space, cut it once to generate two subspaces, and then continue cutting each subspace with a random hyperplane, looping through each subspace in which there is only one data point. Clusters with high densities can be cut many times before they stop cutting, but those with low density can easily be parked in a subspace early. The algorithm implementation step is generally divided into two steps:

(1) Train Dataset is sampled from the training set and built an iTree tree. The dataset works as a training sample for this tree is uniformly sampled. In the sample, randomly select a feature, and within the range of all values of this feature, select a random value between the minimum and maximum values. Divide the sample by a binary division, and divide the sample less than the value into the nodes. On the left, greater than or equal to the value is divided to the right of the node. In this way, a splitting condition and left and right data sets are obtained, and then the above process is repeated on the left and right data sets respectively, and the termination condition is directly reached.

(2) Test Each iTree tree is tested in the iForest forest, the path length is recorded, and each test data anomaly score is calculated based on the anomaly score calculation formula. By building all the iTree trees, you can predict the measured data. The prediction process is to take the test data along the iTree tree along the corresponding conditional branch until it reaches the leaf node, and records the path length h(x) during the process, and finally, calculate the anomaly score of each test data, and its calculation formula is:

$$s(x) = 2^{-\frac{E(h(x))}{c(\varphi)}}$$

h(x) is the depth of the node retrieved by the detected data x in the iTree, E(h(x)) represents the average height of the record x in each tree, which is the average path length of all iTrees in the forest c(φ) constructs the average path length of the binary tree for φ points to normalize the result, where H(k) = ln (k) + ε is estimated and ε is Euler's constant with a value of 0.5772.

$$c(\varphi) = \begin{cases} 2H(\varphi - 1) - \left(\frac{2(\varphi-1)}{n}\right), & \varphi > 2 \\ 1, & \varphi = 2 \\ 0, & \varphi < 2 \end{cases}$$

The anomaly score s(x) is the anomaly index of the iTree that records x in the training data of n samples. The results are analyzed as follows: When $E(h(x)) \rightarrow 1$, $s \rightarrow 1$;

$$E(h(x)) \rightarrow n - 1, \ s \rightarrow 0;$$
$$E(h(x)) \rightarrow c(n), \ s \rightarrow 0.5.$$

Here, s(x, n) is in the range [0, 1]. The closer to 1, the higher the probability of an outlier; the closer to 0 the higher the possibility of a normal. If the s(x, n) of most training samples is close to 0.5, there is no obvious outlier in the entire data set. In summary, the process of the isolated forest algorithm is shown in Fig. 1.

2. Mean interpolation method

Mean interpolation, as the name implies, replaces the missing value of the unanswered data with the mean value of the responding unit in the survey item [10, 11]. Mean interpolation method is to calculate the average value of the answer units in each target variable, and then use the average value of each group as the interpolation value of all missing items in each variable. The interpolation value is

$$y_1 = \frac{\sum_{i=1}^{n} a_i y_i}{n_1},$$

Fig. 1 The flow chart of isolated forest

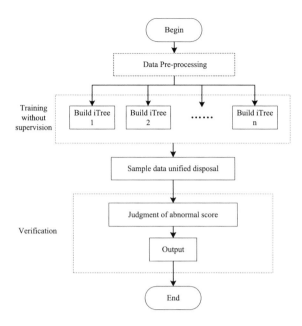

where a_i is an indicative variable, $a_i = 1$ is answered, $a_i = 0$ is no answer, and n_1 is the number of answering units. The overall mean estimate at this time is:

$$Y = \frac{1}{n} \sum_{i=1}^{n} [a_i y_i + (1 - a_i) y_1] = \frac{n_1}{n} y_1 + \frac{n - n_1}{n} y_1 = y_1$$

The sample variance after interpolation is:

$$s^2 = \frac{1}{n-1} \sum_{i=1}^{n} [a_i (y_i - Y) + (1 - a_i)(y_1 - Y)]^2$$

$$= \frac{1}{n-1} \sum a_i (y_i - y_1)^2$$

$$= \frac{n_1 - 1}{n - 1} s_1^2$$

Among them, n_1 is the answer unit number, and s_1^2 is the sample variance of the answer unit.

Although this is a simple and easy method of interpolation, the missing values in the same variable are all replaced by the same mean, which can seriously distort the distribution of the samples. Generally, when the variable obeys or approximately obeys the normal distribution, the average value of this variable can be used as the interpolated value of all its missing values; when the variable obeys the skewed distribution, the median or mode can be considered as the interpolating value.

3   Experimental Setup and Results

This paper evaluates the operation management of a green three-star office building in a cold area in the north. The project has built an intelligent system integration platform which is based on the integration of subsystems in an equal manner. It is a highly integrated management platform that can meet the requirements of comprehensive analysis and decision support. All data in this paper is taken from the intelligent system integration platform. The calculation of non-heating energy consumption is taken from the three electric meters of the main cabinet of the substation and the two electric meters of the information room.

3.1   *Operation Analysis and Processing of Green Building Energy Consumption Data*

The related energy consumption data are processed according to the isolated forest method and the mean interpolation method. Dataset is selected one hourly from January 1 to December 31 of 2017, and the abnormal data of the 26,280 data is received by the head-end and perform interpolation processing. The difficulty in using isolated forest law to monitor is to determine the number of trees and how much they are sampled. After repeated tests and tests, the number of selected trees is 100. The height of the tree is 8 and the training sample is 256. The results of this project are shown in Figs. 2, 3, 4 and 5.

Figure 2, Fig. 3 and Fig. 4 are respectively the result of outlier monitoring and interpolation total active energy in D101, D201, and D301. While Fig. 5 is the

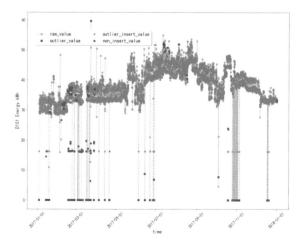

Fig. 2 Monitoring and interpolation of total active energy outlier in D101

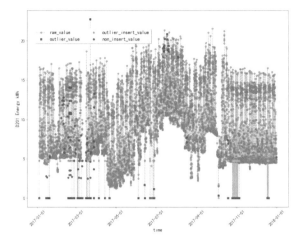

Fig. 3 Monitoring and interpolation of total active energy outlier in D201

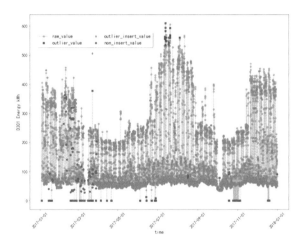

Fig. 4 Monitoring and interpolation of total active energy outlier in D301

result of outlier monitoring and interpolation considering total active energy in D101, D201, and D301. We can apply the above results to follow up green building energy consumption operation evaluation to ensure the accuracy of the operation evaluation results.

The same algorithm is used to monitor and process the abnormal values of the two electricity meters and other data in the information room, and then used to evaluate the energy consumption of the building operation.

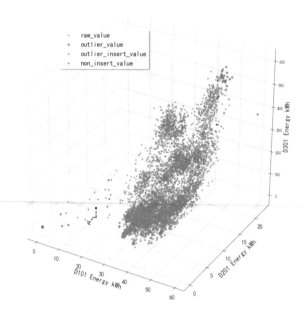

Fig. 5 Monitoring and interpolation of total active energy outlier in D101, D201 and D301

3.2 Operation Evaluation of Green Building Energy Consumption Data

The analysis and evaluation of the green building energy consumption data is achieved by using the total energy consumption per square meter per year of the building as an evaluation index. According to the specification, the evaluation of the energy consumption of this project is based on the calculation of the three electricity meters of the main cabinet of the substation and the two electricity meters of the information room. After applying the data of outliers and missing values to process the above data, the total energy consumption of the building for the whole year of 2017 is calculated (Fig. 6).

According to Clause 5.3.1 and Clause 5.3.2 of the "Standard for Energy Consumption of Civil Buildings" GB/T 51161-2016, when the actual use of public buildings exceeds the specified annual use time and per capita floor area, the actual value of the energy consumption index can be made corrections. According to investigations, the total number of daily work in the building is 2,050, and the average daily use time is 11 h. It is calculated according to the national legal working day of 250 days, so that the whole year's use time is 2750 h. The correction coefficient is calculated according to the above parameters and it is

Correction parameter by time: $\gamma_1 = 0.3 + 0.7\frac{T_0}{T} = 0.3 + 0.7\frac{2500}{2750} = 0.936$

Correction parameter by users: $\gamma_2 = 0.7 + 0.3\frac{S}{S_0} = 0.7 + 0.3\frac{22000/1450}{10} = 1.155$

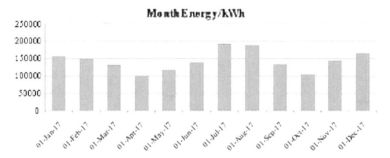

Fig. 6 The total energy consumption of the building for the whole year of 2017

Therefore, the revised energy consumption indicator for the building is:

$$E_{0C} = E_0 * \gamma_1 * \gamma_2 = \frac{1724465}{22000} * 0.936 * 1.155 = 84.74 \, kWh/m^2 * a$$

According to China's "Environmental Energy Consumption Standard for Civil Buildings" GB/T 51161-2016, the above-mentioned building energy consumption index constraint value is 80 kWh/m^2 * a, and the measured value exceeds the constraint value by nearly 6%. It can be judged that there is a wasteful or unreasonable place in the building, and it is necessary to conduct further analysis on the specifics of building energy use. Compared with research projects, most of the existing public buildings use energy at 90 kWh/m^2 * a or more. Although they also do not meet the national standards, the energy use of this building is relatively satisfactory.

4   Conclusions

This paper analyzes the energy consumption data of green buildings, adopts isolated forest algorithm and mean value interpolation method to process abnormal data and missing data, and uses the obtained data to more accurately evaluate the total energy consumption of green buildings. Based on the above results, it can be concluded that the building operation is basically good, but there is also some room for energy saving. Therefore, the analysis and evaluation of the green building energy consumption operation, on the one hand, strengthen the energy-saving consciousness of the building staff, making the building energy use further reduced; on the other hand, through the operation and maintenance management personnel centralized management and management to reduce building energy, especially for the air conditioning heating power and lighting socket. It can fundamentally promote the conservation and rational use of energy resources, ease the contradiction between China's energy resources supply and economic and social development.

Acknowledgements This paper comes from the Ministry of Housing and Urban-Rural Development, Tianjin Science and Technology Commission, and the Tianjin Urban and Rural Construction Committee Fund Projects: 17YFZCSF01280, 2016-24, K82017150.

References

1. Wu, G., Chong, Y., Zhang, C.: On the control of Chinese total energy consumption. In: The 18th International Conference on Industrial Engineering and Engineering Management, vol. 2, pp. 938–942
2. ZhichaoSun, BoWang: An integrated method for building energy consumption analysis based on data mining technology. Comput. Appl. Softw. **34**(11), 103–108 (2017)
3. Zhang, X., Li, M., Zhang, K.: Study on life cycle energy consumption control of green buildings. Sichuan Constr. **34**(01), 28–29 (2014)
4. Civil building energy consumption standards, GB/T 51161-2016, 2016-12-1
5. Wu, W.: Anomaly identification and repair method of building energy consumption monitoring data based on machine learning algorithm. Constr. Sci. Technol. (09), 60–62 (2017)
6. Zhang, S., Wang, P., Xu, Z.: Research on data processing of missing values based on statistical correlation. Stat. Decis. (12), 13–16 (2016)
7. Abe, N., Zadrozny, B., Langford, J.: Outlier detection byactive learning. In: Proceedings of the 12th ACM SIGKDD International Conference on Knowledge Discovery and Datamining. ACM Press, pp. 504–509 (2006)
8. Angelos, E.W.S., Saavedra, O.R., Cortes, O.A.C., et al.: Detection and identification of abnormalities in customer consumptions in power distribution systems. IEEE Trans. Power Deliv. **26**(4), 2436–2442 (2011)
9. Liu, F.T., Kai, M.T., Zhou, Z.H.: Isolation forest. In: The 8th IEEE International Conference on Data Mining (2008)
10. Pang, X.: Discussion on related problems in missing data processing. Stat. Inf. Forum **19**(5) (2004)
11. Jin, Y.J., Shao, J.: Statistical Processing of Missing Data. China Statistics Press, Beijing (2009)

The Method of Random Generation of Electronic Patrol Path Based on Artificial Intelligence

Wen-Xia Liu, Dong-Lin Wang, Wen-Jing Wu and Chen-Fei Qu

1 Introduction

Urban security has become an important issue in the development of cities today [1]. Complete machine monitoring measures do not solve all problems [2]. The way of man–machine integration is still one of the important means for the current building safety precautions. As one of the important human–machine security systems in buildings, the electronic inspection system is still the focus of current security research.

At present, most of the electronic inspection systems in construction projects do not have this function which is to generate patrol routes automatically and randomly. Within the stipulated time range, property personnel must strictly follow the inspection route to complete inspections of the corresponding inspection sites.

The electronic inspection system in a very few engineering projects has the function of automatic generation of inspection routes. However, the inspection route is entered in advance, and the route is fixed. It is static and cannot be automatically generated randomly based on inspection points and corresponding building parameters. The fixing of the inspection route makes it easy for criminals to grasp the law of inspection. Therefore, property personnel must often change the inspection route.

W.-X. Liu (✉) · D.-L. Wang · W.-J. Wu · C.-F. Qu
Tianjin Intelligent Building Design and Operation and Maintenance and Management Integration Engineering Technology Center, Tian Jin Architecture Design Institute, Tianjin, China
e-mail: liuwenxia2018@163.com

D.-L. Wang
e-mail: wangdonglin3898@126.com

W.-J. Wu
e-mail: wuwenjing-2002@163.com

C.-F. Qu
e-mail: chinaqchf@163.com

© Springer Nature Singapore Pte Ltd. 2019   147
S. Tiwari et al. (eds.), *Smart Innovations in Communication and Computational Sciences*, Advances in Intelligent Systems and Computing 851,
https://doi.org/10.1007/978-981-13-2414-7_15

This article proposes a method. This method has caused trouble for property personnel. The method of randomly generating inspection routes can effectively prevent criminals from mastering the inspection rules, ensure the safety of inspectors, and improve the inspection results.

Ant colony algorithm is a bionic algorithm proposed by Dorigo M, Maniezzo V and Colornia A in 1990s [3–6]. At present, the algorithm is well applied in fields such as secondary allocation [7, 8], graph coloring, vehicle scheduling [9–11], integrated circuit design, and multi-objective combination optimization [12].

The ant colony algorithm is inspired by the search of food by ant colony in nature, the algorithm uses the principle of positive feedback to simulate the method that ant colony find the optimal path between food and nests [13, 14]. If the ant encounters an untraveled path, it will select the path randomly while releasing pheromones on the path [15]. The pheromones concentration evaporates, so the longer the path, the lower the pheromones concentration. When other ants encounter the same path again, the selection probability of a short path and a high pheromone concentration will be increased. Therefore, the concentration of pheromone on the optimal path will become higher and higher. Finally, most of ants will choose the optimal path, and the optimal path is funded.

This article describes a method for automatic random generation of electronic inspection routes. The inspection points are converted accordingly, according to the absolute coordinate system in the figure. A three-dimensional path model was established. The inspection route problem was analyzed and simplified. The objective function of path planning is established. Artificial intelligence algorithm is introduced into the model and verified by experimental simulation.

2   Electronic Patrol Problems Simplify and Establish Mathematical Models

Most buildings are multi-layered structures. Between the floors of the building is connected by stairs. Random generation of multi-storey building inspection routes can be superimposed on the basis of a random optimal path on a single floor. The connection between floors is taken into account and the patrol route for the entire building is generated. The inspection process in a single floor can be described as follows: the stairway is used as the starting point and end point, and the inspection point is randomly selected. The optimal path planning between the starting point, inspection point, and end point is accomplished by using the ant colony algorithm. Multiple floors are connected by a stairwell. After random selection of floors and inspection points using a random algorithm, the inspection routes of selected floors are connected in ascending or descending order to form a random inspection route for the entire building. The schematic diagram of the building coordinate system and the electronic inspection path is shown in Fig. 1.

Fig. 1 Sketch of building coordinates and inspection route

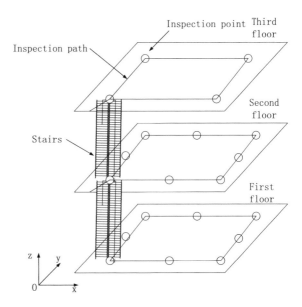

This problem can be simplified to solve the problem of the shortest optimal inspection path with the shortest distance to the selected inspection point. That is, a path P = {p_1, p_2, p_3, …, p_n} is found so that the following objective function is minimized.

$$f(p) = \sum_{i=1}^{n-1} d(p_i, p_{i+1}) + d(p_n, p_1) \qquad (1)$$

In the above formula, p_i randomly selected patrol point number, (p_i, p_{i+1}) the distance between two adjacent inspection points.

In order to optimize the path of building inspection points, the inspection points need to be parameterized so that the inspection path can be combined with the ant colony algorithm to automatically generate patrol routes.

In the intelligent design of a building, the coordinates of the corresponding inspection point can be acquired by using absolute coordinates. At the same time, the coordinates of different inspection points are converted by increasing the length of the staircase path. Building 3D path model is built. The location of the building inspection point is different. And the floor spacing is constant, that is, the height between floors is a fixed constant. As shown in Fig. 1, the Cartesian coordinate system O-xyz is determined. The location of inspection points is denoted by $P_i = (x_i, y_i, z_i)$. Where x_i, y_i is the relative position of the patrol point in the floor. z_i indicates the height of the floor where the patrol is located and the floor. As a result, the inspection route of a building can be represented as a directed graph of M-level N_k inspection points, that is

$$\begin{cases} G_k = (N_k, \ A_k) \, k = 1, 2, \ldots, M \\ N_k = \{1, 2, \ldots, n_k\} \, A_k = \{(i \ , j) \, | i, \ j \in N_k\} \end{cases} \tag{2}$$

Among them, M is the number of building floors.

The coordinates of the point of departure of the inspector are $(0, 0, 0)$. The distance of the inspectors from the inspection point i to the inspection point j is:

$$d_{ij} = |x_j - x_i| + |y_j - y_i| + |z_j - z_i| \tag{3}$$

Among them, (x_i, y_i, z_i) is the coordinates of the inspection point i, and (x_j, y_j, z_j) is the coordinates of the inspection point j.

Therefore, the distance between inspection points on the same floor can be expressed as

$$d_{ij} = |x_j - x_i| + |y_j - y_i| \tag{4}$$

Based on the above settings, the objective function of the inspection route for the selected inspection point of the entire building can be expressed as

$$f(w_{kl}) = \sum_{k=1}^{M} \sum_{l=1}^{n_k} d_{i_l i_{l+1}} + \sum_{k=1}^{M-1} 2^* H_k \tag{5}$$

Among them: $f(w_{kl})$ is the path taken by inspectors during a patrol, which is generated by the selected inspection point.

$w_{kl} = (i_1, i_2, \ldots, i_{n_k})$ represents a permutation combination of k-level inspection points 1, 2, ..., $i_{n_k+1} = i_1$; H_k Indicates the floor height between floor k and floor $k - 1$.

In response to the above problems, building a three-dimensional model of the building and parameterizing the inspection point. For the randomly generated problem, two problems need to be solved: the one is the random selection of floors and patrol points, and the other is the optimization of path planning. Therefore, the problem of random generation of patrol routes is to solve the problems of random selection and optimal path planning.

3   Proposed Approach

3.1   The Floor Random

The floor random algorithm is based on t linear congruential method. First, each floor is assigned a random number, and then, we use the floor random number to judge whether the floor is selected. Assuming random numbers for each floor are random, a random selection of floors can be achieved. When all floors are selected randomly, all floors may not be selected in the case. This situation is not satisfied the requirements of the patrol route. Therefore, the floor random algorithm must increase a condition like the minimum selection number of floors to meet the actual requirements of the patrol route. The steps of the floor random algorithm are as follows:

(1) Get the floor number based on building information, and assign a random probability P_f for all floors: $P_f \in (0, 1)$
(2) Judge the random probability P_f, and when $P_f \geq 0.5$, the number of the floor S_f is recorded. When all floors have been judged, calculate the cumulative number of the selected floors:

$$S_f = \begin{cases} P_f \geq 0.5 & S_f = 1 \\ otherwise & S_f = 0 \end{cases} \tag{6}$$

$$F = \sum_{f=1}^{n} S_f \tag{7}$$

(3) Judge whether the cumulative number is greater than the minimum number of the selected floors F_{max}. If $F < F_{max}$, repeat the first and second steps until $F \geq F_{max}$.

3.2   The Patrol Point Random

The floor random algorithm determines which floors in the building are selected. The patrol point random algorithm is to obtain the random patrol points in the selected floor. The principle of the patrol point random algorithm is similar to the floor random algorithm, and the steps of the patrol point random algorithm are as follows:

(1) Get the patrol points number based on the selected floor, and assign a random probability P_p for all points. Assume that the stairway is the starting and the ending point of the patrol route, and the random probability of the starting point and the ending point is always 100%:

$$\begin{cases} P_p = 1 & p = 1 \\ P_p \in (0, 1) & otherwise \end{cases} \tag{8}$$

(2) Judge the random probability P_p, and when P_p is greater than 0.5, the number of the floor S_p is recorded. When all floors have been judged, calculate the cumulative number of the patrol points:

$$S_p = \begin{cases} P_p \geq 0.5 & S_p = 1 \\ otherwise & S_p = 0 \end{cases} \tag{9}$$

$$P = \sum_{p=1}^{n} S_p \tag{10}$$

(3) Judge whether the cumulative number is greater than the minimum number of the selected floors P_{max}. If $P < P_{max}$, repeat the first and second steps until $P \geq P_{max}$.

3.3 Route Plan

The patrol route for a single floor can be expressed as a TSP issue. Based on the patrol route planning for a single floor, the patrol route planning problem for 3D building can be realized, and the steps can be shown as follows:

(1) Initialize the ant colony size and the maximum number of interactions.
(2) Place m ants on the same starting point. The probability that the k-th ant at the patrol point I selects the next patrol point j is expressed as

$$P^k(i, j) = \begin{cases} \dfrac{[\tau(i,j)]^\alpha \cdot [\eta(i,j)]^\beta}{\sum\limits_{s \notin tabu_k} [\tau(i,s)]^\alpha \cdot [\eta(i,s)]^\beta}, & if\ j \notin tabu_k \\ 0, & otherwise \end{cases} \tag{11}$$

where: $\tau(i, j)$ indicates thepheromone concentration on the path between inspection point i and inspection point j;
$\eta(i, j) = 1/d_{ij}$ represents as stimulating factor, d_{ij} indicates the distance between the petrol point i and the petrol point j.
α, β denotes the importance of pheromone and stimulating factor.
$tabu_k$ denotes as taboo list, the taboo list is the list of patrol points visited by ants.
(3) Record the best path for this iteration and update the pheromone globally.
(4) Update all pheromones when all of ants traverse the selected patrol point in the same layer. The update algorithm is expressed as:

$$\tau_{ij}(t + n) = \rho \cdot \tau_{ij}(t) + \Delta\tau_{ij}, \tag{12}$$

where: ρ is the pheromone's persistent constant, $\rho < 1$;
$\Delta\tau_{ij} = \sum_{k=1}^{m} \Delta\tau_{ij}^k$ denotes as the changes of the pheromone. $\Delta\tau_{ij}^k$ denotes as the pheromone update value of the k-th ant.

$$\Delta \tau_{ij}^k = \begin{cases} \frac{Q}{L_k} & ij \in l_k \\ 0 & otherwise \end{cases} \tag{13}$$

where: $L_k = \sum_{k=1}^{n_k} l_k$ denotes as the total path that the k-th ant has walked through. Q is constant.

(5) The end condition

When the number of iterations reaches the maximum number or reaches the design accuracy, the algorithm is converged. Repeat step 1 and step 2 to complete the optimal route planning for the remaining floors.

3.4 The Overall Process

Combining with the floor random algorithm, the patrol point random algorithm and the ant colony algorithm, the method of random generation of patrol path is completed. The flow of the algorithm is shown in Fig. 2.

4 Experimental Setup and Results

In order to verify the effectiveness of algorithm, this paper uses matlab software to simulate the algorithm. Assuming that the building consists of 5 floors and 13 patrol points in each floor, $F_{max} = 3$ and $P_{max} = 10$. Assuming that the patrol points in each floor are in the same position, the coordinates of the patrol points are ((0, 0), (3, 1), (7, 1), (9, 3), (9, 7), (7, 9), (5, 7), (3, 9), (1, 7), (5, 5), (3, 3), (4, 4), (6, 2), (9, 8)). The simulation results are shown below.

First run the algorithm: The selected floors are $\{1, 2, 3\}$, the number of patrol points in the selected floors is $\{12, 11, 12\}$. The optimal patrol path length is 107.4216. The patrol path of the building are shown in Fig. 3. And the iterative process of the algorithm are shown in Fig. 4.

Second run the algorithm: The selected floors are $\{1, 2, 4, 5\}$, the number of patrol points in the selected floors is $\{12, 11, 12, 11\}$. The optimal patrol path length is 141.7634. The patrol path of the building are shown in Fig. 5. And the iterative process of the algorithm are shown in Fig. 6.

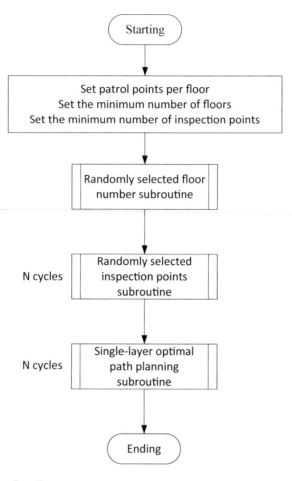

Fig. 2 Algorithm flow diagram

The simulation results show that the floors and the patrol points are selected randomly and the optimal path planning for multiple floors can be achieved. The simulation verifies the feasibility of using the artificial intelligence algorithm to implement the automatic planning algorithm for the random patrol route of three-dimensional building.

Fig. 3 The patrol path of the building

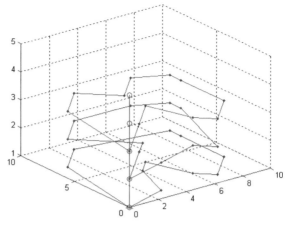

Fig. 4 Algorithm iteration process

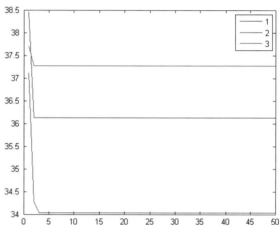

Fig. 5 The patrol path of the building

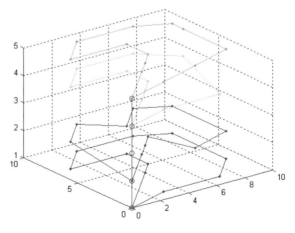

Fig. 6 Algorithm iteration process

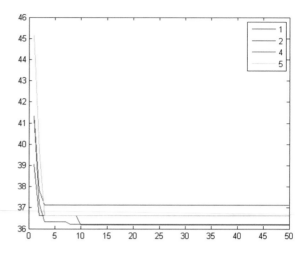

5 Conclusions

Compared with the usual electronic inspection system, this paper realizes the random automatic generation of the electronic inspection path.

This article analyzes and studies the problem of random automatic generation of building inspection routes. Absolute coordinates in intelligent drawings are converted to Cartesian coordinates, and building floors and inspection points are parameterized. Building 3D path data model is built.

At the same time, the patrol path random algorithm flow is established based on the ant colony algorithm and random algorithm in the artificial intelligence algorithm. And it is simulated through experiments. In the experimental simulation, the random automatic generation of the electronic patrol route was implemented, which demonstrated the effectiveness of the artificial intelligence algorithm in randomly generating the patrol route.

References

1. Cai, C.: Smart city development review and outlook 2012–2013. Mod. Telecommun. Technol. **1**, 27–32 (2013)
2. Dong, D.X.: Talking about intelligent building building automatic control system. Urban Constr. Theory Res. (Electronic Version) (7), 58–58 (2016)
3. Hua, S.: Group Intelligence and its Application in Distributed Knowledge Management. Hefei University of Technology, Hefei (2007)
4. Xiu, Z.M., Ke, C., Xia, W.M.: Improved ant colony algorithm in dynamic path planning. Comput. Sci. (1), 314–316 (2013)
5. Su, N.R., Qiang, L., Ji, W.L.: An improved particle swarm optimization algorithm based on bionics. Comput. Eng. Appl. (6), 61–63 (2014)

6. Shu, F., Cheng, Z.Z., Wu, S.Y.: Maximal likelihood doa estimation with bat algorithm. Mod. Electron. Technol. (8), 26–29 (2016)

7. Dorigo, M., Di Caro, G.: Ant colony optimization: a new meta-heuristic. In: Proceedings of the 1999 Congress on Evolutionary Computation, pp. 1470–1477 (1999)

8. Dorigo, M., Maniezzo, V., Colorni, A.: The ant system: optimization by a colony of cooperating agents. IEEE Trans. Syst. Man Cybern. Part B **26**(1), 29–41 (1996)

9. Liuzhibin, Z.: Double heuristic reinforcement learning method based on bp neural network. Comput. Res. Dev. **52**(3), 579–587 (2015)

10. Huang, G.B., Zhu, Q.Y., Siew, C.K.: Extreme learning machine. Theory Appl. Neu-Rocomput. **70**(1/3), 489–501 (2006)

11. Yu, H., Sun, C., Yang, W., et al.: Al-Elm: one uncertainty-based active learning algorithm using extreme learning machine. Neurocomputing **166**, 140–150 (2015)

12. Huang G., Wang D., Lan Y., et al.: Extremelearning machine: a survey. Int. J. Mach. Learn. Cybermetics **2**(2) (2011)

13. Yu, B., Yong, Z., Cheng, C.: Parallel ant colony algorithm applied in optimization of public transit network. J. Dalian Univ. Technol. **2**, 211–214 (2007)

14. Hao, C.: The principle and application of ant colony optimization algorithm. J. Hubei Univ. (Natural Science Edition), (4), 350–352 (2006)

15. Zheng, Z.J.: Research on Load Forecasting Means of Power Network in Low Voltage Platform Based on Cluster Intelligent Means. Shanghai Jiaotong University, Shang Hai (2011)

Part II
Intelligent Communications
& Networking

Channel Power Estimation for DVBRCS to DVBS2 Onboard DSP Payload

Krishna K. Wadiwala, Neeraj Mishra, Deepak Mishra and Hetal Patel

1 Introduction

DVB-S2 [1] is second generation broadband standard for satellite communication. It is used for digital television and for data broadcasting through satellite. It designed to provide flexibility with efficient utilization of the bandwidth with less complex trans-receiver compared to previous standards. Generally, the satellite links are designed with 4–5 dB margins to cater the dynamics of the link. The DVB-S2 provides the tools in terms of choosing modulation and coding parameters to maintain guaranteed QOS for each user even in the worst-case link dynamics. In DVB, only one type of modulation scheme is used where in DVBS2 has variable modulation scheme. It provide a flexibility in terms of modulation. Current standard allows the operator to choose and optimize these parameters on individual basis based on the channel power receiver form the user. In this process there is the requirement to monitor the power in each chain.

The digital subsystems are widely used in satellites to reduce the guard band requirements along with flexibility and re-configurability. Apart from the inherent advantages, digital systems are sensitive to input power. Generally, the requirements for the dynamic range in digital systems varies from 20 to 40 dB range. This poses the high finite words requirement in the digital system. This leads to the complex design of the hardware. Basically, if we can maintain the input power level to digital system at fixed level the design of the digital system will simplified. This can be done by measuring the signal power at the input of the receiver and amplifying or attenuating the signal power in digital domain.

_____

K. K. Wadiwala (✉) · H. Patel
ADIT, New Vallabh Vidyanagar, Gujarat, India
e-mail: Radha.Wadiwala@gmail.com

N. Mishra · D. Mishra
SAC-ISRO, Ahmedabad, Gujarat, India

© Springer Nature Singapore Pte Ltd. 2019
S. Tiwari et al. (eds.), _Smart Innovations in Communication and Computational Sciences_, Advances in Intelligent Systems and Computing 851,
https://doi.org/10.1007/978-981-13-2414-7_16

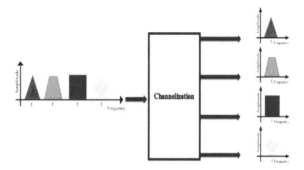

Fig. 1 Basic idea of channelization [2]

In multicarrier signal transmission environment, each channel has a different modulation and coding. To process the each and every channel faithfully, all signals must be in specified signal power range. Apart from this the channel of interest may be of different bandwidths and may be uniformly or non-uniformly, continuously or non-continuously distributed over the input band. This can lead to large variation in received power at the satellite receiver. Thus, there is always a requirement to maintain the constant power for individual channel. The processing in such scenario is through extraction of individual channel form the input band. This can be done by the channelizer device in the satellite. After channelization, power estimation is applied on each narrowband channel. The extracted signal is amplified and attenuated based on the measured power in the channel. Figure 1 shows the basic concept of the channelizer.

In the above requirement there is need to monitor the received power before and after the channelizer in the satellite. The algorithm of the measurement must be independent of the signal parameter. Various author has reported the method of the power measured in the literature.

Power spectrum estimation is important parameter for to check the signal. The heart of the power spectrum estimation is FFT. The FFT segment size is one of the important parameter for power spectrum estimation specifically, small segment size can achieve high detection performance but it concurrently leads to lower frequency resolution and vice versa [3]. Also averaging is important, Filippo Attivissimo et al. presented the effect of averaging techniques under noisy conditions with the view of welch method [4]. This Paper describes the superiority of nonlinear overlapping method under such circumstances. Kurt Barbe et al. presents a PSD with circular overlap for power measurement. The circular overlap imposes a discontinuity so variance is decreases [5].

There are many parameters to control in DVBS2. like number of satellite antenna beams, frequency and polarization reuse. DVBS2 gives a flexibility in terms of modulation. At the receiver side, the demodulation power will reduce. It is a serious issue to solve in future.

The present paper compares the parametric and non-parametric power estimation methods in the view of satellite communication scenario. The proposed algorithm will be used in satellite for monitoring the received power. The Sect. 2 describes the power estimation method in time and frequency domain. The Sect. 3 compares the nonparametric with parametric methods for power estimation. In Sect. 4 describes all nonparametric methods. In Sect. 5 the MATLAB simulation result for all methods are presented. Section 6 provides the details of the implemented algorithm on Xilinx Vertex 5 FPGA. Section 7 provide the conclusion.

2 Theory

This section describes the calculation of signal power using power spectral density [6] in time domain and frequency domain. Time domain analysis shows how a signal power changes over time [7]. The time domain power spectrum estimation defined as is

$$P_1 = \frac{1}{N} \sum_{n=-\infty}^{\infty} |x(n)|^2 \tag{1}$$

where x(n) is an input signal in time domain and N is a number of samples.

Frequency domain analysis shows how the signal's energy is distributed over a range of frequencies [7]. Frequency domain analysis is widely used in communications, geology, remote sensing, and image processing. The frequency domain power spectrum estimation equation is

$$P_2 = \frac{1}{F} \sum_{n=-\infty}^{\infty} |X(f)|^2 \tag{2}$$

where X(f) is an input signal in frequency domain and F is a number of samples in frequency domain.

3 Parametric Versus Nonparametric

There are two classes of methods for power estimation, one is parametric method and other nonparametric method.

In parametric method prior knowledge of the signal structure is required whereas in nonparametric method no prior knowledge is required for power estimation. Hence when the incoming signal characteristics are known then parametric method is used and otherwise nonparametric methods are used. But in parametric method every time the specific power estimation model has to apply for power estimation. Using

Table 1 Table of comparison of parametric and nonparametric method

Parameter	Parametric	Nonparametric method
Methods	AR MA ARMA	Periodogram Modified periodogram Barlett's method Welch's method
Prior knowledge about signal	Needed	Not needed
Window used	No	Yes
Leakage problem	No	Yes
Frequency resolution	Good	Poor
Data length	More	Less
Complexity	High	Low
Dummy frequency peaks	Present	Not present

nonparametric method as no knowledge is required hence only small length of signal structure is used. The comparison between the parametric and nonparametric method is given in Table 1. The nonparametric method is chosen for power estimation as it is being less complex and model independent.

4   Nonparametric Method

There are five common nonparametric PSD available in the literature: the periodogram [8], the modified periodogram [8], Bartlett's method [9], Welch's method [10] and Blackman–Tukey [11]. In this section all methods are describes in detail.

4.1   Periodogram

Periodogram is the simplest method for finding a power spectrum estimation method of the random signal. Periodogram method is described in below Fig. 2.

Consider the $x_N(n)$ is the finite length of the signal. It is produce by multiplying the input signal $x(n)$ with the rectangular window $w_R(n)$. Here N is a window length.

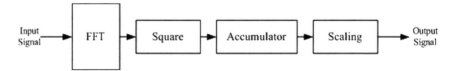

Fig. 2 Periodogram process

Find the autocorrelation of the signal $x_N(n)$. Taking the Fourier Transform of the autocorrelation $\hat{r}_k(k)$, the periodogram becomes

$$\hat{P}_{per}\left(e^{j\omega}\right) = \frac{1}{N}\left|\sum_{n=0}^{N-1} x(n)e^{-jn\omega}\right|^2 \tag{6}$$

4.2 Modified Periodogram

Modified periodogram is slightly different then the periodogram. Triangular window is used for the traction of the input signal. This window provides the smoothing to the power spectrum.

Consider the $x_N(n)$ is the finite length of the signal. It is produce by multiplying the input signal $x(n)$ with the triangular window $w_R(n)$. Find the autocorrelation of the signal $x_N(n)$. Taking the Fourier Transform of the autocorrelation $\hat{r}_k(k)$, the modified periodogram becomes

$$\hat{P}_M\left(e^{j\omega}\right) = \frac{1}{NU}\left|\sum_{n=-\infty}^{\infty} x(n)w(n)e^{-jn\omega}\right|^2 \tag{10}$$

where N is a Window length.
Where,

$$U = \frac{1}{N}\sum_{n=0}^{N-1}|w(n)|^2 \tag{11}$$

4.3 Bartlett's Method

Both the Periodogram Method and Modified Periodogram Method do not give zero variances as the data length approaches in unity. One way to enforce zero variance is by averaging. The bartlett's method is the averaging based method, the steps are listed below.

Consider the N point data input sequence. This signal is subdivided into K overlapping segments. Here each segment has length L. The total number of segment is K. Where L = N/K (Fig. 3).

$$X_i(n) = (n + iL) \tag{12}$$

where $n = 0, 1, \ldots, L-1$ and $i = 0, 1, \ldots, K-1$.

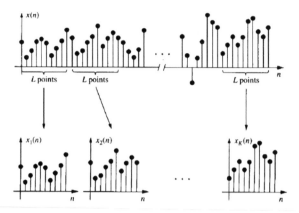

Fig. 3 Partitioning x(n) into nonoverlapping sub sequences

Then find the periodogram of each of the K segment. Take the Averaging of the periodogram for the K segment. The final Bartlett power spectrum estimation equation is

$$\widehat{P}_B\left(e^{j\omega}\right) = \frac{1}{N} \sum_{i=0}^{K-1} \left| \sum_{n=0}^{L-1} x(n+iL)w(n)e^{-jn\omega} \right|^2 \tag{13}$$

4.4 Welch's Method

Welch method allowing the overlapping of the segment in bartlett's method. By applying overlapping, welch method improve the result of the power spectrum estimation.

Consider the N point data input sequence. This signal is subdivided into K overlapping segments. Here each segment has length L. The total number of segment is K. Where $L = N/K$

$$X_i(n) = x(n + iD) \tag{14}$$

where n = 0, 1, ..., L − 1 and i = 0, 1, ..., K − 1.
The amount of overlap is L-D points is shown in Table 2.

Table 2 Table of the amount of overlap between L-D points

D = L	No overlap
D = L/2	50% overlap (Most common)
D = 3L/4	25% overlap

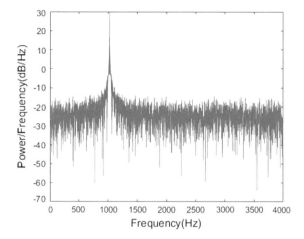

Fig. 4 Periodogram (1000 Hz with noise (SNR = 20))

Then find the periodogram of each of the K segment. Take the Averaging of the periodogram for the K segment. The final Welch power spectrum estimation equation is,

$$\widehat{P}_W\left(e^{j\omega}\right) = \frac{1}{NU} \sum_{i=0}^{K-1} \left| \sum_{n=0}^{L-1} x(n + iL)w(n)e^{-jn\omega} \right|^2 \tag{13}$$

Table 3 shows the comparison of all nonparametric methods.

5 Software Simulation Result

This section shows the MATLAB simulation result. 1 MHz and 2 MHz signal with two different Signal to Noise Ratio (SNR) 20 and 30 are used for performing this experiment. In time domain the power is 0.4900 V/Hz at 1000 and 2000 Hz. In case of frequency domain, the power is 0.4900 V/Hz 1000 and 2000 Hz.

Similarly for nonparametric methods measured PSD is given in Table 4 and also the PSD simulation result is shown in Figs. 4, 5, 6 and 7.

Table 4 shows a MATLAB results for all different nonparametric methods. Here almost the power is constant. All method gives a same power.

Simulation result shows that the amount of change in signal frequency or change in noise variance will not alter the estimated average power. Further due to less complexity for measurement of power of the channel in DVBS2, periodogram is finalized. Periodogram is one of the easiest approach for finding a power. It has lowest complexity among all nonparmetric methods.

Table 3 Comparison of all nonparametric methods

Parameter	Periodogram	Modified periodogram	Barlett's method	Welch's method
Equation	$\frac{1}{N}\left\|\sum_{n=0}^{N-1} x(n)e^{-in\omega}\right\|^2$	$\frac{1}{NU}\left\|\sum_{n=-\infty}^{\infty} x(n)w(n)e^{-in\omega}\right\|^2$	$\frac{1}{N}\sum_{i=0}^{K-1}\left\|\sum_{n=0}^{L-1} x(n+iL)w(n)e^{-in\omega}\right\|^2$	$\frac{1}{NU}\sum_{i=0}^{K-1}\left\|\sum_{n=0}^{L-1} x(n+iL)w(n)e^{-in\omega}\right\|^2$
Bias	$\frac{1}{2\pi}P_x\left(e^{i\omega}\right) * W_B\left(e^{i\omega}\right)$	$\frac{1}{2\pi NU}P_x\left(e^{i\omega}\right) * \left\|W\left(e^{i\omega}\right)\right\|^2$	$\frac{1}{2\pi}P_x\left(e^{i\omega}\right) * W_B\left(e^{i\omega}\right)$	$P_x\left(e^{i\omega}\right) * \left\|W_B\left(e^{i\omega}\right)\right\|^2$
Resolution	$0.8\frac{2\pi}{N}$	Window dependent	$\Delta\omega = 0.8k\frac{2\pi}{N}$	Window dependent
Variance	$P_x^2\left(e^{i\omega}\right)$	$P_x^2\left(e^{i\omega}\right)$	$\frac{1}{k}P_x^2\left(e^{i\omega}\right)$	$\frac{9}{16}\frac{L}{N}P_x^2\left(e^{i\omega}\right)$
Smoothing	Not possible	Good	Improving using increasing segment	Improve using overlapping
Consist parameter	Not	Not	Yes	Yes
Frequency resolution	Decrease with N	Decrease with N	Decrease with N and k	Decrease with overlapping
Leakage	Yes	Yes (Depend on window)	Yes (Depend on window)	Yes (Depend on window)
Computational complexity	Addition = N $\log_2$ N; Multiplication = $\frac{N}{2}\log_2 N$	Addition = N $\log_2$ N + N; Multiplication = $\frac{N}{2}\log_2 N$ + N	$\frac{N}{M}\left(\frac{M}{2}\log_2 M\right)$	$\frac{2N}{M}\left(\frac{M}{2}\log_2 M\right)$
Complexity	Lowest	Moderate	High	Highest

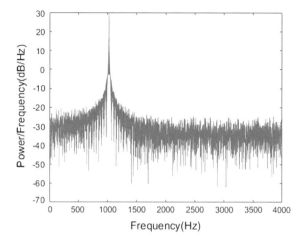

Fig. 5 Periodogram (1000 Hz with noise (SNR = 30))

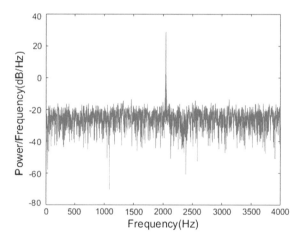

Fig. 6 Periodogram (2000 Hz with noise (SNR = 20))

6   Hardware Simulation Result

The periodogram algorithm is implemented in Xilinx FPGA XC4VSX55FF-1. Figure 8 shows the architecture block diagram. The ADC5463 samples the input signal at 70 ± 4 MHz with 131.072 MHz sampling clock.

This signal is processed by the FPGA and finally the calculated power value is modulated to plane carrier at 16.384 MHz. The DAC5675 converts the digital data to analog form at 131,72 MHz sampling frequency Table 5 shows the values of measured output power in various signal power level at fixed Eb/No. Table 6 shows the output power at various Eb/No at fixed signal power.

Table 4 Simulation result of power spectrum estimation using various nonparametric methods

Method	Periodogram (V/Hz)	Modified periodogram (V/Hz)	Bartlett's method (V/Hz)	Welch's method (V/Hz)
1000 Hz with noise (SNR = 20)	0.5046	0.5048	0.4750	0.4798
1000 Hz with noise (SNR = 30)	0.5007	0.4999	0.4988	0.4924
2000 Hz with noise (SNR = 20)	0.5042	0.5071	0.4798	0.4674
2000 Hz with noise (SNR = 30)	0.4994	0.5002	0.4716	0.4741

Table 5 Measured output power at fixed Eb/No

S. no.	Input power level (dBm)	Output power level	C/N (dB)	Eb/N0 (dB)
1	0	−16.93	10.99	14
2	−1	−16.94	10.99	14
3	−2	−18.01	10.99	14
4	−3	−19.78	10.99	14
5	−4	−21.72	10.99	14
6	−5	−23.43	10.99	14
7	−6	−25.36	10.99	14
8	−7	−27.29	10.99	14
9	−8	−29.31	10.99	14
10	−9	−31.45	10.99	14

Table 6 Measured output power at fixed input power level

S. no.	Input Eb/No (dB)	Input power level	C/N	Output power level
1	3	0	−0.01	−20.83
2	4	0	0.99	−20.35
3	5	0	1.99	−19.98
4	6	0	2.99	−19.64
5	7	0	3.99	−19.37
6	8	0	4.99	−18.99
7	9	0	5.99	−18.65
8	10	0	6.99	−18.19
9	11	0	7.99	−17.95
10	12	0	8.99	−17.45

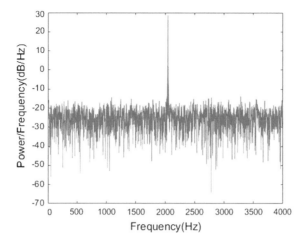

Fig. 7 Periodogram (2000 Hz with noise (SNR = 20))

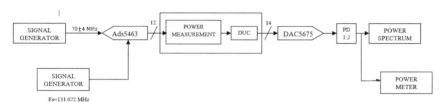

Fig. 8 Architecture diagram

7 Conclusion

In DVBS2, the variable coding and modulations are selected based on the individual channel power. From this experiment, the amount of change in signal frequency or change in noise variance will not alter the estimated average power in chosen algorithm. Hence DVB-S2 periodogram method is choose for power measurement because it has lowest complexity as compare to all other nonparametric methods.

References

1. Alberty, E., et al.: Adaptive coding and modulation for the DVB-S2 standard interactive applications: capacity assessment and key system issues. IEEE Wirel. Commun. **14**(4), 61–69 (2007)
2. Vaidyanathan, P.P.: Multirate Signal Processing Filter Banks. Prentice-Hall, Englewood Cliffs, NJ (1993)
3. Iwata, H., Umebayashi, K., Tiiro, S., Suzuki, Y., Lehtomäki, J.J.: A study on Welch FFT segment size selection method for spectrum awareness. In: 2016 IEEE Wireless Communications and Networking Conference, Doha, pp. 1–6 (2016)

4. Attivissimo, F., Savino, M., Trotta, A.: A study on nonlinear averagings to perform the charac-
terization of power spectral density estimation algorithms. IEEE Trans. Instrum. Meas. **49**(5),
1036–1042 (2000)
5. Barbe, K., Schoukens, J., Pintelon, R.: Non-parametric power spectrum estimation with circular
overlap. In: 2008 IEEE Instrumentation and Measurement Technology Conference, Victoria,
BC, pp. 336–341 (2008)
6. Hayes, M.H.: Statistical Digital Signal Processing and Modeling. Wiley (1996)
7. Prokis, J.G., Manolakis, D.G.: Digital Signal Processing Principles, Algorithms Applications.
Prentice-Hall, Inc (1996)
8. Schuster, A.: On the investigation of hidden periodicities. Terrestr. Magn. **3**, 13–41 (1898)
9. Bartlett, M.S.: Smoothing periodograms from time series with continuous spectra. Nature **161**,
686–687 (1948)
10. Welch, P.: The use of fast fourier transform for the estimation of power spectra: a method
based on time averaging over short, modified periodograms. IEEE Trans. Audio Electroacoust.
AE-15, 70–73 (1967)
11. Stoica, P., Moses, R.: Introduction to Spectral Analysis. Prentice-Hall, Upper Saddle River, NJ
(1997)

A Slotted Microstrip Patch Antenna for 5G Mobile Phone Applications

Kausar Parveen, Mohammad Sabir, Manju Kumari and Vishal Goar

1 Introduction

Nowadays because of quick change in communication field to strength of mobile speed there is an increase in data rate. Most of challenges are rising day by day [1, 2]. To overcome this challenge technological advancement has taken place in the field of new mobile communication network and subscribers increased up to 150 million and more [3]. 4G network did not fulfill the need of customer. To resolve this problem many researchers have worked on millimeter wave, i.e., high frequency band spectrum. Millimeter wave band called as 5th generation mobile network that provides gigabit communication services [4].

The designed microstrip patch antenna has a compact size with dimensions 6 mm × 6 mm × 1.6 mm. The substrate material used for patch antenna also affects the result efficiency and gain, i.e., RT-duroid is one of them. It is used for millimeter wave band and gives good result [3, 5].

In Sect. 2, we discussed the literature survey of the previous work. In Sect. 3, the designed antenna is discussed. Section 4 consists of simulation result such as VSWR

K. Parveen (✉)
Geetanjali Institute of Technical Studies, Udaipur, India
e-mail: kausarparveen24@gmail.com

M. Sabir
ECE Department, GITS, Udaipur, India
e-mail: sabii.sankhla@gmail.com

M. Kumari
CS Department, Government Polytechnic College, Bikaner, India
e-mail: manjusunia08@gmail.com

V. Goar
ECB, Bikaner, India
e-mail: dr.vishalgoar@gmail.com

© Springer Nature Singapore Pte Ltd. 2019   173
S. Tiwari et al. (eds.), *Smart Innovations in Communication and Computational Sciences*, Advances in Intelligent Systems and Computing 851,
https://doi.org/10.1007/978-981-13-2414-7_17

plot, S11 (return loss plot), 3D-gain plot and Directivity. And finally conclusion is described in Sect. 5.

2 Literature Survey

5G networks provide number of services for the user by improving bandwidth and data rate [6]. But the main reason is to increase the need of 5G technology as it reduces the size of the antenna, it is light in weight so it reduces size of antenna [7, 8]. Verma et al. [1] proposed a microstrip patch antenna has a compact size of 20 mm × 20 mm × 1.6 mm resonate at 10.15 with 4.46 dBi gain. Jandi et al. [9] proposed microstrip patch antenna has a compact size of 19 mm × 19 mm × 0.787 mm and provide a gain of 5.51 dB at 10.15 GHz and 8.03 dB at 28 GHz. Rafique et al. [10] presented an antenna has compact structure of 16 mm × 16 mm resonate at 28 and 38 GHz. Adedotun et al. [11] This paper present a simple antenna having square –shaped port patch. Antenna performance is carried out from HFSS antenna software and proposed antenna resonate at 28.3 GHz frequency.

3 Proposed Antenna Dimension and Formula

A small microstrip patch antenna has inset feeding techniques. The parameter of this antenna is 6 mm × 6 mm with substrate thickness 1.6 mm and FR4-material is used for substrate whose dielectric constant is 4.4. A proposed antenna has rectangular radiating patch i.e. 2.69 mm × 4.55 mm. And one slot is etched on ground plane.

These are the steps for antenna design: [12]

- Calculate Width (a practical width that lead to good radiation efficiency is)

$$W = \frac{1}{2f_r\sqrt{\mu_0\epsilon_0}}\sqrt{\frac{2}{\epsilon_r + 1}} = \frac{\upsilon_0}{2f_r}\sqrt{\frac{2}{\epsilon_r + 1}} \tag{1}$$

where υ_0 is the free-space velocity of light.

- Calculate effective dielectric parameters

$$\epsilon_{reff} = \frac{\epsilon_r+1}{2} + \frac{\epsilon_r-1}{2}\left[1 + 12\frac{h_t}{W}\right]^{-1/2}$$
$$\frac{W}{h_t} > 1, \tag{2}$$

$$\frac{\Delta L}{h_t} = 0.412 \frac{\left(\epsilon_{reff} + 0.3\right)\left(\frac{W}{h_t} + 0.264\right)}{\left(\epsilon_{reff} - 0.258\right)\left(\frac{W}{h_t} + 0.8\right)} \tag{3}$$

- Calculate actual length (L), such that, [12]

$$L = \frac{1}{2f_r\sqrt{\epsilon_{reff}}\sqrt{\mu_0\epsilon_0}} - 2\Delta L \tag{4}$$

- Effective Length.

$$L_{eff} = L + 2\Delta L \tag{5}$$

The dimension of proposed antenna are substrate length = 6 mm, substrate width = 6 mm, substrate thickness 1.6 mm, Ground length and width is 6 mm × 6 mm, patch length = 2.69 mm, patch width = 4.55 mm, inset feed line = 0.78 mm, slot length = 04 mm, slot width = 0.4 mm.

4 Simulation and Results

To check the performance of proposed antenna with HFSS software (Fig. 1).

The return loss and VSWR of this proposed antenna is −28.32 and 0.66 dB respectively. This proposed antenna work at resonating frequency is 31.62 GHz with 5.17 dB gain and 5.49 dB directivity as shown in Fig. 2.

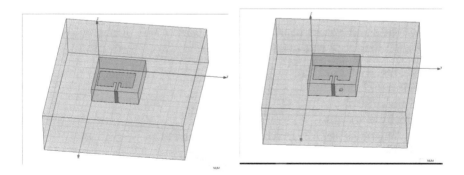

Fig. 1 3D view of proposed antenna without and with ground slot

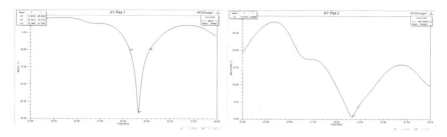

Fig. 2 Return loss and VSWR of proposed microstrip antenna without slot

Fig. 3 3-D gain plot and directivity of proposed antenna without slot

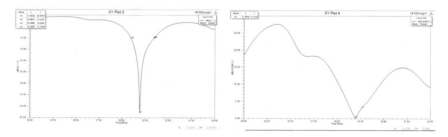

Fig. 4 Return loss and VSWR of proposed antenna with ground slot

After etched on ground plain of proposed antenna with dimension 0.4 mm × 0.4 mm, the simulation results are as follows:

- Return loss = −42.97 dB.
- VSWR = 0.1234 dB.
- Gain plot = 5.2 dB.
- Directivity = 5.55 dB.
- Resonating frequency = 31.90 GHz (Figs. 3, 4 and 5).

From this above result we understood that by cutting slot on ground we have better result and good efficiency of patch antenna.

The performance of the proposed microstrip patch antenna (with and without slot) in terms of VSWR, return loss, gain and directivity is shown in Table 1.

Fig. 5 3D gain plot and directivity of proposed antenna with ground slot

Table 1 Comparison between reference paper and proposed antenna (with and without slot)

Parmeter	Reference paper	Proposed antenna without slot	Proposed antenna with ground slot
Return loss (dB)	−18.27	−28.32	−42.97
VSWR (dB)	2.13	0.66	0.12
Gain (dB)	4.46	5.17	5.2
Directivity (dB)	–	5.49 dB	5.55 dB
Resonant frequency (GHz)	10.15	31.62	31.90
Size of antenna	20 mm ×20 mm × 1.6 mm	6 mm ×6 mm × 1.6 mm	6 mm ×6 mm × 1.6 mm Slot—0.4 mm × 0.4 mm

5 Conclusions

In this paper, a compact size antenna has been simulated for 5 G wireless communication. The proposed patch antenna without slot resonates at 31.62 GHz with a return loss of −28.32 dB while patch antenna with ground slot resonates at 31.90 GHz with return loss of −42.97 dB. From the results it is clear that the return loss of antenna with ground slot has improved over same antenna (without slot) and gives good efficiency and gain. Because of compact size (6 mm × 6 mm × 1.6 mm) it can be used where space is most important issue.

References

1. Verma, S., Mahajant, L., Kumar, R., Singh, H.S., Kumar, N.: A small microstrip patch antenna for future 5G application. In: IEEE 5th International Conference on Reliability, Infocom Technologies and Optimization (ICRITO) (Trends and future Directions), 7–9 Sept 2016, pp. 460–463
2. Sam, C.M., Mokayef, M.: A wide band slotted microstrip patch antenna for future 5G. EPH-Int. J. Sci. Eng. **2**(7), 19–23 (2016). ISSN: 2454-2016

3. Annalakshmi, E., Prabakaran, D.: A patch array antenna for 5G mobile phone applications. Asian J. Appl. Sci. Technol. (AJAST) **1**(3), 48–51 (2017)
4. Mohammed Jajere, A.: Millimeter wave patch antenna design antenna for future 5G applications. Int. J. Eng. Res. Technol. (IJERT) **6**(02), 289–291 (2017). http://www.ijert.org. ISSN: 2278–0181
5. Outerelo, D.A., Alejos, A.V., Sanchez, M.G., Isasa, M.V.: Microstrip antenna for 5G broadband communications: overview of design issues. IEEE Commun. Mag. 2443–2444 (2015)
6. Mohan, G.P., Patil, S.: Design and development of fractal microstrip patch antenna for 5G communication. Adv. Wirel. Mob. Commun. **10**(1), 13–25 (2017). ISSN 0973–6972
7. Verma, M., Sundriyal, N., Chauhan, J.: 5G mobile wireless technology. Int. J. Res. (IJR) **1**(9) (2014)
8. Gupta, A., Gupta, A., Gupta, S.: 5G: the future mobile wireless technology by 2020. Int. J. Eng. Res. Technol. (IJERT) **2**(9) (2013)
9. Jandi, Y., Gharnati, F., Said, A. O.: Design of a compact dual bands patch antenna for 5G applications. In: 2017 International Conference on Wireless Technologies, Embedded and Intelligent System (WITS). IEEE 19–20 Apr 2017
10. Rafique, U., Khalil, H., Rehman, S.U.: Dual-band micristrip patch antenna array for 5G mobile communications. In: 2017 Progress in Electromagnetic Research Symposium-Fall (PIERS-FALL), pp. 55–59, 19–22 Nov 2017
11. Adedotun, O.K., Johnbosco, A.I.E.: Design and optimization of a square patch antenna for millimeter wave applications. Int. J. Sci. Res. (IJSR), **6**(4), (2017). ISSN (Online): 2319-7064. www.ijsr.netpg-2192-2194
12. A balannis Constantine, antenna theory. ISBN 987-0-471-66782-7

Reliability Study of Sensor Node Monitoring Unattended Environment

**V. Mahima, G. R. Kanagachidambaresan, M. Balaji
and Jagannath Das**

1 Introduction

The sensor nodes in a sensor network works on a single objective of sensing and reporting an event to the sink. The sensor node is a serial block having sensing, processing and transceiving subsystem. The overall sensor node senses an event and process the data and transmit the same to the nearby node (i.e.) for further processing. The sensing unit of the sensor node may be analog or digital in nature [1, 2]. The sensor provides data to the processing unit, the processing unit process as per the system requirement and transceive to the nearby node through available communication modules [3, 4]. The communication protocols commonly used are Bluetooth, ZigBee, LowPAN, etc., [5, 6]. The sensor nodes monitoring wide area are prone to failure due to environmental issues, aging and other external disturbances [7]. The sensors are normally resistive, capacitive, and inductive and even touch based sensors. The passive element sensors on impact with varying temperature and pressure face stresses and provide faulty data to the user [8–10]. The fault data from the sensor provides false predictions and catastrophic conditions. Figure 1 represents the sensor node monitoring unattended environment.

V. Mahima
Department of ECE, Vel Tech Rangarajan Dr. Sagunthala R&D Institute
of Science and Technology, Avadi, Chennai 600062, India

G. R. Kanagachidambaresan (✉)
Department of CSE, Vel Tech Rangarajan Dr. Sagunthala R&D Institute
of Science and Technology, Avadi, Chennai 600062, India
e-mail: kanagachidambaresan@gmail.com

M. Balaji · J. Das
Control Systems Group, Onboard Software Section,
ISRO Satellite Centre, Bangalore 560017, India

© Springer Nature Singapore Pte Ltd. 2019
S. Tiwari et al. (eds.), *Smart Innovations in Communication and Computational
Sciences*, Advances in Intelligent Systems and Computing 851,
https://doi.org/10.1007/978-981-13-2414-7_18

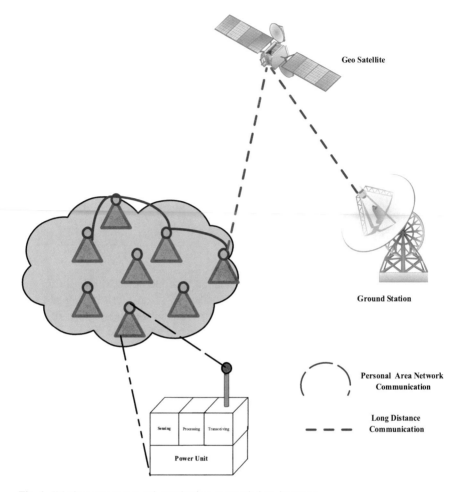

Fig. 1 Wireless sensor network monitoring unattended environment

Table 1 Major types of sensors used

Type	Sensors		
Agricultural	Temperature	Humidity	pH
Outer space	Temperature	Earth	Infrared
Industrial	Temperature	Humidity	Infrared
Sea side	Temperature	Humidity	Seismic

Table 1 illustrates the types of unmanned monitoring and the types of sensor nodes used for the same. The unattended monitoring would be agricultural, Industrial, Aerospace, and sea-side monitoring.

About all the unattended applications utilizes temperature sensors for monitoring and taking decisions. These sensors are normally resistive in nature. The unattended

Table 2 Protocols for unattended monitoring

Protocol	IEEE spec	Frequency band	Nominal range (m)	Base topology	Data rate
Bluetooth	802.15.1	2.4 GHz	10	Piconet	1 Mb/s
ZigBee	802.15.4	2.4 GHz; 5 GHz	10–100	Star	250 kb/s
Wi-Fi	802.11a/b/g	868/915 MHz; 2.4 GHz	100	BSS	54 Mb/s
UWB	802.15.3a	3.1–10.6 GHz	10	Piconet	110 Mb/s

monitoring encourages back up sensors for uninterrupted monitoring of event. The unattended monitoring data is communicated to ground station via satellite communication system [11]. The availability of energy to communicate mainly determines the data transmission [12]. The geo and orbital satellite sends the data received from the sensor nodes directly to the ground station. The sink is connected with the high power oscillator that transmits the sensor data to the ground station through satellite communication system. The power consumption required is satisfied through the renewable energy sources [13, 14]. Table 2 illustrates the different communication modules and their signal range.

The protocols mentioned in the Table 2 got its position due to low power consumption, reliability and ease of access in dynamic environment. The protocol also provides high coverage during line of sight. Normally in the unattended environments, line sight is occasional in nature.

2 Related Works

Many routing algorithms and sensor nodes study has been discussed in the recent research [15, 16]. The sensor nodes also facilitate patient monitoring, ambulatory monitoring, and industrial monitoring purposes [17, 18]. The reliability study of these systems is very meagerly discussed. The objective of such monitoring mainly depends on reliability of the system [19]. The data communicated (or) computed inside the sensor node is in tandem fashion. The failure of any system would result in combination failure and would not provide reliable data [1, 3–5]. The Battery Recovery based Lifetime Enhancement algorithm [12] addresses the network lifetime enhancement due to recovery of the battery. The recovery time is modeled and an increase in network lifetime is observed. The Performance-based analysis of homogenous clustering concept introduced in [13] discusses on queuing approach. The monitoring methodology is classified with the respect to states and store and forward technique is proceeded to have a better lifetime. The unattended monitoring, with the sensor nodes are not always time sensitive in nature. The data become time sensitive during critical durations and normal during normal occasions. The data classification proposed in the Fail Safe Fault Tolerant algorithm for WBSN in

[15, 17] provides better classification of the data with respect to the status of the subject. The subject status with respect to physiological sensors is modeled and an increase in the network lifetime is observed. The markov model based reliability analysis discussed in [6], here the mode of operation is memory less in nature. The present state of the system does not depend on memory. Most of the sensor network models are memory less in nature. However the Monte Carlo based simulation of reliability analysis provides a clear view on all possible fault and failure analysis. The reliability analysis discussed in [7, 8] provides detailed view on the linear sensor network. However the network reliability mainly depends on the individual subsystem reliability. The individual subsystem reliability is subjected to various conditions such as temperature, pressure, atmospheric variations, and external disturbances. The same can be simulated and the overall reliability can be studied using Monte Carlo analysis. In this paper, the real-time sensors are subjected to monitor the event continuously and their expected reliability after particular duration is computed. This paper provides the reliability study of the real-time sensor node on the whole and its individual subsystem.

3   Reliability Analysis

The sensor node subsystem meets different type of faults. The faults would be temporary (or) permanent. The permanent faults are not recoverable. During those conditions a HOT standby would help the sensor to extend the working of the sensors. The sensor system faults are inaccurate data due to wind, cloud coverage, vicinity issues, and other problems. The temporary fault may persist to be a permanent fault in some conditions. The reliability of working of a particular node is demanded by the user.

3.1   Mathematical Model

Figure 2 illustrates the reliability model of the sensor node considering, sensing, processing, transceiving, and power unit. The power unit serves all the other three units and it is connected in series with all the three units.

Table 3 illustrates the necessary system setting. The system parameter requirement is based on the need of the user and the operating environment conditions. The available sensors are calibrated before deployment inside the field to ensure its novel working operation.

From Table 3, the user requirement of sensor node after thirty days is given as 0.98. Equation 1 illustrates the fault rate predicted for a sensor node after 30 day interval. A take care is necessary after every 30 days.

$$\lambda r = -\ln(0.98)/(30 \times 12) \tag{1}$$

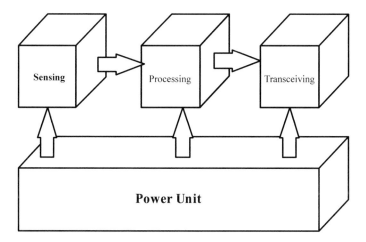

Fig. 2 Reliability relation of the sensor node

Table 3 System parameter requirements

Parameter	Range
Required reliability	0.98 for 30 days
Temperature sensor	−10° to 40°
Humidity range	0–30 g per cubic meter
pH	0–14

λ_r is the fault rate of the sensor.

Equation 2 illustrates the overall reliability of the sensor unit with its fault rate. Since the operating node is tandem in nature. The overall reliability value is the product of the individual subsystem.

$$\omega_k^* = sys_{1k}^* \times sys_{2k}^* \times sys_{3k}^* \times (sys_{4k}^*)^3 \tag{2}$$

sys_{1k}^* Individual sensing subsystem fault rate
ω_k^* Overall sensor fault rate including adc conversion, sensing element, etc.

The sensing system may include one or more sensors. The data from the sensing unit is aggregated making a series operation. Equation 3 illustrates the overall sensing unit fault rate value.

$$\omega^* = \sum_{k=1}^{N} \omega_k^* \tag{3}$$

ω^* Sensing unit fault rate.

Table 4 Sensor parameters

S. no.	Sensor	Number
1.	LM35	40
2.	Humidity (DHT11)	20
3.	Arduino Nano	25
4.	XBee S2	40

Equation 4 provides the complexity factors of the system with respect to each element in a particular subsystem.

$$C_k^* = \frac{\omega_k^*}{\omega^*} \tag{4}$$

C_k^* Complexity factor.

The overall fault rate of the subsystem given in Eq. 5 is analyzed with respect to complexity factor and expected fault rate as given in Eq. 1.

$$\lambda_* = C_k^* \times \lambda r \tag{5}$$

λ_r Fault rate of the subsystem.

Equation 6 illustrates the general reliability equation with respect to fault rate.

$$R_* = e^{\lambda_* t} \tag{6}$$

The total reliability of the tandem system is given in Eq. 7.

$$R_{\text{Sensor Node}} = R_{\text{SU}} \times R_{\text{PU}} \times R_{\text{TU}} \times (R_{\text{Power Unit}})^3 \tag{7}$$

4   Results and Discussion

The proposed system reliability is validated with the Real-time sensor with 98% reliability expectation after a month of deployment. Table 4 illustrates the sensor parameters and number of sensors used to study the reliability sf the sensors.

Figure 3 illustrates the sensor subsystem used for validating the predicted fault rate.

The system is continuously made to work for 48 h and their fault is revised. Table 5 illustrates the Sensor failure rates and reliability values after 30 days. The reliability for a particular system is checked after 30 days. By increasing the number of HOT-STAND sensors the total system reliability is increased and continues uninterrupted monitoring is availed.

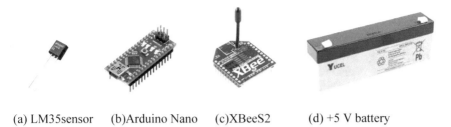

(a) LM35sensor   (b)Arduino Nano   (c)XBeeS2   (d) +5 V battery

Fig. 3 Subsystems used for reliability analysis

Table 5 Sensor failure rates

Sensor	Failure predicted	Number	Total failure (λ_*)	Reliability after 30 days
LM35	0.02	40	0.8	0.999424165856154
Humidity (DHT11)	0.032	20	0.64	0.999539306152014
Arduino Nano	0.01	25	0.25	0.999820016199028
XBee S2	0.003	40	0.12	0.999913603732372
Battery (5 V) lead acid (homogenous)	0.00001	40	0.0004	0.999999712000041

Fig. 4 Operational effectiveness of tandem system

The total reliability as per the Eq. 7 is

$$0.999424166 \times 0.999539306 \times 0.999820016 \times 0.999913604 \times 0.999999712 \times 0.998697^3 = \mathbf{0.998696786}.$$

Figure 4 illustrates the operational effectiveness of the system with respect to hours. The reliability of the system over a period of time reduced and needs replacement. In case of non-replaceable systems HOTSTAND is required to enhance the reliability of the system.

5 Conclusion

The reliability of the system working in tandem operation is studied and its operation effectiveness is also verified. The real-time results provide an idea on the need of HOTSTAND sensors to ensure the continuous monitoring. The extension of this work is planned for more number of sensors and long term system reliability study with communication perspective. The future work includes measuring reliability in real-time environment.

Acknowledgements We would like to thank Mr. S. Sudhakar, Group director, Control Digital and Electronics Group, ISRO Satellite Centre, Bangalore and Vel Tech Nano Satellite team for supporting this work.

References

1. Ammari, H.M., Das, S.K.: Integrated coverage and connectivity in wireless sensor networks: a two-dimensional percolation problem. IEEE Trans. Comput. **57**(10), 1423–1434 (2008)
2. Boudriga, N., Hamdi, M., Iyengar, S.: Coverage assessment and target tracking in 3D domains. Sensors **11**(10), 9904–9927 (2011)
3. Fan, G., Wang, R., Huang, H., Sun, L., Sha, C.: Coverage guaranteed sensor node deployment strategies for wireless sensor networks. Sensors **10**(3), 2064–2087 (2010)
4. Habib, S.J.: Modeling and simulating coverage in sensor networks. Comput. Commun. **30**(5), 1029–1035 (2007)
5. Pashazadeh, S., Sharifi, M.: A geometric modelling approach to determining the best sensing coverage for 3-dimensional acoustic target tracking in wireless sensor networks. Sensors **9**(9), 6764–6794 (2009)
6. Aziz, N.A.B.A., Mohemmed, A.W., Alias, M.Y.: A wireless sensor network coverage optimization algorithm based on particle swarm optimization and Voronoi diagram. In: International Conference on Networking, Sensing and Control (ICNSC'09), pp. 602–607, Mar 2009. IEEE
7. Ammari, H.M., Das, S.K.: A study of k-coverage and measures of connectivity in 3D wireless sensor networks. IEEE Trans. Comput. **59**(2), 243–257 (2010)
8. Mahmood, M.A., Winston, K.G., Welch, S.I.: Reliability in wireless sensor networks: a survey and challenges ahead. Comput. Netw. **79**, 166–187 (2015)
9. Katiyar, M., Sinha, H.P., Gupta, D.: On reliability modelling in wireless sensor networks: a review. Int. J. Comput. Sci. Issues **9**(6), 99–105 (2012)
10. D'amaso, A., Rosa, N., Maciel, P.: Reliability of wireless sensor networks. Sensors **14**(9), 15760–15785 (2014)
11. Xing, L.D.: An efficient binary-decision-diagram-based approach for network reliability and sensitivity analysis. IEEE Trans. Syst. Man Cybern. Part A Syst. Hum. **38**(1), 105–115 (2008)
12. Mahima, V., Chitra, A.: Battery recovery based lifetime enhancement (BRLE) algorithm for wireless sensor network. Wirel. Pers. Commun. **97**(4), 6541–6557 (2017)
13. Nageswari, D., Maheswar, R., Kanagachidambaresan, G.R.: Performance analysis of cluster based homogeneous sensor network using energy efficient N-policy (EENP) model. Clust. Comput. (2018). https://doi.org/10.1007/s10586-017-1603-z
14. Jayarajan, P., Maheswar, R., Kanagachidambaresan, G.R.: Modified energy minimization scheme using queue threshold based on priority queueing model. Clust. Comput. (2017). https://doi.org/10.1007/s10586-017-1564-2
15. Kanagachidambaresan, G.R., Chitra, A.: TA-FSFT thermal aware fail safe fault tolerant algorithm for wireless body sensor network. Wirel. Pers. Commun. **90**(4), 1935–1950 (2016)

16. Kanagachidambaresan, G.R., SarmaDhulipala, V.R., Vanusha, D., Udhaya, M.S.: Matlab based modeling of body sensor network using ZigBee protocol. In: Proceedings of CIIT 2011, CCIS 250, pp. 773–776 (2011)
17. Kanagachidambaresan, G.R., Chitra, A.: Fail safe fault tolerant mechanism for wireless body sensor network (WBSN). Wirel. Pers. Commun. Int. J. **80**(1), 247–260 (2015). https://doi.org/10.1007/s11277-014-2006-6
18. SarmaDhulipala, V.R., Kanagachidambaresan, G.R., Chandrasekaran, R.M.: Lack of power avoidance: a fault classification based fault tolerant framework solution for lifetime enhancement and reliable communication in wireless sensor network. Inf. Technol. J. **11**(6) (2012)
19. Kanagachidambaresan, G.R., SarmaDhulipala, V.R., Udhaya, M.S.: Markovian model based trustworthy architecture. In: Procedia Engineering, ICCTSD. Elseiver (2011)

RSOM-Based Clustering and Routing in WSNs

G. R. Asha⊙ and Gowrishankar Subrahmanyam⊙

1 Introduction

The WSN composed of several SNs for specific purpose. SNs can detect their region and transmit the vital information to the BS. It may use multi-hop and single-hop communication. The better features of the WSNs include low cost, computation power multifunctional, small size, and easy communication with short distances [1]. In the WSN, most of the SNs are prepared with self-supported battery power, which perform suitable operation and communicate between neighboring nodes. An increasing lifetime of the WSN, energy consumption is at most important and increase the lifetime and productivity of the WSNs [2]. Nowadays, most of the protocols and algorithms proposed to improve the throughput and performance of the WSNs such as Hybrid Energy-efficient Distributed Clustering approach (HEED), Low Energy Adaptive Clustering Hierarchy (LEACH) and Improved Multi-Objective Weighted Clustering Algorithm (IMOWCA) [3–5]. The significant difference between WSNs and other Wireless Networks are the resource constraints and operation condition. The resource constraints in terms of energy consumption and primarily arises due to the size of the SNs and their batteries [6]. In WSN, the nodes require energy in terms of battery power to sense the surrounding environment and to transfer the same to the BS in the form of data packet. The energy requirement is proportional to the distance between the SN and BS. Hence, multi-hop communication is recommended

---

G. R. Asha · G. Subrahmanyam (✉)
Department of Computer Science and Engineering, B M S College
of Engineering, Bangalore, India
e-mail: gowrishankar.cse@bmsce.ac.in

G. R. Asha
e-mail: asha.cse@bmsce.ac.in

G. R. Asha
Department of Computer Science and Engineering, Jain University, Bangalore, India

© Springer Nature Singapore Pte Ltd. 2019   189
S. Tiwari et al. (eds.), *Smart Innovations in Communication and Computational
Sciences*, Advances in Intelligent Systems and Computing 851,
https://doi.org/10.1007/978-981-13-2414-7_19

to transmit data from SN to BS which leads to the concept of hierarchical routing [7]. This process maximizes the lifetime of the sensor network by grouping a several SNs into clusters [8]. The Head node is chosen for each cluster called as Cluster Head (CH) to accumulate a Data/Information from its members and transfer to the BS with a minimum energy. In the recent past, many conventional methods incorporated diverse mechanisms in achieving energy efficiency which includes optimization in network layer routing [9, 10]. But these mechanisms are affected lifetime SN and throughput of the WSN. In order to address these issues the RSOM-WSN method is introduced. The RSOM algorithm incorporates neural computing method for energy conservation and the same is absent in the conventional methods. The RSOM structure is Self-Organizing, fault tolerant and recurrent compared to the counterparts. The simple Self-Organizing Map (EBC-S) considers current condition while achieving energy efficiency in WSN [11]. The RSOM takes current and past network condition for clustering and routing, this leads to the better performance compare to the LEACH-WSN, PSO-PSO-WSN, and FCM-PSO-GSO and EBC-S methods.

2   Related Work

The life time of WSN is determined by the number of alive nodes and the energy-hole problem decreases the number of alive nodes. The issue of energy-hole problem is addressed by using EADUC (energy-aware distributed unequal clustering) rule. Here, uneven competition radius forms the clusters of unequal size; the clusters which are near to the BS are smaller in size. The EAUDC protocol forms the clusters based on two major parameters: The position of the SN and residual energy. Thus the clustering overhead is reduced and also the lifetime WSN is improved [12].

The Network topology plays a pivotal role in energy conservation. The topology based routing: Improved Routing Protocol for low power and Lossy network (IRPL). Here, the topology has been segmented into rings of equal area and each SN determines the best route depending on association of rings. The performance of the WSN improved considerably compare to the traditional WSN routing Mechanisms. However delay was not the criteria of optimization [13].

The combination of Clustering and multicast routing is one of the popular techniques of conserving energy in WSN. An improved multicast approach based clustering algorithm allows CH to have any number of multicast sessions. The multicasting provides an opportunity to send data to a much larger area, which in turn increase packet delivery ratio, lower latency and less energy consumption thereby increase the life time of WSN [14].

A Heuristic Algorithm for Clustering Hierarchy (HACH) is an approach for energy conservation. Here, the algorithm schedules the sleeping pattern of SNs, such that the SN having higher energy as CH which in turn increase the life time of the WSN [15].

The soft computing approach, Adaptive Self Organizing, Neural Wireless Sensor Network (RSOM_WSN) treats each SN as a neuron and uses the concept of recurrence to update the input weights of neurons at each iteration by considering the aspect of mobility of the SNs [16].

These works either adopt intelligent clustering techniques to consolidate information from group/set of SNs or adaptive routing techniques to transmit the in-formation from CH to BS. There is no comprehensive approach for clustering and routing. Hence Recurrent Self organizing Map for Wireless Sensor Network (RSOM-WSN) addresses these issues. Here, four time varying parameters of WSN: Received Signal Strength (RSS), Cluster size, Residual Energy and distance between CH & BS are considered to provide comprehensive solution for energy conservation in WSN. The results of RSOM-WSN were promising compared to the well-known energy conservation techniques such as LEACH-WSN, PSO-PSO-WSN, and FCM-PSO-GSO and EBC-S.

3 RSOM-WSN

The RSOM-WSN algorithm comprises of systematic process and the same can be enumerated by steps.

Step I: Node Deployment: SNs are positioned at strategic locations to gather the environmental parameters such as Pressure, Temperature and Humidity. The position and vital parameters such as Initial Energy, Distance from SN to BS, RSS and ID are communicated to the BS.

Step II: Cluster Formation: The SNs forms clusters based on Residual Energy, RSS, Distance and Type. The CH is elected through assigning appropriate weights to the current value of the parameters and ID of recent past CH of the cluster.

Step III: Routing process: Between the CHs and BS the optimum route is computed by considering current network parameter and previous routing information to achieve Load Balancing (LB) and energy conservation through RSOM network and by assigning appropriate weights to the current parameters and previous routing results.

Step IV: Identification of Dead and Standby SNs: The Dead SNs are automatically eliminated from system through algorithm by considering the residual energy of SNs. The SN falls in the critical path are forcefully eliminated to become CH so as to enhance the life time of WSN. Repeat Step II and III till the lifetime of WSN.

3.1 System Model

In RSOM-WSN, SNs are directly analogous to RSOM neurons. Hence, the number of neurons in the RSOM-WSN O/P layer is equal to the number of sensor nodes in the RSOM-WSN network. A typical arrangement of the network is shown in Fig. 1. The arrangement of a single Neuron of RSOM is shown in Fig. 2.

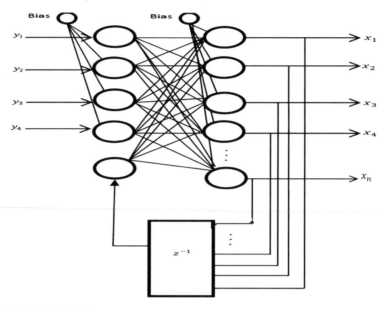

Fig. 1 RSOM-WSN architecture

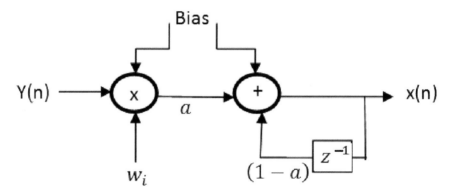

Fig. 2 Single neuron of RSOM

The RSOM is a powerful variant of Kohonen's SOM (K-SOM). The RSOM has an ability to process Temporal Sequences (TSs) through the recurrent connection with the bias. The ability of processing the temporal sequence by RSOM achieves Load balancing and Energy Conservation in WSN.

3.2 RSOM I/P Function

The RSOM is a Two-Layer Neural Network such as one I/P Layer and other one is an O/P layer. The neuron associated with the input layer is fully connected to the O/P layer with the appropriate weight vector (or Synapses) and suitable bias.

3.3 RSOM O/P Function

The neurons in the O/P layer receives the Input from the I/P layer along with the suitable bias values. The result of output layer depended on current input, previous output and bias. The RSOM O/P can be computed from Eq. (1),

$$x(n) = \sigma((1-a)x(n-1) + awy(n) + b) \qquad (1)$$

$$x(0) = 0$$

From Eq. (1)

$x(n)$ Neuron's cumulative output at the nth element of output vector.
$y(n)$ nth term of Input vector.
b Bias Vector.
$0 < a < 1$ the learning constant.
x, y and w vectors all are of same dimension.

Equation 1 can be rewritten to overtly show the contribution that each and every input term has based on cumulative output. A sequence with N-terms, the RSOM $(y, w, \alpha) = x(N-1)$, here

$$\begin{aligned} x(n-1) = \alpha[y(N-1) - w + (1-\alpha)^1(y(N-2) - w) \\ + (1-\alpha)^2(y(N-3) - w) + \cdots \\ + (1-\alpha)^{N-1}(y(0) - w)] \end{aligned} \qquad (2)$$

Generally, the RSOM output function computed as weighted sum of the biased inputs and the same can be expressed as exponential series. In Eq. 2, each input vector is normalized through biased vector w, the result of the O/P layer is normalized through Quashing function (σ).

3.4 Best Matching Unit

The input sequence, each neuron evaluates an absolute output, which value is similar to the magnitude of its output vector. The neurons contend by associating their total output with that of the rest of the system (Network), which competition results are referred to as the Best Matching Unit (BMU), which denotes the neuron that weights are well trained with respect to order of Input, the neuron weights resulted in the low quantity of Output. Through the BMU the input weights of individual neuron with respect to input parameter are adjusted in each iteration.

3.5 Heat Dissipation of SNs

Heat dissipation at SN is directly proportional to the energy consumed during its operation. "Higher the energy consumption: larger the dissipation of heat". In order to decrease the energy consumption SNs the heat dissipation need to be reduced. This can be achieved through a quality cluster formation and optimized routing process. The careful selection of CH and optimized routing path will increase the efficiency and life time of the WSN.

4 Results and Discussion

MATLAB is used to simulate and compare the proposed algorithm (RSOM-WSN) with previous works such as LEACH-WSN, PSO-PSO-WSN, and FCM-PSO-GSO and EBC-S. The Table 1 provides simulation parameters for RSOM-WSN.

The simulation setup considers 300 sensor nodes are deployed in the area of 250 * 250 m² and each sensor nodes of RSOM-WSN has the initial energy of 0.5 J. These sensor nodes are varied with cluster heads from 10 to 50 to generate heat map of the proposed and existing work.

It is evident from Figs. 3, 4, 5, 6 and 7 Substantial reduction in the heat dissipation of network compare to the classical approaches: LEACH-WSN, PSO-PSO-WSN, FCM-PSO-GSO and EBC-S.

Energy dissipation of RSOM-WSN is determined by varying the amount of CHs with respect to the various iteration levels. Heat dissipation of entire network can be identified by the energy dissipation graph. The lifetime of the network is improved carefully choosing the CHs and route during the data transmission. The dead nodes can be avoided at each iteration and the same will be removed while forming the optimized path from the SN to BS.

Table 1 Simulation parameters

Parameter	Value
Area	$250 * 250 \ \text{m}^2$
Sensor nodes	300
Cluster heads	10–50
The initial energy of sensor nodes	0.5 J
Number of simulation iterations	300
Communication range	150 nm
Eelec	50 PJ/bit
εfs	$10 \ \text{PJ/bit/m}^2$
εmp	$0.0013 \ \text{PJ/bit/m}^4$
d0	87.0 m
EDA	5 nJ/bit
Packet size	4000 bits
Message size	200 bits

Fig. 3 Heat map for RSOM-WSN

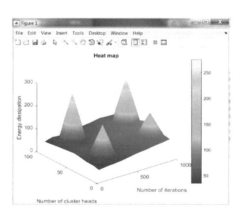

Fig. 4 Heat map for FCM-PSO-GSO

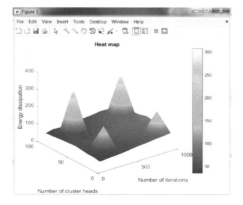

Fig. 5 Heat map for PSO-PSO-WSN

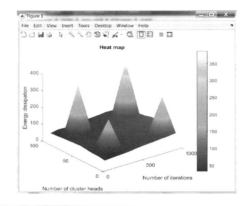

Fig. 6 Heat map for LEACH-WSN

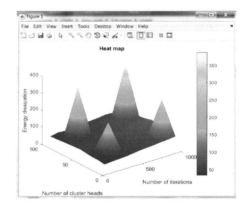

Fig. 7 Heat map for EBC-S

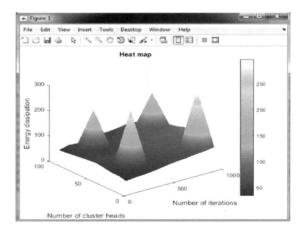

Fig. 8 Comparison of alive

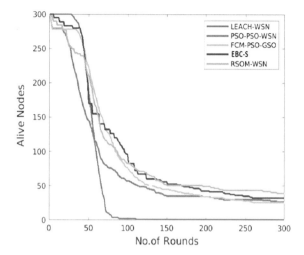

Fig. 9 Comparison of dead
nodes

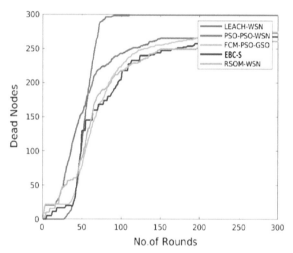

The results of the alive nodes, dead nodes and number of packets received to the BS are simulated by considering the 300 sensor nodes with 25 cluster heads. Figures 8, 9 and 10 shows the alive nodes, dead nodes comparison of RSOM-WSN with LEACH-WSN, PSO-PSO-WSN, FCM-PSO-GSO, and EBC-S.

These comparisons are made in terms of number of rounds. It is evident from Fig. 8, that the number of alive nodes of the RSOM-WSN is more than compare to the other classical methods. Figure 9 depicts the life time of RSOM-WSN is more due to the number dead nodes is lesser than the other classical methods. Figure 10 shows the number of packet transmitted in the RSOM-WSN is considerably higher than other method due to the intelligent routing by carefully eliminating the dead nodes.

Fig. 10 Total packet sent

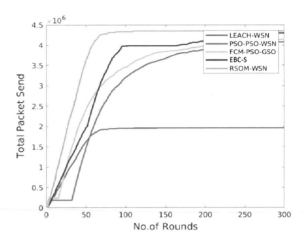

5   Conclusion

The RSOM-WSN technique makes use of RSOM Neural Network for both the clustering and routing. In RSOM clustering the M set of nodes is partitioned into the c clusters. This RSOM clustering and routing mainly depends on the performance parameters which comprises of four parameters such as residual energy, distance, cluster size and received signal strength. These parameters are used to find the optimum CH and shortest path between the CHs in the clustering and routing respectively. The performance of RSOM-WSN is analyzed by following parameters like energy consumption, number of alive nodes and number of dead nodes. RSOM-WSN gives better performance compared to existing methods such as LEACH-WSN, PSO-PSO-WSN, FCM-PSO-GSO, and EBC-S.

References

1. Vimalarani, C., Subramanian, R., Sivanandam, S.N.: An enhanced PSO-based clustering energy optimization algorithm for wireless sensor network. Sci. World J. (2016)
2. Liu, F., Wang, Y., Lin, M., Liu, K., Dapeng, W.: A distributed routing algorithm for data collection in low-duty-cycle wireless sensor networks. IEEE Internet Things J. **4**(5), 1420–1433 (2017)
3. Nayyar, A., Gupta, A.: A comprehensive review of cluster-based energy efficient routing protocols in wireless sensor networks. IJRCCT **3**(1), 104–110 (2014)
4. XingGuo, L., Feng, W.J., Lin, B.L.: LEACH protocol and its improved algorithm in wireless sensor network. In: International Conference on Cyber-Enabled Distributed Computing and Knowledge Discovery (CyberC) (2016)
5. Ouchitachen, H., Hair, A., Idrissi, N.: Improved multi-objective weighted clustering algorithm in Wireless Sensor Network. Egypt. Inform. J. **18**, 45–54 (2017)

6. Kashani, M.A.Z., Ziafat, H.: A method for reduction of energy consumption in wireless sensor network with using neural networks. In: 6th International Conference on Computer Sciences and Convergence Information Technology (ICCIT) (2011)
7. Li, J., Hou, X., Su, D., Munyemana, J.D.D.: Fuzzy power-optimised clustering routing algorithm for wireless sensor networks. IET Wirel. Sens. Syst. **7**(5), 130–137 (2017)
8. Mahajan, S., Dhiman, P.K.: Clustering in wireless sensor networks: a review. Int. J. **7**(3) (2016)
9. Younis, M., Senturk, I.F., Akkaya, K., Lee, S., Senel, F.: Topology management techniques for tolerating node failures in wireless sensor networks: a survey. Comput. Netw. **58**, 254–283 (2014)
10. Li, J., Jiang, X., Lu, I.-T.: Energy balance routing algorithm based on virtual MIMO scheme for wireless sensor networks. J. Sens. (2014)
11. Enami, N., Reza, A.M.: Energy based clustering self organizing map protocol for extending wireless sensor networks lifetime and coverage. Can. J. Multimed. Wirel. Netw. **1**(4) (2010)
12. Gupta, V., Pandey, R.: An improved energy aware distributed unequal clustering protocol for heterogeneous wireless sensor networks. Eng. Sci. Technol. Int. J. **19**(2), 1050–1058 (2016)
13. Zhang, W., Han, G., Feng, Y., Lloret, J.: IRPL: an energy efficient routing protocol for wireless sensor networks. J. Syst. Architect. **75**, 35–49 (2017)
14. Sule, C., Shah, P., Doddapaneni, K., Gemikonakli, O., Ever, E.: On demand multicast routing in wireless sensor networks. In: 28th International Conference on Advanced Information Networking and Applications Workshops (WAINA), pp. 233–238 (2014)
15. Oladimeji, M.O., Turkey, M., Dudley, S.: HACH: heuristic algorithm for clustering hierarchy protocol in wireless sensor networks. Appl. Soft Comput. **55**, 452–461 (2017)
16. Ball, M.: An adaptive, self-organizing, neural wireless sensor network. Electronic Theses and Dissertations. 7049 (2007)

Evaluation of Received Signal Strength Indicator (RSSI) for Relay-Based Communication in WBAN

Pulkit Pandey, Arthav S. Patial and Sindhu Hak Gupta

1 Introduction

A Wireless Body Area network is a network of several computing devices which may also be wearable. The fundamental components of these devices is a biosensor which may be implanted, i.e., fit inside the human body or may be attached to the surface of the human body depending upon the application. It is not necessary for a biosensor to be in physical contact with human body for evaluating physiological parameters. These may be in clothes pocket or maybe in the bags. Since miniaturization is trending these days the networks with miniature sensors are preferred, but their battery recharging and replacement is not possible [1]. Human movement identification has gained more attention lately [2, 3]. This information is helpful for usage in the medical profiles and also for the self-awareness tools [4]. Apart from these, movement identification allows the estimation of propagation channel status of WBAN which can be further used to optimize the performance of a WBAN in terms of power consumption and reliability [5, 6].

For the communication among the nodes, relay-based communication is used [7]. Relay-based communication is a technique of the transmission where the nodes involved, do the transmission of their own information, but at the same time, they also perform the repetition of the information of other users during the transmission to a common destination. In case of relay-based communication, a neighbouring node (R), helps the source node (S) to communicate the information to the destination

P. Pandey (✉) · A. S. Patial · S. H. Gupta (✉)
Amity University Uttar Pradesh, Noida, India
e-mail: pulkit.pandey96@yahoo.in

S. H. Gupta
e-mail: shak@amity.edu

A. S. Patial
e-mail: arthavspatial@gmail.com

© Springer Nature Singapore Pte Ltd. 2019
S. Tiwari et al. (eds.), *Smart Innovations in Communication and Computational Sciences*, Advances in Intelligent Systems and Computing 851,
https://doi.org/10.1007/978-981-13-2414-7_20

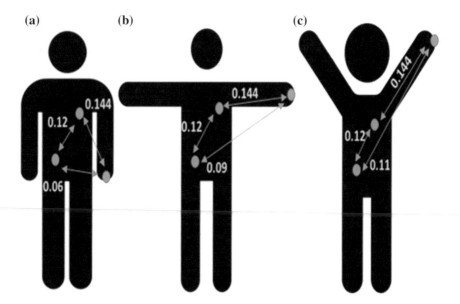

Fig. 1 Network establishment of the proposed method for **a** hand at 0°, **b** hand at 90°, **c** hand at 180°

node (D). Relay-based Communication creates a virtual MIMO as we have effective multiple transmitters and receivers without facing the hardware constraints [8].

This paper aims to investigate the human movement using the RSSI. The major contribution of this research work is the comparison of RSSI with different values of d, which helps to find the pattern of variation as the body moves. Also, the comparison of RSSI with Energy Consumption so as to find the human posture [9, 10] which consumes more energy and the one which consumes less energy.

The rest of the paper is organized as follows, Sect. 2 of the paper illustrates the system model. In Sect. 3, the comparison of simulation results between the relay-based strategies for different values of energy consumption and the path loss model are given. Finally Sect. 4 concludes the paper (Fig. 1).

2 Proposed Approach

The network establishment of the proposed method is in such a way, that we note the distance d, between the Source S, Relay R and Destination D. The path loss among these nodes is found for a reference distance d0. Since, for a link between source and destination, the signals are transmitted based on the Friis formula as described by [11], from S to D, S to R and R to D and are given by (1), (2), and (3),

$$PL_{SD}(d_{SD}) = PL_0 + 10n\log_{10}\left(\frac{d_{SD}}{d_0}\right) \tag{1}$$

$$PL_{SR}(d_{SR}) = PL_0 + 10n\log_{10}\left(\frac{d_{SR}}{d_0}\right) \tag{2}$$

$$PL_{RD}(d_{RD}) = PL_0 + 10n\log_{10}\left(\frac{d_{RD}}{d_0}\right) \tag{3}$$

here the values of PL0 is path loss at the reference d_0 (0.1 m), and n is the path loss exponent which is taken to be unity. The total power consumed by the transmitting end is P total. The power model as described by [12] is given as,

$$P_{total} = P_{POA} + P_{CB} \tag{4}$$

The power consumption of the power amplifiers (PPOA) can be calculated by (5),

$$P_{POA} = (1 + \alpha) P_{out} \tag{5}$$

If de is the drain efficiency and ε is the peak average ratio,

$$\alpha = \left(\frac{\varepsilon}{de} - 1\right) \tag{6}$$

If Rb is bit rate of transmission, $\varepsilon'b$ is energy per bit, d being the transmission distance, and Gt and Gr being the transmitter and receiver antenna gains, respectively, Ml is the link margin, λ is the wavelength of carrier, and Nf is the noise figure of receiver, then,

$$P_{out} = \left(\varepsilon'b\ R_b\right) * \frac{(4\pi)^2 d^k}{G_t G_r \lambda^2} M_l N_f \tag{7}$$

$$N_f = \frac{N_r}{N_0}, \tag{8}$$

where N_r is the PSD of total noise at receiver is input and N_0 is the PSD of total effective noise at room temperature.

Thus, the total energy consumption per bit,

$$\varepsilon_{bt} = \frac{(P_{POA} + P_{CB})}{R_{ber}}, \tag{9}$$

where, R_{ber} is the actual bit error rate.

RSSI can be expressed as,

$$RSSI(d) = Pt(d_0) - PL(d_0) - 10n\log_{10}\left(\frac{d}{d_0}\right) + X_\sigma \tag{10}$$

Table 1 Parameter values for simulation of energy efficiency	RSSI (dB)	Energy (J)		
		0° position	90° position	180° position
	175	414.2	414	414.8
	170	405.5	405.2	406.1
	165	396.8	396.4	397.3
	160	387.9	387.6	388.5
	155	379.2	378.6	379.7
	150	370.5	369.9	370.8

The parameters considered for simulation have been tabulated in Table 1.

3 Experimental Setup and Results

MATLAB has been used for calculation of the total Energy and different parameters which are mentioned above.

Figure 2 illustrates the effect of RSSI on Energy.

From Fig. 2, it is observed that as the RSSI increases, the Energy also increases. As we move towards the right, with an increase in RSSI, the Energy Consumption increases.

Figure 3 highlights the comparison between the RSSI with the Path loss.

It is noted, from Fig. 3, that as the path loss increases, the RSSI also increases with it. As we move towards right, the increase in path loss results in the increase of RSSI.

Figure 4 illustrates the effect of RSSI on Energy.

Fig. 2 Comparison of energy with RSSI

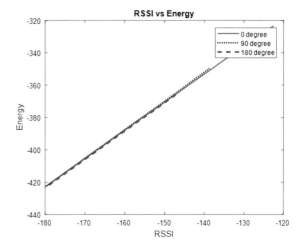

Fig. 3 Comparison of RSSI with path loss

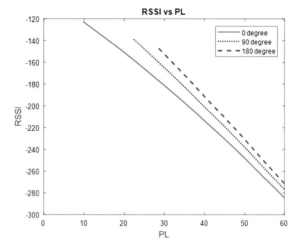

Fig. 4 Comparison of RSSI with distance

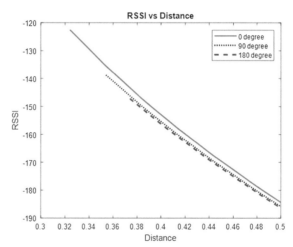

From Fig. 4, it is noted that, as the distance d increases, the RSSI value also increases. As we move towards the right, the increase in distance results in the increase in RSSI.

Figure 5 illustrates the effect of distance on Path loss.

It is noted, from Fig. 5, that as the distance increases, the path loss also increases. As we move towards right, the increase in the distance, results in the increase in the path loss.

The comparison of Energy with the different values of RSSI, is given by Table 2.

The comparison of RSSI with the different values of PL, is given by Table 3.

The comparison of RSSI with the different values of distance (d), is given by Table 4.

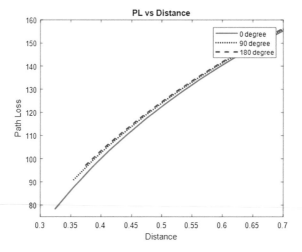

Fig. 5 Comparison of path loss with distance

Table 2 Comparison of energy with RSSI

Parameter	Value
Bit rate (Rb)	1 Mbps
PSD of noise (N0)	−171 dBm/Hz
Reference distance (D0)	0.1 m
Standard deviation (Nσ)	3.8
GtGr	5 dBi
Nf	10
Ml	40

Table 3 Comparison of RSSI with PL

PL (dB)	RSSI (dB)		
	0° position	90° position	180° position
30	181.2	165	152.4
35	197.1	182.7	171.7
40	213.6	200.8	191.2
45	230.7	219.4	211.1
50	248.2	238.4	231.0
55	266.1	257.5	251.2
60	284.6	276.8	271.5

Table 4 Comparison of RSSI with d

d (m)	RSSI (dB)		
	0° position	90° position	180° position
0.38	145.6	148.4	149.3
0.40	152.9	155.4	156.2
0.42	159.9	162.1	162.8
0.44	166.4	168.4	169.1
0.46	172.7	174.4	175.1
0.48	178.6	180.3	180.8

4 Conclusions

In the proposed work, the RSSI is evaluated for WBAN, when the human body is under three different arm postures: arms in downward direction (0°), arms in direction parallel to ground (90°), and arms in upward direction (180°). The critical comparative analysis of the WBAN, concludes that different movements of the human arm varies the values of RSSI, energy efficiency, path loss, and the reliability of the WBAN. It was also noted that the different arm postures resulted in highly varied RSSI. This fact can be utilized in the future for the detection of patient's arm position and any movement related to it.

References

1. Gungor, V., Hancke, G.P.: Industrial sensor networks challenges, design principles, and technical approaches. IEEE Trans. Ind. Electron. **56**(10), 4258–4265 (2009)
2. Archasantisuk, S., Aoyagi, T.: Human movement classification for body area networks using signal level fluctuation via neural network. In: Proceedings of IEICE Society Conference (2013)
3. Archasantisuk, S., Aoyagi, T., Uusitupa, T.: Human movement classification using signal level fluctuation in WBAN at 403.5 MHz and 2.45 GHz. IEICE Tech. Rep. **114**(61), 73–78 (2014)
4. Archasantisuk, S., Aoyagi, T.: The human movement identification using the radio signal strength in WBAN. In: 2015 9th International Symposium on Medical Information and Communication Technology (ISMICT), pp. 59–63. IEEE (2015)
5. Smith, D., Lamahewa, T., Hanlen, L., Miniutti, D.: Simple prediction-based power control for the on-body area communications channel. In: 2011 IEEE International Conference on Communications (ICC), pp. 1–5, June 2011
6. Smith, D., Hanlen, L., Miniutti, D.: Transmit power control for wireless body area networks using novel channel prediction. In: Wireless Communications and Networking Conference (WCNC), 2012 IEEE, pp. 684–688, Apr 2012
7. Pabst, R., Walke, B.H., Schultz, D.C., Herhold, Patrick, Yanikomeroglu, H., Mukherjee, S., Viswanathan, H., et al.: Relay-based deployment concepts for wireless and mobile broadband radio. IEEE Commun. Mag. **42**(9), 80–89 (2004)
8. Jayaweera, S.K.: Virtual MIMO-based cooperative communication for energy-constrained wireless sensor networks. IEEE Trans. Wirel. Commun. **5**(5), 984–989 (2006)
9. Jameel, F., Haider, M.A.A., Butt, A.A.: Robust localization in wireless sensor networks using RSSI. In: 2017 13th International Conference on Emerging Technologies (ICET), pp. 1–6. IEEE (2017)

10. Mounir, T.A., Mohamed, P.S., Cherif, B., Amar, B: Positioning system for emergency situation based on RSSI measurements for WSN. In: 2017 International Conference on Performance Evaluation and Modeling in Wired and Wireless Networks (PEMWN), pp. 1–6. IEEE (2017)
11. Yazdandoost, K.Y.: Channel model for body area network (BAN) (2007). IEEE802. 15-07-0943-00-0ban
12. Murali, V., Gupta, S.H.: Analysis of energy efficient, LEACH-based cooperative wireless sensor network. In: Proceedings of the Second International Conference on Computer and Communication Technologies, pp. 353–363. Springer, New Delhi (2016)

Side-Channel Attacks on Cryptographic Devices and Their Countermeasures—A Review

M. M. Sravani and S. Ananiah Durai

1 Introduction

A cryptographic device/system must be viable to secure confidential data/information. Further, efficient crypto systems must provide data integrity, non repudiation, data confidentiality and authentication. Techniques of securing data in cryptography may include secret key cryptography (SKC), public key cryptography (PKC) and hash function [1]. SKC involves algorithms such as AES, DES and Rivests Ciphers. In the PKC method of data encryption, RSA and Diffie–Hellman algorithms are usually used [2]. The hash encryption method utilizes algorithms such as message digest, MAC, and SHA (SHA-0, SHA-1, SHA-2 and SHA-3) [3]. The implementation of these methods of encryption depend on the application and the level security required. Eventhough these various encryption standards secure the confidential data, they are vulnerable for leakage of information and may end up to be weak encryption standards due to various attacks.

In recent years, few reviews were reported on various attacks on crypto devices/ systems [4, 5]. The primarily reported attack that misuses the leaked information by the alternative means is the side-channel Attack (SCA). Physical interaction to a secured device/system may be a major source that can be tapped by the enemies for the purpose of data theft. The leaked information is called as side-channel information. SCA misuses this side-channel information and may perform further process for data retrieval.

Over the years, few surveys on this issue were reported as in [4–6]. In [4], SCA impacts on physical cryptographic devices were surveyed. These reported attacks

M. M. Sravani (✉) · S. Ananiah Durai
Vellore Institute of Technology, Chennai, India
e-mail: mannemuddusravani@gmail.com

S. Ananiah Durai
e-mail: ananiahdurai.s@vit.ac.in

© Springer Nature Singapore Pte Ltd. 2019
S. Tiwari et al. (eds.), *Smart Innovations in Communication and Computational Sciences*, Advances in Intelligent Systems and Computing 851,
https://doi.org/10.1007/978-981-13-2414-7_21

were primarily on FIPS140-3 encryption standard. Also, the types of post-sensed analysis process for data retrieval and their countermeasures were discussed. The review reported in [5] explores the online application leakage of the side-channel information and the mitigation techniques. Competent and efficient countermeasure techniques for both hardware and software attacks on crypto devices/systems are well detailed in [6]. As an Extension to the above surveys, this paper aims to report on recent attacks that may break the encryption algorithms. Developments on few efficient countermeasures for these attacks were also presented.

The rest of the paper is organized in five sections. The next section gives a comparison of the traditional model of attack and side-channel model of attack. In Sect. 3, the types of attacks are discussed in brief. Section 4 provides a review on the various SCA attacks and their countermeasures. Finally, a conclusion is formulated.

2 Cryptanalysis of Attacks

In cryptographic point of view [7], the models of attacks can be classified into two ways, Classical (Traditional) cryptanalysis and Physical security (Side-channel cryptanalysis).

In this Classical (Traditional) cryptanalysis, the secret key can be retrieved with the help of mathematical models of analysis. The observer actually knows few protocol of the algorithm and may be aware of the transmission data between sender and receiver. Figure 1 shows the Classical (Traditional) model of Cryptanalysis.

In recent years, many researchers are exploring the possibilities of techniques to break the algorithm by analysing the protocol and properties in terms of implementation steps and operating environment. The cryptologist may likewise have the capacity to observe the device process when a fault is introduced during the attack trail. Figure 2 depicts the physical security (side- channel cryptanalysis).

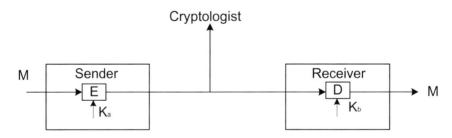

Fig. 1 Traditional model of cryptanalysis [5]

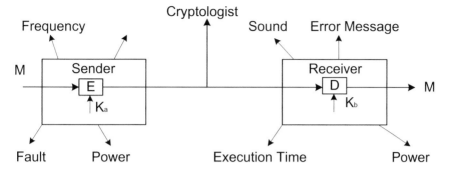

Fig. 2 Side-channel cryptanalysis [5]

3 Classification of Attacks

3.1 Natural Means of Attacks (Direct and Indirect)

In direct approach, the devices/system especially related to business framework can be tampered physically to attack the secure information. Such tampered resistant hardware may include smartcards, electronics wallets and dongles. Hardware security is an essential part in many security frameworks. In an indirect way without disturbing the hardware, attacks are still possible. Further, devices/system properties are analysed and the information to extract the secret key is obtained. Power analysis, EM analysis and Timing analysis are such type of attacks.

3.2 Classification of Attacks Based on Device Behaviour

Based on device performance, SCA attacks can be classified as active attack and passive attack [7]. In active attack, the hacker alters the internal data that can induce a fault message to the receiver. The internal data may be altered by providing a voltage or clock glitches (Fig. 3).

Alternatively, the cryptosystem can be continuously monitored for an open port and whenever an open port is available, the data may be accessed by the observer. This type of attack does not disturb the actual process of any device and it is hence termed as passive attack (Fig. 4). The observer's main task in this attack is to steal the information without leaving any trace that can be tracked down by the owner. Thus, the observer can identify the owner's actual message. Time, power or EM measurements are few possible means of the passive method of message theft.

Fig. 3 Active attack [36]

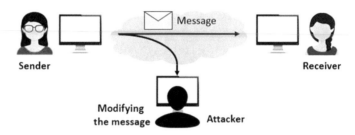

Fig. 4 Passive attack [36]

3.3 Classification Based on Systematic Review of the Attack Surface

According to Anderson et al. [9], attacks can be classified into three basic types, invasive attack, semi-invasive attack and non-invasive attack. Invasive type is a direct access of the internal components of the cryptographic modules without repacking the device. Stressed glass components and custom pressure air gaps are of these types of attacks. Non-invasive means of attacks get the externally leaked information from the system, without interfering in the actual process of the physical device. This attack is completely undetectable. Such attacks are timing analysis, power analysis and EM analysis.

Both invasive and non-invasive attack [6] falls on the extreme side where the former intervenes into the systems process, whereas the latter do not. However, an attack that may fall in between these two extremes is also possible which is termed as a semi-invasive type [6]. This class of attack, access the device but without interfering into the passivation layer of a device.

3.4 Classification Based on Analysis Process

The analysis-based classification has two techniques, simple and differential. In simple analysis, the secret key information can be accessed directly from the operations performed by the system.

Differential analysis can be done by correlating the processed data and leaked information of the cryptographic device. These analyses can break many cryptographic algorithms easily [10].

4 SCA Attacks and Their Countermeasures

4.1 Timing Attack

Kocher [11] in San Francisco, first demonstrated this attack during 1996. He collected all the timing samples from various I/O. Then these samples are correlated with the side channel Information to retrieve the secret key. These timing attacks will occur due to non-constant timings during the device operations. Figure 5 shows the timing analysis process.

Figure 6 depicts the retrieval time of password for various trials with different possible combinations. This analysis is based on support vector machine (SVM), which computes the password retrieval time from the time delay between the clicks of the keys in the keypad of ATM machine/phone unlock [13].

Therefore, the various timings in the process of any device can be helpful for the attackers to retrieve the secret key components. This is due to the fact that the operations performed in the cryptographic device depends on secret key and plain text parameters.

Countermeasures for this timing attack can be based on the time management of various operations. The time maintained for every optimized operation should be constant whatever the data may be. This could be achieved by branch equalization.

Branch equalization is the adding of extra milliseconds to the shorter transaction which will equalize it to the maximum operation time in the devices. However, this

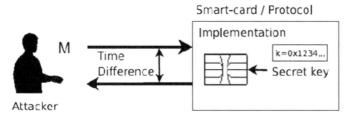

Fig. 5 Timing analysis [12]

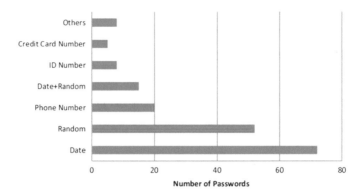

Fig. 6 Experimental data of PIN class distribution [13]

countermeasure adds delay for each process thus affecting the overall computational time.

In RSA [11], countermeasures are done during Montgomery multiplication. The intermediate values during this multiplication steps are masked such that the secret key will not be revealed during the correlation operation. Blinding technique is also a powerful countermeasure during the decryption process. This technique introduces random number during the decryption, so that the attacker will not have valuable information to retrieve the data.

4.2 Power Analysis Attack

The present and the future technologies are all of digital circuits in which CMOS plays a prominent role. In the CMOS circuit, power dissipation occurs during the circuit operation. Power dissipation can be broadly classified into three parts, static, dynamic and short circuit power dissipation [12]. These types of power dissipation can be explored with the operation of a simple CMOS inverter shown in Fig. 7. Static dissipation occurs when the leakage currents appear continuously from the power supply. The static power equation can be given as [12];

$$P_{Static} = I_{leak} \cdot V_{dd} \tag{1}$$

If the dissipation occurs while charging and discharging of the load capacitor present in the circuit, then it is no longer static and is dynamic. This depends on the load capacitor; thus the power dissipation may vary according to the load capacitor value. It can be given by [12];

$$P_{Dynamic} = C_L \cdot V_{dd}^2 P_{0 \rightarrow 1} f \tag{2}$$

Fig. 7 CMOS inverter [7]

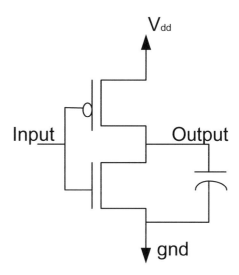

Both NMOS and PMOS transistors switch alternatively from 0 to 1 or from 1 to 0 according to the input levels. At certain instance, both the transistor will be in ON condition that may lead to a short circuit between Vdd and Vss. The dissipation occurring during this short circuit is referred to as short-circuit power dissipation [12];

$$P_{SC} = I_{mean} \cdot V_{dd} \tag{3}$$

The total dissipation can be given by [12];

$$P_{total} = P_{Dynamic} + P_{Static} + P_{SC} \tag{4}$$

This Ptotal depends on, Vdd that causes the vulnerabilities of side-channel attack. The power analysis attack was first observed by Kocher et al. [10]. This attack is well explained by Paul Kocher in his correlation between the power consumed by the device and side-channel information. Using the different cryptanalysis techniques, the secret key of the algorithm can be recovered [10]. Power analysis can be classified in two ways, simple power analysis (SPA) and differential power analysis (DPA).

4.2.1 Simple Power Analysis (SPA)

This is a direct and simple method of interpreting the power consumption of the device for retrieving the secret key. The power consumption that occurs during the cryptographic operation is the source for the attacks that utilize this analysis. Power traces which are a set of sample data of power consumption during the cryptographic operation play a major role to detect the secret key in SPA. This is due to the fact that the traces indirectly reveal the actual operation progressing inside the device.

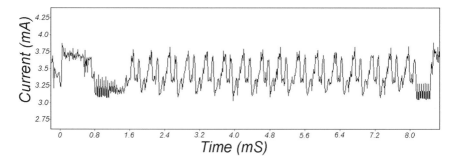

Fig. 8 SPA of DES operation [10]

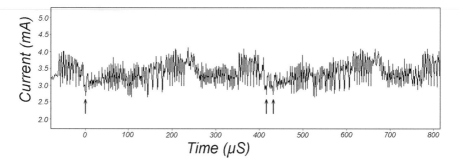

Fig. 9 DES, round 2 and 3 operations [10] of SPA

Thus from SPA analysed traces, the secret key can be retrieved which eventually can break the algorithm. Figures 8 and 9 shows the traces of SPA analysis during a specific operation of DES Encryption algorithm. The countermeasures for this direct method of SPA analysis is simple to implement. Elimination of the operations which has the possibility to reveal the secret key could be a feasible method to counter the attack.

In devices like microprocessors, masking method of countermeasures can be effectively implemented. The microcode in such device largely depends on operands which utilize large amount of power that could be a gateway for attacks. By masking the intermediate values of certain device operations, the effective retrieval of the data using SPA can be mitigated [10]. Alternatively, in extreme cases by reducing the average code, cycles such as minimizing conditional loops, subroutines and instantiation and average power consumption can be reduced thereby protecting the secret key. Further, usage of symmetric cryptographic algorithms (hardwired hardware) may consume less power, which masks the secret key.

4.2.2 Differential Power Analysis (DPA)

DPA utilizes the statistical model to retrieve the secret key. The power variation is usually minimal during any internal operations of the crypto device; therefore SPA technique has less probability to retrieve the secret key. On the other hand, DPA that employs mathematical analysis is proven to have greater potential to break the crypto devices/systems. Figure 10 shows current traces of different samples. Different DPA traces are analysed with the known voltages to read the secret key. In [14], a new type of rigid-based DPA attack is reported. This is found to be prominent in nano-scale chips for retrieval of the secret key.

The number of power traces analysed in DPA can be single or multiple. The former is referred to as horizontal DPA wherein single traces are analysed, whereas the later termed as vertical DPA uses more than one traces. One such attack utilizing horizontal DPA in elliptical curve cryptography (ECC) algorithm is reported in [15]. It is to be noted that ECC can be affected in less time.

The countermeasures for DPA can be achieved by conventional techniques such as reduced signal size, low-noise injection, balancing of hamming code and state transition. However, better shielding can be achieved by physical shielding of the device under attack [10]. In public key cryptography such as ECC, combination of randomization of processed data and blinding the ECC point may resist the DPA attack [15]. Further, an on-chip CORe [16] technique can also be implemented as a one of the countermeasure for DPA attack in AES algorithm. This method creates confusion to the attacker by inducing power noise, especially during the s-box operation.

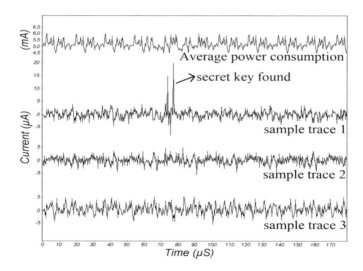

Fig. 10 DPA traces [10]

4.3   Electromagnetic Attack

This attack was first reported by Gandolfi et al. [17] and later explored by Quisquater and Samyde [18].

When a coil is placed nearer to the chip under attack, due to the electromagnetic radiation emf is induced in the coil, which causes a current flow. This current, which is the side-channel information, will be the source for the attackers for data retrieval. A simple analysis in this type of attack can be done with single power trace that directly depends on induced current. On the other hand, as in DPA, a set of sample power traces from the device can be analysed using the statistical process to retrieve the secret key. Figure 11 shows the experiment setup to measure the EMA analysis using oscilloscope. Figure 12 depicts the step by step procedure to find the secret key using the side-channel information. Quisquater and Samyde [18] proposed a few more countermeasures for EM attack such as introducing noise to the data, reducing the EM field, signal information reduction, blockage of EM radiation and low-power processor design.

4.4   Scan-Based Attack

The information leakage through scan chain IOs is called scan attacks. The side-channel information provides the information on circuits internal state. The scan chains designed for hardware testability, acts as a backdoor for certain potential attacks [20].

HMAC-SHA-256 [21] based circuitry is vulnerable to this scan-based attack. This attack can be successful even without the knowledge on architecture parameters such as processing time, key and register information of SHA-256. A hash generator circuit

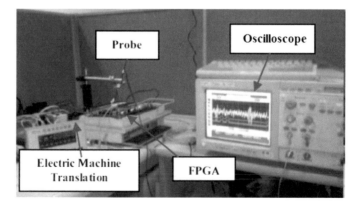

Fig. 11 Measurements setup of EMA [19]

Fig. 12 Flowchart of EMA attack [19]

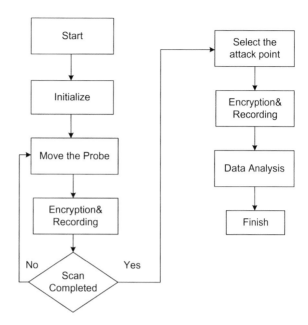

Table 1 Comparison between different types of attacks

Author	Type of attack	Device under attack	Target algorithm
Fleming et al. [13]	Timing attack	Keypad of the ATM)	DES
Kabin [15]	Horizontal DPA attack	Point multiplication of montgomery	ECC
Gu [19]	Electromagnetic analysis	Cryptographic FPGA	3DES, AES
Oku [21]	Scan-based attack	Hash generator circuit (registers)	HAMC-SHA-256

in this architecture is the source for the side-channel information. The registers are attacked to retrieve the keys.

In order to counter these types of attacks, scan chains are physically disconnected after the performance tests are completed. design for testability (DFT) is an approach where disconnecting the scan chains are not necessary at the end of every process. The test procedure includes providing the test vectors and the expected test responses to the device-under-test (DUT) for an on-chip comparison [22] (Table 1).

The procedure for DUT is depicted in Fig. 13. The RST bit determines the test condition. When RST is 1, the output signal 1 of the AND gate A1 is feedback to the counter. The output level 1 of A1 will enable the counter to start counting. When the MSB bit is 1, the inverter disables the state mask and thereby the key mask is enabled which puts the DUT under test mode as the flip-flop is in the feedback loop. Once the RST is 0, the key mask gets disabled. Even if the attacker monitors during

Fig. 13 Representation of scan test controller

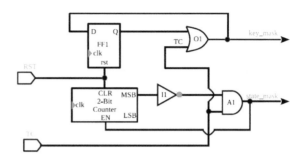

both test and functional mode, the intermediate data cannot be determined. However, this countermeasure is not much resistant to other algorithms. Alternatively, different approaches such as shift registers based secure design [23] and built-in self-test [24] can be employed to resist the scan-based side-channel attack.

4.5   Cache-Based Attack

In recent years, as high data rate is in demand, speed requirement between the CPU and main memory are increasing.

In order to achieve this, cache CPU is introduced. If there is any cache miss, it can lead to delays in loading the data from main memory to cache memory. This timing access may vary depending upon the device operations. The time-dependent operation can lead to leak of side-channel information when a timing measurement of cache operation is done. These types of attacks are referred to cache-based attack. At first, Kelsey et al. [25] identified these types of attacks. Later, the possibility of such attacks in software-based implementation of AES algorithm was well explored by Gajrani et al. [26].

The countermeasure for this attack is done by replacing the cache with cached S-box. Performance depends mostly on the choice of OS, which can efficiently lead to debilitating cache flushing, utilize application particular algorithmic concealing and to take up a division of cache hardware designs accordingly. Compared to all other public and private key algorithms, SHA family is found to be the best choice that does not utilize the lookup tables.

4.6   Acoustic Attack

In the year 1965, the information related to one of the SCA attacks was introduced by Wright [27]. During the world war, the British government tried to break the secret code conversation between the Egyptian embassies located in Egypt and London.

P. Wright suggested a simple technique by placing a microphone close to the rotator machine used by the Egyptians. The conversation was traced from the default click sounds produced by the rotators. This sort of attacks can be categorized as an acoustic type side-channel attack. The acoustic SCA is prevalent in the modern day 3D printing [28], a booming technology in additive systems. In this, for 3D designs, stepper motors are primarily used. By tapping the unequal noise of the motor during the printing process, the complete product can be redesigned by the observer.

Placement of soundproof boxes to attenuate all sound frequencies could be the best countermeasure to shield the acoustic attack. In certain cases as in 3D printing [28], external sound noise can be introduced during the device operation without affecting the design process to combat the attack.

4.7 Visible Light Attack

Kuhn in [29] demonstrated the vulnerability due to light reflection and their possibility of attack in a light-based devices. The average luminosity from a CRT diffuse reflection can be utilized to build back the original signal. The attack due to this sort of side-channel information is coined as a visible light attack. Similar type of attack can also be noticed in LED-based devices. A high level transmitted from an LED on a device can be broken down to derive the data processed [30].

Countermeasure for such visible light attack includes reduction in the dark phase and essentially increasing the signal frequency. Multiple fluorescent-operated lamps that are not synchronized in phase may be another option to combat this attack. In display screens, LCD backlit will be a good choice than phosphorous [29] as it can simultaneously refresh all the pixels in a row. Thus, the leakage of side-channel information can be prevented.

4.8 Photonic Attack

In the year 2008, the photonic side channel was introduced. The phenomena of photonic emission when the transistors are in saturation region causes a leak in the side-channel information. The emission rate of the transistors is proportional to the switching frequency of the circuit. If the saturation voltage increases, the emission of photons also increases. In IC, photonic emissions are prevalent at the backside of the chip [31]. Although, the silicon substrate has high rate of absorption for wavelengths shorter than the band gap energy, the silicon substrate is transparent to near-infrared (NIR) emissions. Hence, any NIR photons emitted by the device will pass through the silicon substrate and can be sensed from the backside of the IC. Two types of PEA photonic emission attack is noticed, simple phonic emission analysis (SPEA) and differential photonic emission analysis (DPEA). SPEA is a power technique similar to the power analysis discussed in Sect. 4.2. This can successfully break the AES 128 secret key with much lower cost [32]. DPES analysis is found to be powerful than SPEA, which can effectively reveal the secret key of DES, AES and as well as RSA [32]. In air-gapped computers by the use of photonic emission analysis, the data can easily be retrieved.

Countermeasure of this attack was discussed in [32]. To protect RSA against the photogenic side channel, it is needed that the secret-key-dependent branches must be removed from the implementation. Moreover, general countermeasures against photogenic side-channel attacks include varying the memory locations of central variables used by the decryption process and adding dummy operations involving the same memory space access patterns [32].

4.9   Fault Attack

This attack is based on the tampering of the cryptographic device; certain faults will automatically leak the information of the secret key. Faults can be generalized into [33] permanent fault attack, transient fault attacks, location accessing fault and faults in timing process.

4.9.1   Permanent Fault Attack

If the cryptographic device gets damaged permanently, the fault will be introduced that leads to the display of incorrect results. By disconnecting the data wire, the fault attack can be eliminated.

4.9.2   Transient Fault Attack

During the device operation, there is a possibility that faults can be introduced at a certain stage of the process. This can affect the device operation. Such attacks, especially occur in power supply and signal frequency generators.

4.9.3   Location Accessing Fault

Some attacks have the potential to induce fault at a very specific location such as memory cell. Such attacks are location based.

4.9.4   Faults in Timing Process

Attacks that have the ability to induce faults at a specific time during the process are prominent in digital circuits.

These faults are random and a frequent check on such attacks during device operation is unnecessary. If any of these faults occur during the device operation, restarting of the device operation will clear the attack. Randomization is the best option for these fault attacks. Especially, public key algorithm is more vulnerable to such attacks; hence, protection at the end of the signature computation is necessary, moreover integrity needed to be checked at every computational time [33].

4.10   Frequency Based Attacks

Frequency-based side-channel attack is much prevalent in mobile devices such as PDAs, cell phones and pagers. It has been proved that this attack is powerful even when the traces are misaligned, whereas the previously researched DEMA are weak [4]. The first order frequency attack proposed by Lui is found to defeat the Desynchronization technique of countermeasure. The effective method of countering this attack is to randomly introduce delays. However, no adequate measures to shield the device from such frequency-based attack is reported yet.

4.11   Error Message

In the communication process, sender sends a requesting signal to receiver before sending the actual message. The receiver responds to the requested signal by sending an acknowledgment, and then the communication is established between the sender and the receiver. The return message to establish such communication is the source of the side-channel information called as an error message attack. This error message attack was well described by Vaudenay [34] in the symmetric encryption technique.

Countermeasure for the error message attack is done by Padding of bits. The extra bits added at the end of information bits is called as padding. This is done before the encryption of the original message in the communication protocol. On the contrary, if padding is done after the encryption of the original message, the error message attack appears again during communication.

4.12   Pressure Attack

The digital doors are usually protected by strong passwords. Eventhough the level of password is strong, it can be easily cracked through the pressure. The raw vibration signals generated when an authorized user presses the password is the source of side-channel information. This side-channel information due to the applied pressure while keying in the password can be extracted through the vibration signals. Thus, the password can be retrieved. The pickoff of this vibration signal [35] could be achieved by simply attaching a small sensor to a door or door lock, thus an attacker reads the side-channel information. This information then can be analysed by the feature extraction and a classifier to extract the password.

Similar to acoustic attack, this attack can also be shielded by introducing random noise to the buttons, so that the press of a buttons yield an indistinguishable signal to the attacker. Further, additional authentication such as an ID card system adds another layer of security [35].

4.13 Combination of Side-Channel Attacks

Combinations of two or three means of side-channel information can be exploited by the attacker for breaking the algorithm. Such attacks can tap into at least two combinations such as timing and power attacks, and electromagnetic and phonetic attacks. There is no specific countermeasure for such combination attacks; the earlier discussed countermeasures for individual attacks can be adapted in combinations to combat.

5 Conclusion

Cryptographic algorithms are vulnerable to attacks eventhough how much intense it may be built. Much effort on the cryptography calculations while disregarding different parts of security threat may be futile. In an effort to guard your phone not by building a high range of security locking system, but rather by putting an enormous stake of encryption algorithm makes it vulnerable for a data theft. This paper aims to present the various avenues of side- channel attacks on crypto devices/systems. The comprehensive counter measures on such attacks were also reviewed based on various authors point of view. A complete review regarding sources of side channel attacks and the deployment of countermeasure to truly protect the systems were also presented.

References

1. Willam, S.: Cryptography and Network Security: Principles and Practices. Pearson Education, India (2006)
2. Galbraith, S.D.: Mathematics of Public Key Cryptography. Cambridge University Press (2012)
3. Grah, J.S.: Hash functions in cryptography. Master thesis, Institute of Informatics, Department of Mathematics, University of Bergen, June 2008
4. Zhou, Y., Feng, D.: Side-channel attacks: ten years after its publication and the impacts on cryptographic module security testing. In: Information Security Seminar (2005)

5. Khan, A.K., Mahanta, H.J.: Side channel attacks and their mitigation techniques. In: First International Conference on Automation, Control, Energy and Systems (ACES) (2014)
6. Kang, Y.J., Bruce, N., Park, S., Lee, H.: A study on information security attack based side-channel attacks. In: 18th International Conference on Advanced Communication Technology (ICACT), Pyeongchang (2016)
7. Standaert, F.X.: Introduction to side-channel attacks. In: Verbauwhede, I.M.R. (ed.) Secure Integrated Circuits and Systems, pp. 27-42. Springer (2010)
8. Schneier, B.: Security pitfalls in cryptography. http://www.schneier.com/essay-pitfalls.html
9. Anderson, R., Bond, M., Clulow, J., Skorobogatov, S.: Cryptographic processors—a survey. Proc. IEEE **94**(2), 357–369 (2006)
10. Kocher, P., Jaffe, J., Jun, B.: Differential power analysis. In: Proceedings of the 19th Annual International Cryptology Conference on Advances in Cryptology, CRYPTO 1999, pp. 388–397. Springer, London (1999)
11. Kocher, P.C.: Timing attacks on implementations of Diffie-Hellman, RSA, DSS and other systems. In: Koblitz, N. (ed.) The Proceedings of the 16th Annual International Conference on Advances in Cryptology (CRYPTO96), vol. 1109, pp. 104–113. Springer (1996)
12. Spadavecchia, L.: A network-based asynchronous architecture for cryptographic devices, (Thesis Submission) (2005)
13. Fleming, C., Cui, N., Liu, D., Liang, H.: Attacking random keypads through click timing analysis. In: International Conference on Cyber-Enabled Distributed Computing and Knowledge Discovery, pp. 118–121 (2014)
14. Wang, W., Yu, Y., Standaert, F.X., Liu, J., Guo, Z., Gu, D.: Ridge-based DPA: improvement of differential power analysis for nanoscale chips. IEEE Trans. Inf. Forensics Secur. **13**(5), 1301–1316 (2018)
15. Kabin, I., Dyka, Z., Kreiser, D., Langendoerfer, P.: Evaluation of resistance of ECC designs protected by different randomisation countermeasures against horizontal DPA attacks. In: IEEE East West design and Test Symposium (EWDTS) (2017)
16. Yu, W., Kse, S.: A voltage regulator-assisted lightweight AES implementation against DPA attacks. IEEE Trans. Circ. Syst. I Regular Papers **63**(8), 1152–1163 (2016)
17. Gandolfi, K., Mourtel, C., Olivier, F.: Electromagnetic analysis: concrete results. In: CHES 2001, LNCS 2162, pp. 251–261 (2001)
18. Quisquater, J.J., Samyde, D.: Electromagnetic analysis (EMA): measures and countermeasures for smart cards. In: The Proceedings of the International Conference on Research in Smart Cards (E-smart 2001), vol. 2140-LNCS, pp. 200–210. Springer (2001)
19. Gu, K., Wu, L., Li, X., Zhang, X.M.: Design and implementation of an electromagnetic analysis system for smart cards. In: 2011 Seventh International Conference on Computational Intelligence and Security, Hainan, pp. 653–656 (2011)
20. Yang, B., Wu, K., Karri, R.: Scan-based side-channel attack on dedicated hardware implementations of data encryption standard. In: Proceedings of International Test Conference 2004 (ITC 2004), Charlotte, pp. 339–344 (2004)
21. Oku, D., Yanagisawa, M., Togawa, N.: A robust scan-based side-channel attack method against HMAC-SHA-256 circuits. In: 2017 IEEE 7th International Conference on Consumer Electronics-Berlin (ICCE-Berlin), Berlin, pp. 79–84 (2017)
22. Rolt, J.D., Di Natale, G., Flottes, M.L., Rouzeyre, B.: Thwarting scan-based attacks on secure-ICs with on-chip comparison. IEEE Trans. Very Large-Scale Integr. (VLSI) Syst. **22**(4), 947–951 (2014)
23. Luo, Y., Cui, A., Qu, G., Li, H.: A new countermeasure against scan-based side-channel attacks. In: 2016 IEEE International Symposium on Circuits and Systems (ISCAS), Montreal, QC, pp. 1722–1725 (2016)
24. Namin, S.H., Mehta, A., Namin, P.H., Rashidzadeh, R., Ahmadi, M.: A secure test solution for sensor nodes containing crypto-cores. In: 2017 IEEE International Symposium on Circuits and Systems (ISCAS), Baltimore, MD, pp. 1–4 (2017)
25. Kelsey, J., Schneier, B., Wagner, D., Hall, C.: Side channel cryptanalysis of product ciphers. In: Proceedings of the 5th European Symposium on Research in Computer Security, LNCS 1485, pp. 97110 (1998)

26. Gajrani, J., Mazumdar, P., Sharma, S., Menezes, B.: Challenges in implementing cache-based side channel attacks on modern processors. In: 2014 27th International Conference on VLSI Design and 2014 13th International Conference on Embedded Systems, Mumbai, pp. 222–227 (2014)
27. Wright, P.: Spy Catcher: The Candid Autobiography of a Senior Intelligence Officer. Viking Press (1987)
28. Faruque, A., Abdullah, M., Chhetri, S.R., Canedo, A., Wan, J.: Acoustic side-channel attacks on additive manufacturing systems. In: 2016 ACM/IEEE 7th International Conference on Cyber-Physical Systems (ICCPS), Vienna, pp. 1–10 (2016)
29. Kuhn, M.G.: Optical time-domain eavesdropping risks of CRT displays. In: Proceedings 2002 IEEE Symposium on Security and Privacy, pp. 3–18 (2002)
30. Loughry, J., Umphress, D.: Information leakage from optical emanations. ACM Trans. Inf. Syst. Secur. **5**, 262–289 (2002)
31. Tajik, S., Dietz, E., Frohmann, S., Dittrich, H., Nedospasov, D., Helfmeier, C., Seifert, J.P., Boit, C., Hübers, H.W.: Photonic Side Channel Analysis of Arbiter PUFs (2016)
32. Carmon, E., Seifert, J.P., Wool, A.: Photonic side channel attacks against RSA. In: 2017 IEEE International Symposium on Hardware Oriented Security and Trust (HOST), McLean, VA, pp. 74–78 (2017)
33. Clavier, C., Feix, B., Gagnerot, G., Roussellet, M.: Passive and active combined attacks on AES combining fault attacks and side channel analysis. In: 2010 Workshop on Fault Diagnosis and Tolerance in Cryptography, Santa Barbara, CA, pp. 10–19 (2010)
34. Vaudenay, S.: Security flaws induced by CBC padding applications to SSL, IPSEC, WTLS. In: EUROCRYPT 2002, LNCS 2332, pp. 534–545 (2002)
35. Ha, Y., Jang, S.H., Kim, K.W., Yoon, J.W.: Side channel attack on digital door lock with vibration signal analysis: longer password does not guarantee higher security level. In: 2017 IEEE International Conference on Multisensor Fusion and Integration for Intelligent Systems (MFI), Daegu, pp. 103–110 (2017)
36. TechDifferences.: Difference Between Active and Passive Attacks. (2008). https://techdifferences.com/difference-between-active-and-passive-attacks.html

Implementation and Analysis of Different Path Loss Models for Cooperative Communication in a Wireless Sensor Network

Niveditha Devarajan and Sindhu Hak Gupta

1 Introduction

An optimization Wireless Sensor Network (WSN) is being utilized commercially and industrially due to its increasing technical advancements. WSN consists of numerous miniature-sized, battery-operated, low-power devices called nodes. Nodes are small-sized, low storage, mini computers having the capabilities of collecting, processing and computing data. They work jointly to form the network. The parameters that can be sensed by these sensors include temperature, pressure, humidity, intensity, vibrations, motions, pollutants and much more [1]. The architecture of these nodes is such that they can be placed in harsh environmental conditions, uneven terrains like the snowy environment where other networks might fail to communicate [2]. This has enabled WSN's to be used for various critical applications including environmental surveillance, health monitoring, military, and so on. The sensor nodes may be deployed to obtain critical data and to study the external factors that affect the functioning and reliability of a sensor network. Once the nodes have been deployed, they work collectively as an independent structure for the transfer of information through them. WSN is a memory and resource constrained network in which energy efficiency is the main and crucial issue as limitations of hardware makes replacement of batteries a difficult task [3]. The successful operation of WSN depends on the life span of the sensor node. Limiting the energy consumption to improve or enhance the energy efficiency becomes an alternative way to solve the energy limitation problem. Also, only finite amount of information transfer can take place due to limited storage capacity of a sensor node. However, multipath channel fading has a strong impact on

N. Devarajan · S. H. Gupta (✉)
Department of Electronics and Communication Engineering, Amity University
Uttar Pradesh, Noida, Uttar Pradesh, India
e-mail: shak@amity.edu

N. Devarajan
e-mail: nivedith.dev@gmail.com

© Springer Nature Singapore Pte Ltd. 2019
S. Tiwari et al. (eds.), *Smart Innovations in Communication and Computational Sciences*, Advances in Intelligent Systems and Computing 851,
https://doi.org/10.1007/978-981-13-2414-7_22

227

the transfer of information as it is capable of increasing the path loss and reducing the reliability of data transfer [4]. Therefore, deploying multiple nodes in harsh environments with direct transmission between nodes will not be effective and reliable. As a solution, the advantages that cooperative communication provides can be utilized in the energy limited WSN to increase the network lifetime. The collaboration of sensors can create a situation for reliable communication links to save energy and reduce the network lifetime. In cooperative communication, single antenna nodes at a proximate distance with each other can share antennas in such a way that create a virtual MIMO (V-MIMO) framework, with the aim of sharing resources and creating multiple independent fading paths. It inculcates the idea of multiple sensor nodes being physically gathered or grouped together to transmit and receive information in tandem. This technique has the capability of increasing the network capacity, and diversity order [5]. Cooperative communication exploits multihop communication to transfer the data from source node to destination node via neighboring nodes. Neighboring nodes act as relay nodes. There is a need to be more aware about the cooperative relaying strategies that will help in taking the information through the nodes to ultimately reach the destination node. The relaying signaling strategies can be broadly classified into Amplify and Forward signaling technique, Decode and Forward signaling technique and Coded Cooperative communication technique. Cooperative relaying strategies may be implemented in a network to increase its lifetime. In this work Amplify and Forward signaling technique will be considered, it being relatively uncomplicated to implement [5, 6].

If the signal transfer was wired, the communication in WSN would have been reliable with a predictable attenuation in the signal transferred. But since the communication channel is wireless in nature, the behavior of the channel is stochastic and random, there are several processes like interference, scattering, reflection that happens due to objects that are present in the atmosphere which will attenuate the transfer of information, reduce the signal power and increase the errors. So, to predict the path loss that is taking place and to evaluate the performance of the system, mathematical path loss models must be considered [7]. Path loss models are physical models that predict path loss suffered by the signal from transmitter to receiver due to the wireless nature of the channel. A prediction method that is accurate and reliable aids to enhance, optimize and improve the coverage area, transmitted power and removing problems due to interference. In this work, Free Space path loss model, Empirical Log-distance model, Two-ray model, and Ray tracing model are being considered [2].

This work aims to show that compared to the non-cooperative relaying or direct transmission scheme, cooperative relaying scheme is more effective in terms of received Energy and reliability. For the analysis, a simple WSN consisting of three nodes with adjustable node antenna heights have been considered, and Energy received at the destination has been evaluated taking the aid of the path loss models. Further, Amplify and Forward single relay cooperative relaying scheme has been implemented in this WSN and Energy received at the destination node has been evaluated again. Comparative analysis shall reveal that cooperative relaying scheme

Table 1 Notations and terminology

R	Received signal at the relay node
e_{s1}	Transmission symbol power in phase 1
β	Amplification factor
α	Log normal fading channel constant
e_{s2}	Transmitted symbol power in phase 2
v	Relay's information
r	Channel capacity in bits per second
B	Bandwidth
Y	Received signal at the destination node
n	Additive white gaussian noise
PL	Path loss
d	Distance between the transmitter node and the destination node/Distance between relay and the destination node

shows better performance than non-cooperative relaying scheme in terms of the received Energy at the destination node.

Analysis of proposed model

- Energy values have been evaluated with the help of path loss values and mathematical equations for Cooperative as well as Non-Cooperative scenario.
- Using the mathematical equations evaluated, the effect of applying Amplify and forward Cooperative communication is observed (Table 1).

2 Proposed Approach

The WSN setup has been depicted in Fig. 1. With the source node (S), relay node (R), and destination node (D) having adjustable antenna heights. The nodes have been adjusted for three heights (0.25, 1, and 1.5 m) from the ground. The position of the source node is fixed but the destination node is mobile in nature and is moved along a straight path along which the real time measured values of path loss are taken. The relay node taken between source node and the destination node will Amplify the data received from source node and further the data to the destination node. This has been depicted in Fig. 1. The destination node is varied from 5 m to a maximum of 30 m with a varied step size of 5 m and the values were calculated at these distances. The heights of antennas on the nodes are adjustable and are adjusted at equal heights (0.25, 1, and 1.5 m) above the ground to find out values of path loss according to different heights of source, relay and the destination node.

Amplify and Forward (AF) relay Cooperative protocol has been considered at the relay. In first phase of Amplify and Forward relay protocol, the relay receives the

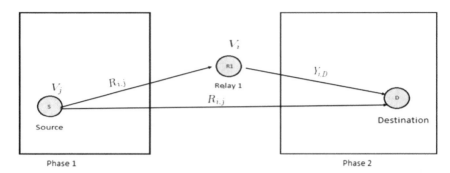

Fig. 1 Proposed wireless sensor network model

information transmitted by the source node. In second phase, this received information is amplified by the relay node and is retransmitted to the destination node. It is
assumed that no loss of signal power takes place during the transmission [8]. Considering Amplify and Forward protocol and Shannon Hartley theorem, the expression
for the transmitted energy in the second phase has been derived and is given in (1).

$$e_{s2} = \frac{(Y_{i,D} - n_{i,D})^2 \left[e_{s1} |\beta_{i,j}|^2 + \frac{e_{s1}}{\left(2^{\frac{r}{B}} - 1\right) B} \right]}{\alpha_{i,D}^2 \left[\sqrt{e_{s1}} \beta_{i,j} v_i + n_{i,j} \right]^2} \tag{1}$$

Equation (1) is the relation that relates the transmitted symbol power e_{s2} in terms
of bandwidth B, channel capacity r, and the amplification factor $\beta_{i,j}$.

Path loss models have been explored being free space path loss model, Ray tracing
path loss model, Two-ray path loss model and Log-distance path loss model.

Free space path loss model: Free space path loss is the loss in signal strength of an
electromagnetic wave resulting from a line-of-sight path through free space, without
any obstacles close by to cause reflection or diffraction. This model is utilized predict
path loss and to find and determine signal strength where there is clear line-of-sight
or pathway between the transmitter node and the receiver node. The expression to
predict path loss in dB is computed in [2] and by (2).

$$PL_{free} = 20 \log_{10} \left(\frac{4\pi f d}{c} \right) \tag{2}$$

f denotes frequency in GHz and d denotes distance between the transmitter node
and the destination node. It is computed in meters. The above equation can also be
expressed as given in (3).

$$PL_{free} = -27.56 + 20 \log_{10}(f) + 20 \log_{10}(d) \tag{3}$$

Equation (3) can be expressed in terms of energy by using the Plank's equation given by $(E = hf)$. Using this relation, this equation can be expressed as given in (4).

$$PL = 636.01 + 20 \log_{10}(d) + 20 \log_{10}(d) \tag{4}$$

Ray tracing path loss model: The ray tracing model is used to calculate path loss for larger distances. The expression to predict the path loss in dB is computed in [2] is given by (5).

$$RL_{free} = 20 \log_{10} \left(\frac{d^2}{h_T h_R} \right) \tag{5}$$

d denotes distance between in meters between transmitter node and receiver node h_T denotes height of source node and h_R denotes height of the destination node in meters.

Two-ray path loss model: The two-ray model is similar to plane earth path loss model. The received signal is composed of two components one being the direct line of sight component and the other being the ray reflected from the ground. The expression to predict the path loss in dB is computed in [2] is given by (6).

$$TR_{free} = \begin{cases} PL_{free}, d < d_c \\ RL_{free}, d \geq d_c \end{cases}, \tag{6}$$

where parameter d_c is the cross over distance defined as

$$d_c = \frac{4\pi h_T h_R}{\lambda} \tag{7}$$

Here, λ is the wavelength, h_T denotes height of source node and h_R denotes height of the destination node in meters.

Log-distance path loss model: The path loss values that is calculated when multipath effects, terrain variations are considered. The expression to predict the path loss in dB is computed in [2] is given by (8).

$$LD_{free} = L(d_o) + 10 n \log_{10} \left(\frac{d}{d_o} \right) + \chi_\sigma, \tag{8}$$

where n is path loss exponent that shows rate at which the received signal decreases with distance, $L(d_o)$ is path loss in dB at a reference distance, d_o. χ_σ is a parameter that represents the mean Gaussian random variable having standard deviation σ (dB) and describes the shadowing effects.

The equation for the received power P_r in the second phase has been derived from [9] and is given by (9).

$$P_r = \frac{e_{s2} \times C}{d^{Pathloss}}, \tag{9}$$

where d stands for distance between the relay node and the receiver node, e_{s2} is transmitted power to the destination node, C is a constant mentioned in (9) has been derived from [9] and has been given by Eq. (10).

$$C = G_t G_r \frac{c^2}{4\pi f_c} \tag{10}$$

Here, G_t, G_r are the antenna gains at transmitter and receiver node respectively, f_c stands for carrier frequency, c is the speed of light.

3 Experimental Setup and Results

Simulation tests have been done using MATLAB. Non-Homogenous WSN scenario has been assumed. Analysis has been accomplished under Log Normal Shadow Fading channel scenario. Additive White Gaussian Noise (AWGN) and Bipolar Phase Shift Keying (BPSK) modulation is considered. Measured path loss values have been taken from [2] to calculate the received Energy for plotting the graphs and these values have been taken for a snowy environment. The graphs have been shown, considering the antenna height of nodes to be 0.25 m. Here, all the nodes are assumed to be fully cooperating. Ideal values of the variables have been assumed and the graphs have been plotted accordingly.

4 Results

Figure 2 illustrates the effect of path loss on Energy. It is observed that as path loss increases, received Energy also increase. It is observed that received Energy of Empirical log-distance model is more compared to other path loss models. Therefore, the Empirical log-distance model exhibits a better performance compared to other path loss models. Also, the plots reveal that the received Energy has improved in comparison to the non-cooperative scenario. Therefore, implementing cooperative communication improves the reliability of the network by increasing the values of the received Energy.

Figure 3 illustrates the effect of distance on Energy. It is observed that as distance increases, received Energy also increase. The received Energy of Empirical log-distance model is more compared to other path loss models. Therefore, the Empirical log-distance model exhibits a better performance compared to other path loss models. Also, the plots reveal that the received Energy has improved in comparison to the non-cooperative scenario. Therefore, implementing cooperative communica-

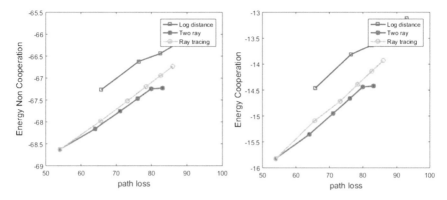

Fig. 2 Graphs for path versus energy for cooperative and non-cooperative scenario

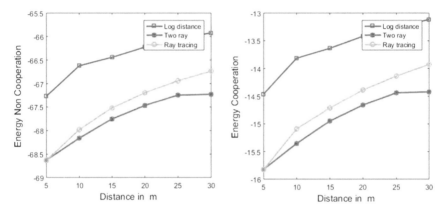

Fig. 3 Graphs for distance versus energy for cooperative and non-cooperative scenario

tion improves the reliability of the network by increasing the values of the received Energy.

Figure 4 illustrates the effect of path loss on received Energy for log-distance path loss model, considering cooperative and non-cooperative scenario. The plots reveal that the received energy has improved in comparison to the non-cooperative scenario. Therefore, implementing cooperative communication improves the reliability of the network by increasing the values of the received Energy.

Figure 5 illustrates the effect of path loss on received Energy for Two-ray path loss model, considering cooperative and non-cooperative scenario. The plots reveal that the received Energy has improved in comparison to the non-cooperative scenario. Therefore, implementing cooperative communication improves the reliability of the network by increasing the values of the received Energy.

Figure 6 illustrates the effect of path loss on received Energy for Ray tracing path loss model, considering cooperative and non-cooperative scenario. The plots reveal that the received Energy has improved in comparison to the non-cooperative scenario.

Fig. 4 Path loss versus
received energy for
log-distance path loss model

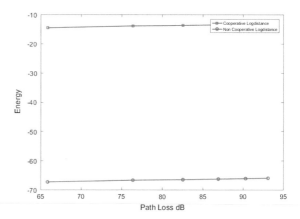

Fig. 5 Path loss versus
received energy for two-ray
path loss model

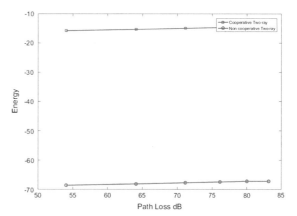

Therefore, implementing cooperative communication improves the reliability of the
network by increasing the values of the received Energy.

Table 2 is a comparative compilation of the values of Received Energy for Cooperative and Non-Cooperative scenario for the considered path loss models. The results
have been compiled taking three distances (d = 10, 20 and 30 m) between the source
node and the destination nodes. The above results been compiled considering taking
the height of nodes to be 0.25 m. Varying the heights of nodes reveal similar results.
Based on the above table and the plots depicted, some important conclusions can be
derived.

Fig. 6 Path loss versus received energy for ray tracing path loss model

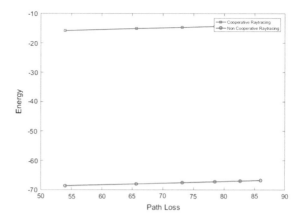

Table 2 Values of received energy at different distances and path loss models

S. no.	Path loss models	Energy (dB) at distance d in (m)					
		Non-cooperative scenario			Cooperative scenario		
1.		10	20	30	10	20	30
	Log-distance	−66.6212	−66.2362	−65.9213	−13.8162	−13.4212	−13.1163
	Two-ray	−68.1607	−67.4675	−67.2271	−15.3558	−14.6626	−14.4421
	Ray tracing	−67.9788	−67.1935	−66.7343	−15.0906	−14.3885	−13.9239

5 Conclusions

Performance of WSN has been analyzed for cooperative and non-cooperative (Direct Transmission) scenario. Mathematical modeling has been done to the find out the received Energy in the second phase for the cooperative scenario and it is observed that the gain offered by the relay in the second phase depends upon Amplification factor, Rate and Bandwidth. It has been observed that Empirical Log-distance model exhibits a better performance in terms of received Energy. The simulation and analytical results reveal that when cooperative communication is implemented, there is a significant improvement in the Received Energy by approximately 79.74% in the Log-distance path loss model, 77.94% in Two-ray model and 77.23% in Ray tracing model when compared to the non-cooperative scenario. This result has been calculated for three different node heights and reveals similar results.

References

1. Kumar, D., Aseri, T.C., Patel, R.B.: EEHC: energy efficient heterogeneous clustered scheme for wireless sensor networks. Comput. Commun. **32**(4), 662–667 (2009)
2. Cheffena, M., Mohamed, M.: Empirical path loss models for wireless sensor network deployment in snowy environments. IEEE Antennas Wirel. Propag. Lett. **16**, 2877–2880 (2017)
3. Rani, S., Malhotra, J., Talwar, R.: Energy efficient chain based cooperative routing protocol for WSN. Appl. Soft Comput. **35**, 386–397 (2015)
4. Butcharoen, S., Pirak, C.: An adaptive cooperative protocol for multi-hop wirelessnetworks. In: 2015 17th International Conference on Advanced Communication Technology (ICACT), Seoul, 2015, pp. 620–635
5. Katiyar, H., Rastogi, A., Agarwal, R.: Cooperative communication: a review. IETE Tech. Rev. **28**(5), 409–417 (2011)
6. Duy, T.T., Kong, H.Y.: On performance evaluation of hybrid decode-amplify-forward relaying protocol with partial relay selection in underlay cognitive networks. J. Commun. Netw. **16**(5), 502–511 (2014)
7. Kurt, S., Tavli, B.: Path-loss modeling for wireless sensor networks: a review of models and comparative evaluations. IEEE Antennas Propag. Mag. **59**(1), 18–37 (2017)
8. Iqbal, Z., Kim, K., Lee, H.N.: A cooperative wireless sensor network for indoor industrial monitoring. IEEE Trans. Ind. Inf. **13**(2), 482–491 (2017)
9. Bharati, S., Zhuang, W., Thanayankizil, L.V., Bai, F.: Link-layer cooperation based on distributed TDMA MAC for vehicular networks. IEEE Trans. Veh. Technol. **66**(7), 6415–6427 (2017)

Parallel Approach for Sub-graph Isomorphism on Multicore System Using OpenMP

Rachna Somkunwar and Vinod M. Vaze

1 Introduction

The Graph isomorphism issue is worried about deciding when two graphs are isomorphic. This is a troublesome issue and in the general case there is no known productive calculation for doing it. Although, the graph isomorphism problem is polynomial time solvable if graphs are restricted to trees, planar graphs, random graphs, mesh graphs, interval graphs, permutation graphs, bounded valance graphs, etc. In Randomly Connected graphs, nodes are randomly connected with other nodes. In mesh graph, structure is regular, e.g., 2D, 3D, 4D. The 2D mesh is graphs in which each node is connected with its four neighborhood nodes. In Bounded Valence graphs, every node has a number of edges lower than a given threshold, called Valence. For, e.g., Valance (3) graph, every node should have three neighbors. Sub-graph can be used to recognize functional and non-functional characteristics in various graph applications includes social, chemical [1], bioinformatics [2], computer vision [3, 4], pattern recognition [5]. With the current developing enthusiasm on graph databases, it has turned out to be more critical to have a quick answer for the issue of sub-graph isomorphism, or finding related patterns in large graphs. The sub-graph isomorphism is the problem of detection of input graph inside the target graph. The term isomorphism is nothing but one to one mapping of two graphs: number of edges and nodes should be same [6].

In general the implementation of serial-based sub-graph algorithm is time-consuming; if the size of graph grows exponentially and it also requires rapid calculation for isomorphism algorithms. Because of the serial-based algorithms which are impractical and its complexity constraints, parallel approach is proposed for the serial-based algorithms.

R. Somkunwar (✉) · V. M. Vaze
Shri Jagdishprasad Jhabarmal Tibrewala University, Jhunjhunu, Rajasthan, India
e-mail: rachnasomkunwar12@gmail.com

© Springer Nature Singapore Pte Ltd. 2019
S. Tiwari et al. (eds.), *Smart Innovations in Communication and Computational Sciences*, Advances in Intelligent Systems and Computing 851,
https://doi.org/10.1007/978-981-13-2414-7_23

For large graphs, parallel implementation of isomorphism is fruitful than serial-based implementation. Several isomorphism algorithms on supercomputer have been studied in order to improve the execution time of large graphs, but the use of super-computer has real restrictions due to high cost.

While most sub-graph Isomorphism algorithms are designed for the single processor environment, there are some that have been extended to be suitable for use in a parallel multiprocessor or multicore environment. Schatz et al. [7] parallelized the approach using a master-worker load balancing scheme of several individual queries and also tried to parallelize sub-graph query but shown very limited scalability tests.

In this work [8], an efficient method is proposed for finding and counting the sub-graphs from large graphs using G-tries. G-tries are specialized data structured used for searching sub-graphs.

Authors [9] have extended the work of SP-SUBDUE parallel approach by modifying the evaluation phase. SUBDUE algorithm is used for finding the most compressing sub-graph in a large database and SP-SUBDUE is based on MPI frame-work.

In this research [10], authors have proposed scalable parallel approach for census computation. Census sub-graph is finding the frequencies of small sub-graphs in large networks. One thread is used for dealing with highly unbalanced search tree.

In this paper [11], a new algorithm is proposed which is based on thread parallel constraint based search method. This method is more suitable for supplemental graphs.

This paper presents [12], divide and conquer method for dividing graphs into smaller blocks. These smaller blocks are proceeded by individual processing unit and then results are merged together using hierarchical procedures.

In [13], parallel approach is produced called subneum. Subneum enumerates sub-graphs using edges instead of vertices. Authors have also used fast heuristic methods for which can efficiently accelerate non isomorphic sub-graph enumeration. Authors have proposed an algorithm for drug discovery application using GPU-based parallel approach [14], two different tools are also discovered to narrow down the large ligand search space and also able to handle the data intensity of the large graphs.

This paper presents parallel and efficient algorithm to the sub-graph isomorphism problem using three different ways. Parallel implementation of all type of algorithm is not possible due to dependency of logical module of algorithm. If the logical part of algorithm depends on other part, then parallel implementation is difficult to implement. If the logic of algorithm is having less dependency on the resources and the other logical part of algorithm, then it might be possible to implement the parallel algorithm for the same.

In case of Ullman algorithm [15], parallel implementation is not possible due to its various dependencies on the previous and the next steps. Also Ullman is based on various permutations and combinations.

In case of VF2 [16], it is backtracking type of algorithm and in backtracking type of algorithms, the step depends on the previous steps, in such cases, parallel implementation is difficult.

2 Proposed Method

2.1 Motivation

The parallel algorithm proposed in this work is divided into three phases: Phase 1 is based on grouping of similar nodes, Phase 2 is used for reducing the size of group and Phase 3 is responsible for generating the path. Execution time of Phase 1 is negligible as compared to total execution time of algorithm. Execution time of Phase 3 depends on the Phase 2. Phase 2 is taking much time as compared to total execution time.

Parallel implementation is required only for the Phase 2 which is known as optimization of cluster. Cluster optimization is an iterative process in which the iteration is performed until the average of cluster size becomes constant, i.e., during the execution, if two continuous iterations are having same cluster size, then iterative process is stopped. So while doing cluster optimization for each cluster, we have to perform these operations by using parallel concept.

2.2 Algorithm

Algorithm is divided into three phases:

Input: Graph G1 (Sub-graph) and G2 (Fullgraph)
Output: To check whether G1 is Isomorphic to G2 or not.

Phase 1: Cluster Formation

Graph clustering is a process in which the clusters of graphs are created on the basis of labels and attributes. Some of the algorithm uses graph clustering techniques to create clusters of different kinds of nodes of graph.

During the execution of phase1, algorithm creates cluster for each node of subgraph in which cluster members are those nodes of full graph having similarity with node of sub-graph and the node of sub-graph is considered as head of cluster. Clusters are created using minimum distance, shown in Eq. 1.

$$Mindist = \sqrt{(Indegree\ of\ fullgraph - Indegree\ of\ subgraph)^2}$$
$$+ (Outdegree\ of\ fullgraph - Outdegree\ of\ subgraph)^2 \quad (1)$$

Phase 2: Cluster Optimization

Phase 2 is known as cluster optimization. We have seen in the phase 1 that clusters are created as per the number of nodes of sub-graph. If there are 10 nodes in sub-graph, then there will be 10 numbers of clusters and each cluster will contain the members of nodes of full graph in such a way, the size of each cluster varies. But in our algorithm, the number of members of class cluster is very important and therefore we will try to decrease the size of each cluster by removing the unwanted nodes of cluster.

Phase 3: Path Finding

In phase 3, we started to calculate the size of smallest cluster and we try to create the path of node of smallest cluster with the node of remaining clusters. But while creating this path, we have to use backtracking technique.

So finally in all the phases, we have used different types of techniques. In phase 1 we have used the concept of node cluster, in phase 2, we have optimized the size of cluster and in phase 3 path is generated by using backtracking technique.

The benefit of this algorithm is that we can use this algorithm for graph isomorphism as well as sub-graph isomorphism and we tested this algorithm for different kinds of graphs such as randomly generated graphs, M2D, M3D and M4D types of graphs and execution time is also calculated.

2.3 System Design of Parallel Sub-graph Isomorphism Using OpenMP

OpenMP is a standard library implemented for FORTRAN, C, and C ++. It is specially designed for multi-processing of any program. To execute more than one part of program by using multiple core processors is called as parallel processing. While using the OpenMP in any C/C++ program, it is necessary that multiple numbers of core processors are required for the program or not, that means if large number of arithmetic and conditional operators are performed in your program, then you can use the power of parallel processing by using OpenMP. For the large graphs, large numbers of operations are performed in the algorithm and to perform all the large number of calculations, if we use multiple core processors, then execution time of program decreases and performance of the system increases.

So to minimize the execution time of program we have to use OpenMP in our algorithm in phase 2. To use the OpenMP library in your program, it is necessary to link the library files of OpenMP with C or C++ program. We have used OpenMP pragma for the parallel processing of cluster optimization in which each core optimize the nodes of cluster, i.e., one core per cluster shown in Fig. 1. This concept of parallel processing is useful for the bigger graphs, for the smaller graphs it is not essential or it is not required. For the bigger graph, parallel processing of phase 2 is required.

In our program, we have used the concept of configuring the number of core processors that means if the machine is having four core processors, but if the user wants to use only two numbers of core processors then user can pass these parameters to the program.

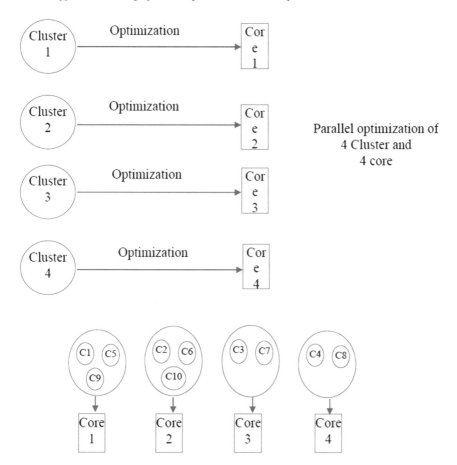

Fig. 1 System design of parallel sub-graph isomorphism

2.4 *PSGIAlgo: Pseudocsode of Parallelization*

```
1.for it = 0 to 10          // 10 clusters
2. Apply parallel operation to each ith cluster using OpenMP
4.for I = 0 to SNC                //SNC: Subgraph Node Count
5.N1 = get ith number of node in G1
6.MNC = get matching node of N1
7.for m = 0 to SNC
8.if I != m then go to step 9
9. N2 = get mth number of node in G1
10. CALl FilterNode(N1,N2,G1,G,i)          // Get all optimized nodes
```

11. end if
12. end for
13. count = 0
14. for j = 0 to MNC
15. if N1.matchStates[j] = false
16. count ++
17. end if
18. end for
19. if count == 0 then
           I = SNC
           it = 10
           status = true
20. end if
21. end for
22. totalClusterSize = 0
23. for i = 0 to SNC
24. N1=get node of subgraph (G1)
25. MNC = get Match Node count of N1
26. size = 0
27. for j = 0 to MNC
28. if N1.matchStates[j] = false
29. size++
30. end if
31. end for
32. totalClusterSize = totalClusterSize + size;
33. end for
34. avgClusterSize = totalClusterSize/SNC
35. if lastAvg = avgClusterSize
           break then go to step 36
36. end for   // iteration completed

3   Experimental Setup

In experiment, program run on an Intel (R) Core (TM) i3 Quad Processor equipped with 8 GB RAM. Algorithm is implemented in C++ language using OPenMP on Linux Suse Operating system for the purpose of faster execution. Dataset and Testset are created using C language and it is developed by http://www.mivia.unisa.it. Dataset contains 10,000 fullgraphs and testset contains 1000 sub-graphs with 6000 nodes.

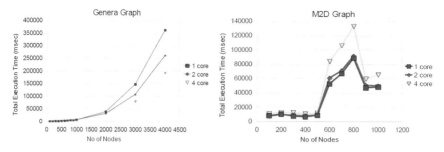

Fig. 2 Comparative result on 1 core, 2 core and 4 core processors for general and M2D graphs

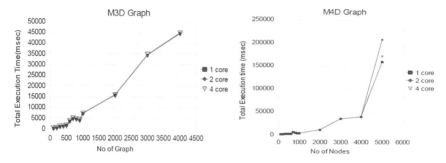

Fig. 3 Comparative result on 1 core, 2 core and 4 core processors for M3D and M4D graphs

4 Results

In this section, we empirically evaluate the performance of Parallel Sub-graph Iso-morphism Algorithm named as PSGIAlgo. Our Experimental evaluation consists of three phases. The PSGIAlgo is tested by using 1 core, 2 cores and 4 cores of proces-sor for different kinds of graphs such as general, M2D, M3D graphs. After testing it is found that performance of PSGIAlgo increases as the number of cores increases. PSGIAlgo performs better in case of large graphs, i.e., we recommend to increase number of cores only if graph size is bigger otherwise for medium and small size of graph 1 core is sufficient for better performance. Figure 2 shows comparative perfor-mance of 1 core, 2 cores and 4 cores for General and M2D graphs. Figure 3 shows comparative performance of 1 core, 2 cores and 4 cores for M3D and M4D graphs. Figure 4 shows the comparative performance of proposed algorithm PSGIAlgo with Ullman and VF2 algorithms. It can be observed that the number of nodes increases, our algorithm outperforms more and more, which is expected.

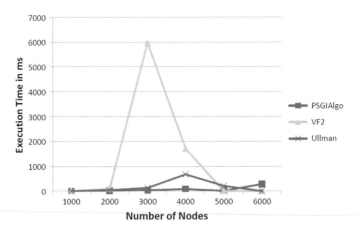

Fig. 4 Comparison between Ullman, VF2 with PSGIAlgo

5   Conclusion

In this paper we have shown how to divide the task into number of cores using PSGIAlgo. This paper introduced and described the PSGIAlgo algorithm. This algorithm is implemented in C++ using OpenMp for finding Sub-graph isomorphism. Sub-graph Isomorphism is a complex problem of graph theory. Many algorithms are designed to solve the problem of sub-graph isomorphism but all the existing algorithms are implemented to work for smaller and medium size of graphs. Design of existing algorithm is not compatible for parallel.

Parallel implementation is required only for the phase 2 which is known as optimization of cluster. By comparing the performance of PSGIAlgo with the existing algorithm it is cleared that the existing algorithms are not implemented for the parallel processing but PSGIAlgo is designed for parallel processing for large graphs. PSGIAlgo gives the better performance by using more than one core processors but in case of VF2 and Ullman we cannot use more number of core processors.

We tested our algorithm for different graphs such as General, M2D, M3D and M4D graphs and we achieved an almost better performance on four cores.

References

1. Régin, J.-C.: Développement d'outils algorithmiques pour l'Intelligence Artificielle. Application à la chimie organique. Ph.D. thesis, Universit_e Montpellier 2,199
2. Bonnici, V., et al.: A subgraph isomorphism algorithm and its application to biochemical data. BMC Bioinform. **14**(7) (2013)
3. Damiand, G., et al.: Polynomial algorithms for subisomorphism of nd open combinatorial maps. Comput. Vis. Image Underst. (2011)

4. Solnon, C., et al.: On the complexity of submap isomorphism and maximum common submap problems. Pattern Recognit. (2015)
5. Conte, D., et al.: Thirty years of graph matching in pattern recognition. Int. J. Pattern Recogn. Artif. Intell. (2004)
6. Bondy, J.A., Murty, U.S.R.: Graph Theory with Applications, vol. 290. Macmillan, London (1976)
7. Schatz, M., Cooper-Balis, E., Bazinet, A.: Parallel network motif finding. Technical report, University of Maryland Insitute for Advanced Computer Studies (2008)
8. Ribeiro, P., Silva, F., Lopes, L.: Efficient parallel subgraph counting using g-tries. In: 2010 IEEE International Conference on Cluster Computing (CLUSTER). IEEE (2010)
9. Ray, A., Holder, L.B.: Efficiency improvements for parallel subgraph miners. In: FLAIRS Conference (2012)
10. Aparicio, D., Paredes, P., Ribeiro, P.: A scalable parallel approach for subgraph census computation. In: European Conference on Parallel Processing. Springer, Cham (2014)
11. McCreesh, C., Prosser, P.: A parallel, backjumping subgraph isomorphism algorithm using supplemental graphs. In: International Conference on Principles and Practice of Constraint Programming. Springer, Cham (2015)
12. Son, M.-Y., Kim, Y.-H., Oh, B.-W.: An efficient parallel algorithm for graph isomorphism on GPU using CUDA (2015)
13. Shahrivari, S., Jalili, S.: Fast parallel all-subgraph enumeration using multicore machines. Sci. Program. (2015)
14. Jayaraj, P.B., Rahamathulla, K., Gopakumar, G.: A GPU based maximum common subgraph algorithm for drug discovery applications. In: 2016 IEEE International Parallel and Distributed Processing Symposium Workshops. IEEE (2016)
15. Augustyniak, P., Ślusarczyk, G.: Graph-based representation of behavior in detection and prediction of daily living activities. Comput. Biol. Med. (2017)
16. Cordella, L.P., et al.: A (sub)graph isomorphism algorithm for matching large graphs. IEEE Trans. Pattern Anal. Mach. Intell. (2004)

Optimized Solution for Employee Transportation Problem Using Linear Programming

Alind and Himanshu Sahu

1 Introduction

With recent advances in world economy, many giant companies are providing mass scale employment where employees tend to work through day and night. Companies have adopted various cab services to facilitate easy transportation. Cabs provide both pickup and drop services. There are various factors that have impact on cost perspective of such service. The possible metrics are total distance traveled by all cabs running, number of cabs that companies need to send, etc. These services are the part of cost to company for an employee, i.e., CTC. Companies try to minimize CTC per employee but there are certain constraints also that prevent easier solution to such problem. The refraining constraints are many such as limited number of employees per cab, upper limit on travel time for employee, etc. The proposed work emphasizes on certain aspects of such problem and tries to reduce total cost incurred to company by minimizing the requirement of cabs to satisfy demand of all employees with some common restrictions.

1.1 Employee Transportation Problem

Employee Transportation Problem (ETP) [1, 2] comes under umbrella of term Vehicle Routing problem (VRP) [3, 4], an area in which intensive research is on for near 50 years. VRP are set of problems that involve finding some optimal travel

Alind (✉) · H. Sahu
University of Petroleum and Energy Studies, Dehradun, India
e-mail: alind@ddn.upes.ac.in

H. Sahu
e-mail: hsahu@ddn.upes.ac.in

© Springer Nature Singapore Pte Ltd. 2019
S. Tiwari et al. (eds.), *Smart Innovations in Communication and Computational Sciences*, Advances in Intelligent Systems and Computing 851,
https://doi.org/10.1007/978-981-13-2414-7_24

path for a group of vehicles that services customers and are also subjected to few constraints. There is a classification of VRP based on these constraints, Capacitated VRP (CVRP) and VRP with time window (VRPTW). CVRP refers to constraint on number of persons that a vehicle can accommodate, whereas VRPTW concerns with time window set by employee for their arrival at destination. ETP has a flavor of both these constraints, where carrying capacity of each vehicle is limited by C and each employee travelling must not travel for a time longer than time proportional to minimum travelling distance from center to their destination.

The remaining of the paper is organized as follows Sect. 2 provides the literature review. Section 3 provides the problem statement. Section 4 describes the proposed method. Section 5 provides the result whereas Sect. 6 provides the conclusion.

2 Literature Review

In this section different research articles that have been studied are summarized. Based on these literatures the theoretical background of VRP and ETP has been established.

Sathyanarayanana and Joseph [4] mention two different classes of optimization for VRP; re-optimization model and a priori optimization. Re-optimization means finding next point to be visited by the vehicle if current location of vehicle is provided along with different current parameters. Whereas, a priori optimization tries to find optimal solutions pro-actively and later tackle the parameter changes with time.

Stochastic VRP [4] is a VRP problem where one or several components of problem are random and may alter with time. These are classified in three categories:

Stochastic Customers: Each customer is present with a probability of p and absent with probability of 1-p.

Stochastic Demand: The demand of each customer is a random variable.

Stochastic Time: Servicing time is dependent on customers and travel time is also variable based on various factors like, type of cab, traffic etc.

VRP with uncertain demands. Secomandi [5] tries to solve a VRP where customer demands are uncertain and are confirmed when vehicle reaches the customer location. They developed optimistic approximate policy iteration (OAPI) and one-step roll out algorithm (ORA). It dynamically routes (in real time) a single vehicle to serve the demands of various customers whose locations are earlier provided. It minimizes the expected distance traveled to serve all dynamic customers' demands. The problem with this approach is its limitedness in problem hypothesis, where only one vehicle in the system is assumed.

VRP with extended resource strategy. Clara and Storer [6] provides a set partitioning-based model for SVRP which involves two-stage stochastic solution and provides a better resource utilization policy where a vehicle, while returning to the depot can service few of the nodes which were not serviced due to path failure of other vehicles. This work proposes rather two-stage roll out algorithm instead of one-step roll out algorithm. Two stages involve more computation but also provide better results.

VRPs in general are the extension of Traveling Salesman Problem, which is a NP Hard problem, and while solving NP hard problem we can only be exponentially efficient. Linear Programing approach has already been well researched and many optimizing techniques have been developed for solving these problems. So by reducing our ETP into Linear Programing we will be able to take advantage of vast literature already available for the subject.

After reviewing literature related to the subject we finally concluded that although all the papers are trying to solve VRPs but these problems are very different problem to ETP so none of their techniques can be applied to solve the ETP efficiently and comparison with these problem will also be futile.

3 Problem Statement

The Employee Transportation Problem can be understood well with the analogy of a call center and its employees. Suppose there exists a manager of a call center who wants to provide cab service to its employees. Manager knows the home locations of its employees and their distances with each other and to call center. Now Manager wants to compute the minimum number of cabs for call center's employees while satisfying certain constraint.

The problem can be generically stated as

- Given:
  - Location of center, and home location of all employees.
  - Time required to travel from center to each location, and from each location to every other location.

- Find a set of paths to run a cab on each path such that:
  - Number of cabs required is minimum.
  - No cab can transport more employees than a fixed constant number (Capacity) [7, 8].
  - All employees will leave the call center at the same time.
  - No employee will remain in the cab more than a limited time. This limited time can be different for different employees and is proportional of the minimum distance duration between call center and employee's home [9].

Now, to solve this problem we mathematically formulate such a problem as a 6-tuple $M = \{V, E, P, v_0, \alpha, C\}$.

1. $V = \{v_0, v_1, v_2, \ldots v_n\}$ is vertex set.
2. v_0 represents the center while remaining vertices denote passengers' home location. E is a matrix of size $|V| \times |V|$, where $E_{i,j}$ represents, travel time duration between v_i and v_j

3. is set of paths, $p_i \forall v \in V, \ni p \in P$ such that v $\in$ p.

- $\forall v_i \exists a p_j : v_i \in p$
- Suppose $p_i = \{v_0, v_x, v_{x+1}, v_{x+2}, \ldots \ldots \ldots v_y, v_{y+1},\}$
- Head (p) = v_z: Head Node.
- $C_{p_i, v_y} = \sum_{n=0}^{y-1} E_{n,n+1}$ and also $C_{p_i} = C_{p_i, v_z}$.
- $|p_i| - 1$: Length of path p_i.

4. $T_{v_i v_j}$: Minimum travelling time between v_i and v_j
5. α: Travel Time Limit Constraint, such that $\forall p_i \in P, \forall v_k \in p_i : C_{p_i, v_k} \leq \alpha T_{v_0 v_k}$ and $\alpha \geq 1$.
6. C represents cab capacity. $\forall p_i$ in P: Length of $p_i \leq C$.
7. $|P|$ is minimum among all possible P.

4   Proposed Algorithm

In our proposed method Linear Programming approach is used to solve ETP. MVRP [6] defines an approach where number of cabs is given, and one has to find which vertex should be traversed by which cab so as to minimize the distance traveled. Whereas, ETP provides location of center and various points to be traversed only. One has to then find out optimal paths so as to minimize the requirement of cabs. One obvious approach as in [10], to solve this problem is reducing it in a Linear programming problem by feeding it various constraints along with all possible paths in the graph to a LP solver. But, with increase in number of nodes, number of path increases exponentially. So, this solution is not scalable with increasing number of locations. However, a closer observation depicts that a large set of these paths could be filtered out, to make this approach scalable.

So, we divide our algorithm into two sub-parts, namely, ETP: Valid Path Generation and ETP: LP Route Selection. Valid path generation tries to reduce the number of path which will be fed to LP Route Selection; which serves as a LP solver with provided constraints. Now, before we mention Valid Path Generation, we state a module Satisfy Constraints (r) which qualifies whether a path r is satisfying various constraints or not.

Satisfy Constraints (r, α, C). This function is responsible to check whether a given path satisfy various constraints. First, a check on number of nodes is required, because the passengers that could be travelling in a cab are limited by C excluding the driver, then we check the travelling time constraint. We need to consider only head node of the path for this constraint, as the loop is called for every new addition of a node

in the paths, and thus all nodes prior to head node were already checked for this
constraint v_0.

```
SatisfyConstraint(r,α,C)
1. If r is valid
2.     z ← Head(r);
3. If(|r| ≤ C) && (C_r ≤αT_{v0,z}) Then
4.     Return true
5. Else
6.     Return false
```

Algorithm 1: ETP: Constraint Satisfaction

4.1 ETP: Valid Path Generation

This function is responsible for discarding paths which already are violating the
constraints and thus could never turn out to be present in solution set. We recursively
add new paths starting from v_0 (the center) by adding neighbor to the paths. Now,
we do not include a path in solution till it is ensured that no vertex could be further
added to this path. We do this because, if a cab has capacity to serve more employees,
it should always try to serve those employees.

```
ValidPathGenerator( V,E,v0,α,C)
1. P constraint satisfying valid paths only, but not minimal
2. Set S ← φ, which stores paths
3. Priority Queue, Q ← φ, keyed on Cp.
4. while Q ≠ φ do
5.    p ← Extract Min(Q);
6.    z ← Head(p);
7.    R ← new paths by adding neighbor of z;
8.    flag ← false;
9.    for each r ∈ R do
10.         if Satisfy Constraints(r, α, C) then
11.             flag ← true;
12.             Q ← Q ∪ {r};
13.    if !flag then
16.       S ← p;
17. return S;
```

Algorithm 2: ETP: Valid Path Generation

4.2   ETP: LP Route Selection

2 reduces the numbers of paths that one needs to feed to LP Solver module; which now makes it scalable, at least for 150 nodes. Now, we describe the objective functions and constraints to solve ETP.

LPRouteSelection(p)
1. for *each v ∈ V* **do**
2. $S(v) \leftarrow \varphi;$
3. for *each p ∈ S* **do**
4.   **if** *v ∈ p* **then**
5. $S(v) \leftarrow S(v) \cup p;$
6. return S

Now, we can define the objective function and constraints which will be fed to LP Solver, in terms of x_p, v_0, V, S & S(v).

We also define a variable x_p for corresponding to each path in S. Its value determines whether a path is selected by LP Solver or not. A non-zero value denotes selection of path and zero denotes its discarding. So, to minimize the number of cabs required, one should minimize the number of time x_i receives a non-zero value such that all vertices are also covered by the paths selected. This could be mathematically denoted as:

$$\textbf{Objective Function} : \sum_{\forall p \in S} \chi\, p$$

$$\textbf{Constraints} : s.t.\, \forall v \in V - \{v0\}, \sum_{\forall p \in S} \chi\, p \geq 1, \text{ and } \chi\, p \geq 0 \qquad (1)$$

5   Results

Implementation of the algorithm is done by using JAVA. The application is tested with travel time constraint $\alpha = 1.5$ and cab capacity of 5 with 100 employee home locations taken all around the city of Kanpur (1600 KM2). Screen shots of application are given below in Fig. 1. We simulate method defined in this paper under multiple circumstances to check the efficiency of it. Table 1 reflects concise result of all these experimentations. First Column of the table represents number of vertices or employees selected for corresponding experiment and first row represents the capacity of corresponding experiment. Completion time of these experiments is noted in milliseconds in corresponding cells. Although location of vertices play a significant role in the experiments it is near to impossible to mention location of each of these vertices but while selecting locations we have exploit some sociological human tendencies. According to these tendencies we have assume that we can select some localities and can group the vertices inside these selected localities. This assumption

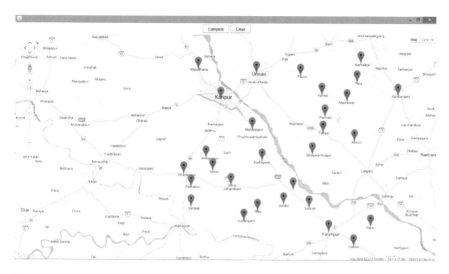

Fig. 1 Selection of locations

Table 1 ETP: performance statistics (in milliseconds)

Vertices/capacity	5	6	7	8	9
50	52	84	91	105	164
75	254	300	481	571	811
100	454	1045	1239	1330	1622
125	640	1298	2485	4887	10210
150	860	1492	3050	7993	10633

is plainly taken for bringing the experiments to real world domain as much as possible (Fig. 2).

6 Conclusion

In conclusion, ETP has a grand scope and a solution to the ETP can be applied in many areas such as Student Transportation, Railways or Military cargo deployment, etc. The properties of ETP indicate that it is applicable in many traffic-related problems. We can state that provided framework with this paper can be used to solve ETP with 150 or more employees within few minutes with standard values of other constraints. Our research about the given problem produces some significant result which could be relevant in many Vehicle Routing Problems.

In future we will develop a feasible solution of extension of a ETP. The extended version will not only demand the minimum number of cabs but also take the distance traveled by all cabs into consideration. Of course we have to prioritized between dis-

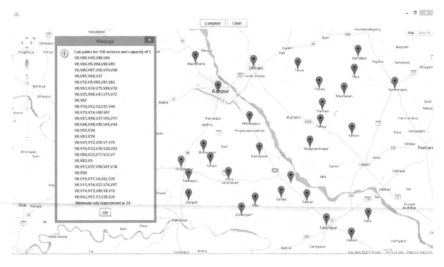

Fig. 2 Representation of Final Cab Locations is on the map and the Routes are represented by the sequence of nodes

tance and number of cabs, to resolve this conflict we can use a linear mathematical expression which will contain both number of cabs sand distance traveled by all the cabs. Extended ETP will now only have to minimize this expression. We can also modify the proposed strategy to be applicable on VRPTW. A solution to such a problem will open gates for many real-world applications where parameter optimization is required.

References

1. Yüceer, Ü.: An employee transporting problem. J. Ind. Eng Int. **9**(1), 31 (2013)
2. Önder, İ.: An employee transporting problem and its heuristic solutions. MS thesis (2007)
3. Laporte, G.: The vehicle routing problem: an overview of exact and approximate algorithms. Eur. J. Oper. Res. **59**(3), 345–358 (1992)
4. Sathyanarayanana, S., Joseph, K.S.: A survey on stochastic vehicle routing problem. Inf. Commun. Embed. Syst. 1–7 (2014)
5. Secomandi, N.: A rollout policy for the vehicle routing problem with stochastic demands. Oper. Res. **49**(5), 796–802 (2001)
6. Novoa, C., Storer, R.: An approximate dynamic programming approach for the vehicle routing problem with stochastic demands. Eur. J. Oper. Res. **196**(2), 509–515 (2009)
7. Laporte, G., Mercure, H., Nobert, Y.: An exact algorithm for the asymmetrical capacitated vehicle routing problem. Networks **16**, 33–46 (1986)
8. Ralphs, T.K. et al.: On the capacitated vehicle routing problem. Math Program. **94**(2–3), 343–359 (2003)
9. Baldacci, R., Mingozzi, A., Roberti, R.: Recent exact algorithms for solving the vehicle routing problem under capacity and time window constraints. Eur. J. Oper. Res. **218**(1), 1–6 (2012)

10. Balinski, M., Quandt, R.: On an integer program for a delivery problem. Oper. Res. **12**, 300–304 (1964)
11. Eilon, S., Watson-Gandy, C.D.T., Christofides, N.: Distribution Management: Mathematical Modelling and Practical Analysis. Griffin, London (1971)

Spectrum Prediction Using Time Delay Neural Network in Cognitive Radio Network

Sweta Jain, Apurva Goel and Prachi Arora

1 Introduction

With the increase in number of users, the need of spectrum has also increased significantly. However the spectrum band available for the communication is limited. Federal Communications Commission (FCC) in US is one of the principle authorities that distribute and manage the spectrum bandwidth. The FCC frequency graph [1] demonstrates the range allocated over all the frequency bands. This graph clarifies the range shortage particularly for the band under 3 GHz. Despite what might be expected, real estimations that were taken in downtown Berkeley [2] uncovers a distinctive utilization of 0.5% in 3–4 GHz frequency band which further drops to 0.3% the 4–5 GHz band. This suggests that the spectrum available for communication is mostly allocated but it is often used in inefficiently. To deal with the problem of spectrum scarcity and inefficient spectrum usage, FCC has approved the unlicensed users to use the spectrum in absence of licensed users. This gives an approach to different innovations such as dynamic spectrum access, cognitive radio networks, software defined radios etc. Cognitive Radio Networks can be primarily used for this purpose due to their ability to adapt to networks conditions.

Cognitive Radio technology enables the secondary user (SU) to discover and use the frequency band or spectrum hole in the licensed frequency band. A SU senses the channel continuously for spectrum holes and maintain a pool of free holes. Then the CR user decides which spectrum hole to use based on various parameters. Sometimes

S. Jain (✉) · A. Goel · P. Arora
Maulana Azad National Institute of Technology, Bhopal, India
e-mail: shweta_j82@yahoo.com

A. Goel
e-mail: apurvagoel18@gmail.com

P. Arora
e-mail: prachiarora0111@gmail.com

© Springer Nature Singapore Pte Ltd. 2019
S. Tiwari et al. (eds.), *Smart Innovations in Communication and Computational Sciences*, Advances in Intelligent Systems and Computing 851,
https://doi.org/10.1007/978-981-13-2414-7_25

Fig. 1 Trade-off between
spectrum sensing time and
transmission time

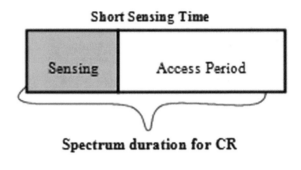

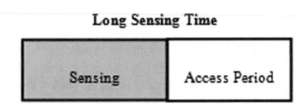

the holes are shared between more than one SU. However if the primary user (PU) arrives, CR users have to leave the band immediately which is known as Spectrum Mobility. But there are several drawbacks or shortcomings in CRN which may hinder the capacity and capability of CRN [3].

- Spectrum Sensing has large time-delays which are non-negotiable in wide band.
- According to real time results, the efficiency of spectrum utilization is undermined.
- It is because of time-delay caused by sensing and thus causing delay in decision.
- Many CR users can join the spectrum at any time. Each user has different QoS specifications as well as bandwidth requirement. So, the assignment of the appropriate band to CR may introduce time-delay, thus causing inefficiency.
- Transmission collision may occur due to traditional CSMA based mobility policy as CR user may not be able to evacuate the band within the allowable range of interference time.

In order to overcome these problems prediction based techniques have to be studied. SU can reduce the sensing time and can make use of that time for transmission. Since the hole is available only for a limited times, there is always a trade-off between sensing time and transmission time (Fig. 1). The channels which are predicted to be busy may be excluded from sensing to save the sensing time. The quality of channel in terms of idle duration, idle probability and other properties can also be predicted. Thus the probability of selecting a high quality channel increases. The time of appearance of PU can also be predicted which improves the efficiency of spectrum mobility algorithm used. However spectrum sharing based on prediction has never been discussed as per our knowledge. Thus prediction methods in CRN can highly improve its capability.

2 Background and Related Work

Spectrum prediction is a challenging problem in cognitive radio networks that involves Primary user activity prediction, channel state prediction, transmission rate prediction and radio environment prediction. Spectrum prediction is supposed to add an additional layer of pro-activeness to CR capacities. It can give decreased sensing time, provide robust and flexible channel choice and build up proactive handover procedure to stay away from impact with authorized client. In this segment we discuss various channel methods and their applications. Spectrum prediction reduces the spectrum sensing time and improves the bandwidth efficiency, reliability and scalability.

Prediction based in Bayesian Inference: It is a method of statistical inference which uses Bayes theorem. In [4], a channel prediction scheme is proposed. Spectrum sensing process is demonstrated as the Non-stationary Hidden Markov Model and model parameters are evaluated through Bayesian Inference utilizing Gibbs sampling.

Prediction based on Hidden Markov model: Hidden Markov model is a statistical model for modeling generative sequences and it can be regarded as an underlying process that generates an observable sequence.HMM can be used for analyzing and modeling sequential data or time series in different fields. In [5], HMM based channel state prediction is proposed. The authors suggest that time delays are introduced during spectrum sensing which reduce the accuracy of sensing. Hence transmission collision can occur between PU and SU. Spectrum utilization efficiency can be improved by selecting a channel that is predicted as well as sensed to be idle.

Prediction based on Artificial Neural Networks: Artificial neural networks (ANN) are identical to biological cells of human brain. Cells consist of a number of interconnected processors known as neurons. Different learning algorithms are used to train a ANN model that depends on the desired output and the type of neural network being used. Weights are updated during the training phase using the learning algorithms. In [6], author proposed a Multilayer Perceptron (MLP) predictor that does not require a priori knowledge of traffic characteristics of the channel used by the PU. The problem of Channel state prediction is considered as binary series prediction problem. Binary series is used to obtain input and output that is generated for each channel by sensing the channel status for duration T in every slot. MLP is trained to predict the channel; batch back propagation algorithm is used to update the parameters of MLP.

In [7], to predict the channel states based on its history two type of neural networks are proposed; TDNN and RNN. SU are assumed to logically divide the PU channel into separate time slots. PU traffic on any channel is assumed to follow Poisson process. The sigmoid function is used for the neurons in hidden layer and output layer. BP algorithm is used for learning and Mean Squared error (MSE) is calculated as a performance measurement. In [8], main benefits of using MFNN is that as a function of measurement they provide a general purpose black box modeling of the performance of Cognitive Radio functions. MFNN is flexible and provides good accuracy. CR radio networks can effectively learn using MFNN even if the number of inputs and outputs are high. The approach makes use of the function approximation of MFNN to obtain the performance and environment measurements.

In [9], four supervised machine learning approach, namely SVM with Gaussian, SVM with Linear Kernel, RNN and Multilayer Perceptron, have been applied and compared for forecasting the primary user activity. Poisson, Interrupted Poisson and Self-comparative traffics are utilized for the examination of authorized client condition. Information generated by each traffic distribution is utilized in the training stage exclusively with the assistance of each learning model after which, the testing is finished for the primary activity prediction. The simulation results recommend the best learning model for different types of primary user activity.

Hence we observe that a number of prediction techniques can be applied to predict the spectrum state of the primary user to reduce the sensing time and improve the efficiency. However there are certain limitations of each technique. HMM provides accurate prediction and can be used widely but it can mostly be applied on statistical models. Neural network based models can provide reduced sensing time, provide robust and flexible channel selection and establish proactive handover strategy to avoid collision with licensed user. Although in this section we present a brief review of some of the channel prediction methods, a detailed survey of some other techniques such as Prediction based on autoregressive model [10, 11], prediction based on moving average [12, 13], etc. can be found in [14].

3 Proposed Work

3.1 Slot Status

Slot status denotes the pattern of the PU on/off time on the spectrum over a time period. The input provided to the neural network for the training is slot history of the channel. The occupancy of the channel by the PU is random activity. A random function is used to generate the history. Poisson distribution is used to generate the data values. The traffic intensity is varied during the generation of data to provide real life scenario. The slot status is a binary values, status is -1 when channel is free and 1 when the channel is occupied. At least 50% occupancy is considered. The algorithm for the generation of data for 10,000 time slots is:

```
BEGIN
DECLARE VARIABLES input1 = 10000 j and λ
FOR λ 1 to 5
r_s1 = Poisson (λ, 1, 2000) FOR i 1 to 2000
IF (r_s1 (i) > (λ − 1))
SET input_1 (j) = 1
ELSE
SET input_1 (j) = −1
END IF
j = j + 1;
END FOR; END FOR; END
```

3.2 Time Delay Neural Network

TDNN is an artificial neural network that works primarily on sequential data. It generally works on larger data patterns and it can recognize those features that are independent of time shift. Delayed copies augments input signal as other input. It does not have any internal states. Here RNN can also be used but TDNN is used because it is simplest, primarily works on sequential data and widely used neural network. And CNN cannot be used here it is generally used for images classification then data need to be convert in image form while in TDNN there is no such requirement. TDNN are implemented generally as Feed Forward neural network rather than Recurrent Neural Network. It consists of various interconnected layers which are made up of clusters. A traditional TDNN has three layers: input layer, output layer and middle layer. Middle layer is responsible for change in input through filters. To achieve the invariance in time shifts and to represent the data at different time points, asset of delay can be added to input. Delays add a temporal dimension to network. Major advantage of TDNN is that it does not require a priori knowledge to setup filter at different layers. It is usually trained using supervised learning. TDNN can allow non-uniform sampling

$$xi(t) = x(t - \omega) \tag{1}$$

where o_i is the delay that is associated to each component i. Input can be a convolution to the original input sequence.

$$xi(t) = \sum_{\tau=1} c_i(t - \tau) * \tau \tag{2}$$

For delay line memories

$$c_i(t - \tau) = 0 \{ 1 \textbf{ if } t = \omega_i, 0 \textbf{ else} \} \tag{3}$$

TDNN is trained using supervised learning algorithm. Usually back propagation algorithm and its variants are used to train the TDNN because of its simplicity but it does not perform so well for the conditions where a priori information is not available. Lavenberg Marquardt can also be used to train this neural network.

$$E = \frac{1}{p} \sum_{p=1}^{p} (E_p) \tag{4}$$

Main objective of any training is to reduce global error E which is defined as:

$$E = \frac{1}{N} \sum_{i=1}^{N} (o_i - t_i)^2 \tag{5}$$

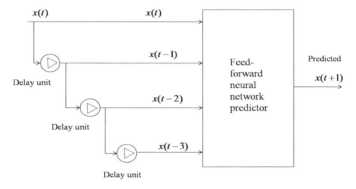

Fig. 2 Tapped delay line neural network (TDNN) with three delay elements [16]

where o_i is output node and N is total number of nodes (Fig. 2).

Here, input signal is augmented as other inputs with delayed copies. This neural network is time-shift invariant as it does not have an internal state. A set of delays are added to input in order to achieve time-shift invariance. Works have been done to create adaptable TDNN to eliminate the need of manually tuning TDNN. Key feature of TDNN is its ability to express the relation between input and time. This feature is used by TDNN to recognize pattern between delayed inputs.

3.3 Lavenberg Marquardt Algorithm

This algorithm was designed to approach second-order training speed by computing Hebbian matrix. If the performance functions has a form of sum of square. Hebbian matrix can be calculated as

$$H = J^T J \tag{6}$$

And gradient can be computed as

$$g = J^T e \tag{7}$$

where J is the Jacobian matrix (JM) that contains first derivatives of the errors of network with respect to biases and weights, and e is vector of network errors. Standard back propagation can be used to compute JM since BP is much less complex than Hebbian learning. Approximation to the Hebbian matrix is used by LM algorithm in Newton like update.

$$x_{k+1} = x_k - \left[J^T J + \mu I\right]^{-1} J^T e \tag{8}$$

Table 1 Implementation details

Properties	Specification	Properties	Specification
Number of layers and min gradient	3 and 1e-07	Network used	Time delay distributed feed forward network
Range of delays	[0, 1, 2]	Name of layers	Input, computational, output layer
Number of neurons	15	Algorithm used	Lavenberg-Marquardt
Max number of epochs	1000	Adaption learning	LEARNGDM (gradient descent with momentum)
Transfer function	TAN-Sigmoid	Performance	Mean square error

If μ is large, this method becomes gradient descent with small step size and if is 0, it is Newton method that uses approximate Hebbian matrix. Thus is decreased after every successful step and if tentative step increase performance function the value of is increased.

4 Implementation and Results

4.1 Toolbox

The proposed work is implemented using MATLAB. A neural Network is designed using Time delay Neural Network which is being used to predict the channel status (Table 1).

4.2 Toolbox

TDNN described in above section is trained using Lavenberg-Marquardt algorithm. The slot history generated using Poisson random function is the input to the neural network. For the generation of slot history, minimum traffic intensity is maintained to be at least 50% i.e. for at least for half of the time the channel is occupied by the PU. The performance of the network during training is measured for different traffic intensities. When the slot history is fed to the neural network it trains itself according to the slot history. Our network was tested by varying the size of input data. The target values are also provided initially. Weights are adjusted according to the training algorithm. The performance of network during training is measured using Mean square method. After the training the network is simulated and the results are generated. Output of the network is the slot status predicted by the neural network. As the output, neural network predicts when the channel or spectrum would be free for next set of time slots. It is possible that the network wrongly predicts the status of the channel. Probability of generation of false positive is also calculated. The false positive rate is measured in terms of error. The proposed work is compared with MLP predictor that uses back propagation algorithm [6] and MLP predictor trained using Lavenberg-Marquardt algorithm [15].

5 Implementation and Results

The result of the experimental evaluation are calculated for three methods: on the basis of performance during training which is measured in terms of mean square error, on the basis of difference in performance when the number of time slots for which network is being trained changes and on the basis of Probability of false positive alarms.

Performance and Comparison: The proposed work is compared with two other methods. In first method Multilayer perceptron is being used and Lavenberg Marquardt algorithm is used for training of the network. In the second method Batch back propagation algorithm is used to train the MLP predictor [6]. The results were generated by simulating the neural networks for different traffic intensities. The performance of the neural networks varies with traffic intensity. The values generated for each traffic intensity are the average values obtained by simulating the neural networks for over 20 times for each value of the traffic intensity. Figures 3, 4 and 5 show the performance comparison in terms of MSE of proposed TDNN trained using LM-algorithm, MLP trained using LM [15] and performance of MLP trained using Back propagation (gradient descent) [6] respectively. Figure 6 shows the comparative analysis of all the three neural networks.

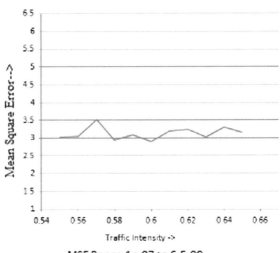

Fig. 3 TDNN-LM performance in terms of MSE

Fig. 4 MLP-LM
performance in terms of
MSE

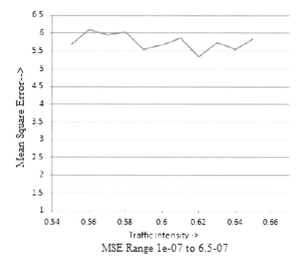

MSE Range 1e-07 to 6.5-07

Fig. 5 MLP-BP
performance in term of MSE

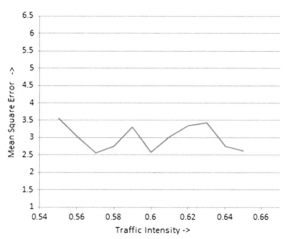

After comparing the results of the proposed work with the other two existing works, it was found out that the proposed work performs the MLP predictor trained using back-propagation algorithm and MLP predictor trained using Lavenberg-Marquardt algorithm during the training phase as the mean square error of the proposed work was much less than the other two works. But due to the precision being too high it is not possible to compare them and calculate the percentage improvement obtained. However the algorithms were also compared using the performance parameter generated by MATLAB simulations.

Average performance of TDNN-LM is 1.57, while for MLP-LM it was 1.36 and for MLP-BP the average performance was measured to be 1.15. So TDNN LM performs 15% better than MLP-LM and 36% better than MLP-BP during the training. After the training phase the neural networks were simulated and the output was calculated.

Fig. 6 Performance graph

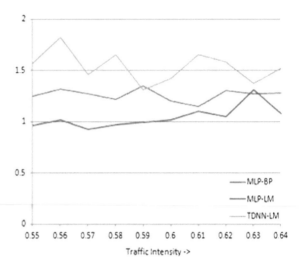

The efficiency or the probability of correctly predicting the status of the spectrum was calculated to determine the performance of the output generated.

The efficiency of the networks is measured in terms of output. Efficiency denotes the probability of predicting the correct output. The probability of wrongly predicting the slot status will be 1-efficiency. Efficiency is measure by the formula:

$$E = \frac{no.\ of\ times\ output\ prediction\ is\ correct}{total\ no.\ of\ slots} \tag{9}$$

Figure 7 represents the difference in efficiency in all the three methods in terms of difference in target data and output.

The efficiency of TDNN-LM, MLP-LM AND MLP-BP is 0.58, 0.46 and 0.48 respectively. The proposed work shows 26% improvement with respect to MLPLM and 20% improvement with respect to MLP-LM.

Change in the mean square error with the increase in number of time slots provided for training was also measured for the proposed work. It is represented in Fig. 8. Initially as the number of slots increased performance was increased but then it became constant. Initially, when less number of slots is provided for training, mean square error was more because the data provided for training was not sufficient for the neural network. The number of iterations was less; hence the neural network was not trained properly. But after 10,000 slots it became constant because the number of iterations required for the training phase was sufficient. Hence further increase in number of slots did not make any difference in the performance.

Fig. 7 Graph between efficiency and traffic intensity

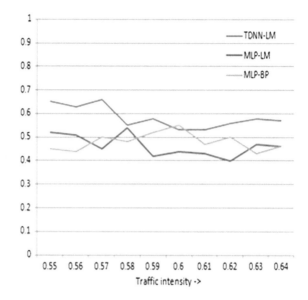

Fig. 8 Change in performance with respect to time slots

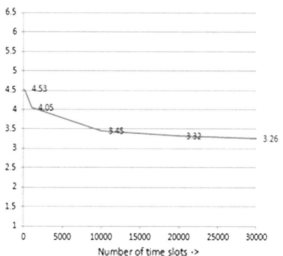

6   Conclusion and Future Work

In the current work, a prediction method based on TDNN has been developed to predict the channel status based on the past history. TDNN has been trained using LM algorithm which is a fast and efficient supervised learning algorithm. Then we have simulated this algorithm on MATLAB and analyzed results in terms of Mean square error and efficiency. The simulation results show 15% improvements when compared to MLP-LM and 36% improvements have been calculated as compared to MLP-BP during the training. The proposed work also shows 26% improvement with respect to MLP-LM and 20% improvement with respect to MLP-LM in terms of efficiency. Introduction part briefly discusses CRN, their function, spectrum sensing and different spectrum sensing methods. We have also discussed the need and importance of spectrum prediction and different methods used for prediction. ANN is most efficient among all the spectrum prediction methods due to its ability to learn and act according to the changing environment which is necessary for CRN.

References

1. http://www.fcc.gov/oet/info/database/spectrum/
2. Cabric, D., Mishra, S.M., Willkomm, D., Brodersen, R.W., Wolisz, A.: A cognitive radio approach for usage of virtual unlicensed spectrum. In: 14th IST Mobile and Wireless Communications Summit, pp. 1–4 (2005)
3. Akyildiz, I.F., Lee, W.-Y., Vuran, M.C., Shantidev, M.: Next generation/dynamic spectrum access/cognitive radio wireless networks: a survey. Comput. Netw. J. (Elsevier) **50**(13), 2127–2159 (2006)
4. Xing, X., Jing, T., Huo, Y., Li, H., Cheng, X.: Channel quality prediction based on Bayesian inference in cognitive radio networks. In: INFOCOM, Proceedings of IEEE, pp. 1465–1473 (2013)
5. Chen, Z., Guo, N., Hu, Z., Qiu, R.: Channel state prediction in cognitive radio, part ii: single-user prediction. In: Proceedings of IEEE Southeastcon, pp. 50–54 (2011)
6. Tumuluru, V.K., Wang, P., Niyato, D.: A neural network based spectrum prediction scheme for cognitive radio. In: IEEE International Conference on Communication (ICC), Cape Town, South Africa, pp. 1–5 (2010)
7. Shamsi, N., Mousavinia, A., Amirpour, H.: A channel state prediction for multi-secondary users in a cognitive radio based on neural network. In: International Conference on Electronics, Computer and Computation (ICECCO), pp. 200–203 (2013)
8. Baldo, N., Zorzi, M.: Learning and adaptation in cognitive radios using neural networks. In: 5th IEEE Consumer Communications and Networking Conference (CCNC 2008), pp. 998–1003 (2008)
9. Agrawal, A., Dubey, S., Khan, M.A., Gangopadhyay, R., Debnath, S.: Learning based user activity prediction in cognitive radio networks for efficient dynamic spectrum access. In: Signal Processing and Communications (SPCOM), pp. 1–5. IEEE (2016)
10. Wen, Z., Luo, T., Xiang, W., Majhi, S., Ma, Y.: Autoregressive spectrum hole prediction model for cognitive radio systems. In: IEEE International Conference on Communications Workshops, pp. 154–157 (2008)
11. Eltholth, A.: Forward backward autoregressive spectrum prediction scheme in cognitive radio systems. In: Signal Processing and Communication Systems (ICSPCS), pp. 1–5. IEEE (2015)
12. Butun, I., Cagatay Talay, A., Turgay Altilar, D., Khalid, M., Sankar, R.: Impact of mobility prediction on the performance of cognitive radio networks. In: Wireless Telecommunications Symposium (WTS), pp. 1–5 (2010)
13. Lin, Z., Jiang, X., Huang, L., Yao, Y.: A energy prediction based spectrum sensing approach for cognitive radio networks. In: 5th International Conference on Wireless Communications, Networking and Mobile Computing, pp. 1–4 (2009)

14. Jain, S., Goe. A.: A survey of spectrum prediction techniques for cognitive radio networks. Int. J. Appl. Eng. Res. **12**(10), 2196–2201 (2017)
15. Yin, L., Yin, S., Hong, W., Li, S.: Spectrum behavior learning in cognitive radio network based on artificial neural network, Military Communication Conference, Milcom, pp. 25–30 (2011)
16. Mehrotra, K., Mohan, C.K., Ranka, S.: Elements of Artificial Neural Networks. MIT press

Reliability Factor Based AODV Protocol: Prevention of Black Hole Attack in MANET

Prakhar Gupta, Pratyaksh Goel, Pranjali Varshney and Nitin Tyagi

1 Introduction

MANET stands for Mobile Ad hoc Network. It is also known as wireless ad hoc network. It is a network of mobile devices connected wirelessly without any infrastructure. Being wireless, it offers advantages such as reduces overhead of wiring and thus saves money, easy to establish and many more. But it also exhibits many disadvantages. It does not have any central administrator to manage all the nodes of the network. The nodes which are part of the network simply work using mutual understanding. Since MANET is self-configuring network, all the nodes which are part of the network possesses the ability to configure themselves. The absence of administrator allows any untrusted node to be the part of the network. Such nodes are known as malicious nodes and they can drop the packets, take the important information, and cause many more damages to the network. The main challenge of the MANET is to reduce attacks caused by malicious nodes or to detect such nodes at early stages. Black hole attack is a type of attack in which the malicious node simply becomes the part of the network and then absorbing everything coming to it. It can be seen as a hole in which packets which are coming are falling. Thus, desired destination node does not receive the packets, therefore, it affects the whole network. We have proposed an algorithm which is based on reliability factor approach. It calculates the reliability factor of the nodes in the path, through which packets are to be forwarded, and if the value of reliability factor is high then packets are forwarded. At some stage, it is possible that the node has a low value of reliability factor and also it is not malicious. Therefore, fake RREQ concept is used to detect the malicious node and if the nodes prove to be non-malicious the packets are forwarded. This algorithm greatly reduces the number of packets dropped.

P. Gupta (✉) · P. Goel · P. Varshney · N. Tyagi
GLA University, Mathura, India
e-mail: gupta.prakhar9596@gmail.com

© Springer Nature Singapore Pte Ltd. 2019
S. Tiwari et al. (eds.), *Smart Innovations in Communication and Computational Sciences*, Advances in Intelligent Systems and Computing 851,
https://doi.org/10.1007/978-981-13-2414-7_26

2   Related Work

Different authors have given different ways for eliminating black hole attack in MANET and here are some:

- In [1], the author used TAODV (Trust based ad hoc network) approach for prevention of black hole attack. All the nodes in the network are initialized with the value 0.7. When source broadcasts RREQ and gets route through RREP, first, the sequence number is checked and then the trust index of the replying nodes. If the value is less than 0.7, then the nodes are said to be the black hole nodes and are blacklisted. The nodes with high trust index is selected for relaying message. If the packets are delivered to the destination, then the trust index value is increased else it is decreased.
- In [2], the author proposed a method in which source node waits for RREP packet from multiple nodes. In this waiting time, sending node puts its packet in the buffer. When RREP arrives from more than one node, the full path to the destination is extracted. Multiple nodes must share some hops. Initial node determines the path to the destination using these hops, if no nodes are visible in this repetitive route it waits for other route reply. This solution guarantees the best route to the destination but suffers from increased time delay.
- In [3], the AODV protocol is embedded with the trust function. The nodes are categorized as reliable, most reliable or unreliable based on trust function value. In this method, a trust table is maintained by every node. This table stores the trust status of particular node with its neighbours. The function used for trust estimation is $T = tanh(r1 + r2)$, where r1 and r2 are the two ratios calculated. Source node checks the trust status to search the route to the destination. To update the trust status, some fake packets are sent.
- In [4], the author proposes a solution in which every node maintains DRI table. This is the table containing entry for every node and whether the particular node has received through this neighbour and node has received packets. The values can be 0 or 1. This table is updated when packets is received from a particular node or through a particular node. This table is also updated while searching reliable path.
- In [5], the author proposed a method which creates N nodes network. te is the encounter time of node with its neighbour and tb is the time at which link breaks between node and its neighbour. A list of neighbour node is maintained for every node. Hint value is calculated using formula $h = tb - te$. If h is less than or equal to threshold value, status is set to a black hole node else connected to fair node. Using this, a list of fair nodes can be maintained and the packets can be sent through these fair nodes.
- In [6] a method uses Ant Colony Optimization (ACO) in which packets follow the shortest route to the destination. The node which receives the backward ant packet checks the sequence number in its routing table. If the sequence number in the backward ant packet is greater than that in the routing table, the node is suspected as malicious. An alarm packet is sent to all the neighbours which contain a list of all such malicious nodes to be avoided.

- In [7], another method tries to make the source node independent of its neighbour. It attaches id of the destination to the RREP and it is termed as MRREP. When RREQ reaches the destination, it replies with MRREP. Since malicious node will always send a fake reply with highest sequence number but due to the fact that it cannot fake the destination id, the malicious node can be detected using this method and black hole attack can be prevented.
- In [8], the author proposed a method in which source node sends fake RREQ and waits for RREP equal to the half of the RREP waiting time. It will be always true that non-malicious node will never react to fake RREQ, so the replying nodes will be malicious and will be added to the black hole node list.
- In [9], the author used a method in which every node listens to its neighbours promiscuously that is each node checks that the packet forwarded to its neighbour id further forwarded. In order to verify that packet is further forwarded, it uses a caching mechanism which stores the packets forwarded to the next node. If the node cannot tap the same packet then it assumes the adjacent node as malicious and that packet is considered dropped. This decreases the threshold value of neighbouring node using $T = 1 - D/F$, where, T is the threshold value between 0 and 1, D represents the packets dropped by a node and F indicates the total packets forwarded to that node.
- In [10], the authors have proposed that the large difference between the sequence number of the source node and the destination sequence number of the node that sent back RREP can be used to detect malicious behaviour. They match the receiving node number with the initial node sequence number, and if the difference is greater between the two, then that node is an attacking node and should be discarded immediately.
- In [11], another author proposed a paper which is based on Intrusion Detection using Anomaly Detection that can be used for black hole prevention.
- In [12], the author proposes to use fidelity table in which node will be assigned a level to measure the reliability of a node. If the fidelity level of any node drops to 0 then it would be considered as a malicious node and is then eliminated.
- In [13], the author uses zone routing protocol which is a hybrid protocol to take advantage remaining two types of routing protocols. In this, every node has its own zone inside which they maintain their connection safely.
- In [14], the author has proposed a method that reduces overhead. He used the concept of ID in which multiple agents in the network communicate with one another or with the server which is responsible for providing information related to monitoring, analysis, and attack.
- In [15], the author has used the concept of static distribution of keys instead of distributing them dynamically, and the key is selected based on value of hop count and this technique minimizes the overhead.

Fig. 1 Route establishment
in AODV

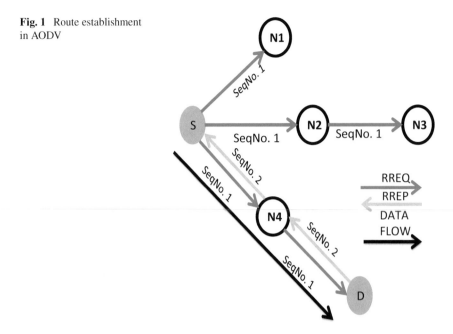

3 AODV Routing Protocol

As shown in Fig. 1, AODV routing protocol has been explained. When a node has
to deliver a packet, first it looks in its table, whether it has a path to the receiving
node or not. If the node does not have a path to the destination, it initiates the route
discovery process. A RREQ (route request) is broadcast to its neighbour nodes to
reach destination. If there is any route error, node sends RRER to source node. The
nodes having route to the destination send RREP with high sequence number to
the source node. In this way route to the destination is established. At the time of
route establishment, a malicious node becomes the part of the network by sending
RREP to the source node with higher sequence number. Then, the source node sends
the packet to destination through the route with highest sequence number and the
malicious node drops the packet instead of passing it to its neighbour. Figure 2, on the
other hand, shows that first RREQ sends them M replies with high sequence number
so route is established to M and packets are forwarded to it which further drops all
the packets.

4 Proposed Framework

This section describes the proposed reliability factor based approach for prevention
of black hole attack.

Fig. 2 Black hole attack in
MANET

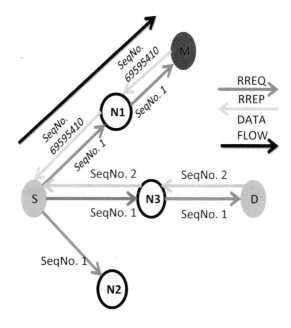

Fig. 3 Algorithm to prevent
black hole attack

1. A mobile ad-hoc network is formed by connecting number of nodes. Initially the value of reliability factor(r) of every node is 0.5.

2. RREQ is transmitted by the source node in order to find the path to the destination.

3. On receiving RREQ, neighbour node checks if it contains the path to the destination else it broadcasts further to its neighbours.

4. Then source node receives RREP, then the sequence number as well as reliability factor value is checked.

5. The node is selected based on the value of reliability factor value as follows:

   I. If r>0.5 then node is selected

   II. If r<=0.5 then it is checked for malicious behaviour

   III. A fake RREQ is sent containing fake destination identifier. If the node is malicious it will reply to fake RREQ and will be caught.

6. The packets are forwarded.

7. The reliability factor(r) of every node is increased (in case packets are forwarded) or decreased (in case packets are dropped) using a formula (value is computed in base2):

$$r = r \pm \frac{1 + \log x}{1 + \log N}$$

x→ Number of packets forwarded   N→ Number of packets received

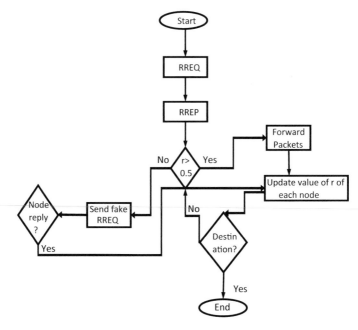

Fig. 4 Flowchart of the proposed algorithm

4.1 Proposed Algorithm

The algorithm of our proposed work that we have implemented is shown in Fig. 3. And the flowchart for the same is shown in Fig. 4.

Initially, the value of reliability factor (r) is 0.5 for all the nodes in the network. When node wants to send packets to the other node, then it sends RREQ for route discovery. When it receives RREP, the value of **r** is also checked along with the sequence number. The value near to zero indicates that node is (or maybe) malicious. If the value of **r** is greater than 0.5, then route will be established and packets will be forwarded. If the value is less than or equal to 0.5, then the malicious behaviour is checked by sending fake RREQ. Since the malicious node will always reply to RREQ, it will also reply to fake RREQ and black hole attack will be prevented. This algorithm uses double verification to declare that node is malicious. When the route is established, if the packets are successfully delivered to the destination, then the formula used in algorithm will give value equal to or more than the previous value of **r** else it will give less value of **r**. The minimum value of r will be zero and maximum one. If value of **r** exceeds more than one, then it will be set to one. If it decreases less than zero, than it will be kept zero.

Table 1 Simulation details

Parameter	Value
Type of attack	Black hole attack
Simulation tool	NS-2.35
Simulation area	$2000 * 2000$
Simulation time	1000 s
MAC type	IEEE 802.11
Mobility model	Random way point
Number of nodes	10, 20, 30, 40, 50
Application traffic	CBR
Routing protocol	AODV
Queue limit	50 packets
Channel	Wireless
Packet size	512 Bytes
Transmission rate	10 Kbps
Pause time	10 s

5 Simulation Results

The proposed algorithm was implemented on NS-2.35 and was tested, and all the information regarding simulations is given in Table 1. It was found that the Packet Delivery Ratio (PDR) and throughput were more and end to end delay was less than the normal AODV with black hole attack. Three parameters are used for the measurement of the algorithm performance:

(a) Packet Delivery Ratio (PDR): It is the ratio of the number of packets received to the number of packets sent.

(b) Throughput: Rate of sending information in a network is throughput.

(c) End to End delay: It is the time taken by the time taken by the packet to reach the destination from the source.

The comparison is done between:

- AODV protocol without any black hole attack.
- AODV protocol with black hole attack.
- Reliability factor based approach with AODV protocol.

Figures 5, 6, and 7 show the various comparisons. Figure 5 compares the throughput of the network. It can be seen that throughput for the R-AODV in all the cases is more than the normal AODV with black hole attack. Whereas, it can be seen in Fig. 6 that the ratio of number of packets forwarded and number of packets received (PDR) is more for R-AODV than normal AODV with black hole attack. Finally, it can be also observed in Fig. 7 that time taken by the packet from source to the destination in case of R-AODV is always less than the normal AODV with black hole attack.

Fig. 5 Throughput versus number of nodes

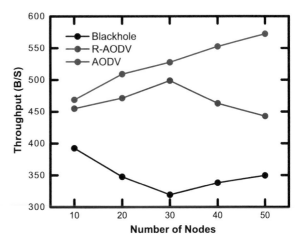

Fig. 6 Packet delivery ratio versus number of nodes

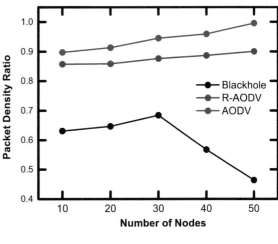

Fig. 7 End to end delay versus number of nodes

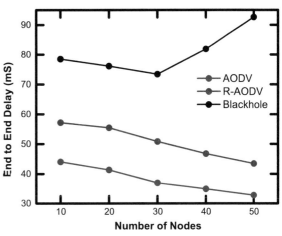

6 Conclusion

In this paper, we propose reliability factor based approach which can be used for the detecting and preventing black hole attack in MANET. It is found that better results are found using the proposed method. So, a little modification in AODV makes it more secure against black hole attack.

References

1. Jain, A., Prajapati, U., Chouhan, P.: Trust based mechanism with AODV protocol for prevention of black-hole attack in manet scenario. In: Symposium on Colossal Data Analysis and Networking (CDAN), pp. 1–4. IEEE (2016)
2. Sharma, N., Sharma, A.: The black-hole node attack in manet. In: 2012 Second International Conference on Advanced Computing & Communication Technologies (ACCT), pp. 546–550. IEEE (2012)
3. Singh, S., Mishra, A., Singh, U.: Detecting and avoiding of collaborative black hole attack on manet using trusted aodv routing algorithm. In: Symposium on Colossal Data Analysis and Networking (CDAN), pp. 1–6. IEEE (2016)
4. Devassy, A., Jayanthi, K.: Prevention of black hole attack in mobile ad-hoc networks using mn-id broadcasting. Int. J. Mod. Eng. Res. 2(3) (2012)
5. Chauhan, R., et al.: An assessment based approach to detect black hole attack in MANET. In: 2015 International Conference on Computing, Communication & Automation (ICCCA), pp. 552–557. IEEE (2015)
6. Sowmya, K., Rakesh, T., Hudedagaddi, D.P.: Detection and prevention of blackhole attack in manet using aco. Int. J. Comput. Sci. Netw. Secur. (IJCSNS) 12(5), 21 (2012)
7. Narayanan, S.S., Radhakrishnan, S.: Secure aodv to combat black hole attack in MANET. In: 2013 International Conference on Recent Trends in Information Technology (ICRTIT), pp. 447–452. IEEE (2013)
8. Sathish, M., Arumugam, K., Pari, S.N., Harikrishnan, V.: Detection of single and collaborative black hole attack in MANET. In: International Conference on Wireless Communications, Signal Processing and Networking (WiSPNET), pp. 2040–2044. IEEE (2016)
9. Thachil, F., Shet, K.: A trust based approach for aodv protocol to mitigate black hole attack in manet. In: 2012 International Conference on Computing Sciences (ICCS), pp. 281–285. IEEE (2012)
10. Jaiswal, P., Kumar, D.R.: Prevention of black hole attack in MANET. IRACST–Int. J. Comput. Netw. Wirel. Commun. (IJCNWC), ISSN, pp. 2250–3501 (2012)
11. Alem, Y.F., Xuan, Z.C.: Preventing black hole attack in mobile ad-hoc networks using anomaly detection. In: 2010 2nd International Conference on Future Computer and Communication (ICFCC), vol. 3, pp. V3–672. IEEE (2010)
12. Tamilselvan, L., Sankaranarayanan, V.: Prevention of co-operative black hole attack in manet. JNW 3(5), 13–20 (2008)
13. Rajput, S.S., Trivedi, M.C.: Securing zone routing protocol in manet using authentication technique. In: 2014 International Conference on Computational Intelligence and Communication Networks, pp. 872–877, Nov 2014
14. Arya, K.V., Rajput, S.S.: Securing AODV routing protocol in MANET using NMAC with HBKS technique. In: 2014 International Conference on Signal Processing and Integrated Networks (SPIN), pp. 281–285, Feb 2014
15. Roy, D.B., Chaki, R.: Baids: detection of blackhole attack in manet by specialized mobile agent. Int. J. Comput. Appl. 40(13), 1–6 (2012)

Part III
Web & Informatics

A Survey of Lightweight Cryptographic Algorithms for IoT-Based Applications

Ankit Shah and Margi Engineer

1 Introduction

Lightweight cryptography is a developing term which secures the information in an improved way utilizing low assets and giving higher throughput, conservativeness and having low power utilization. Likewise, cryptographic algorithms, the lightweight cryptographic algorithms are additionally isolated into two sections: Symmetric figures and Asymmetric figures. Pervasive figuring prevalently utilizes lightweight symmetric block ciphers. Symmetric ciphers contain Block and Stream Ciphers. They are intentionally utilized with gadgets and furthermore there are no strict limitations to get classified into lightweight. Security, Cost, and Performance are three noteworthy parts to deal with for each lightweight cryptographic architect. It is very difficult to the three major design goals: security and cost, security and performance, or cost and performance at once, while it is easy to optimize any one of them [1]. Elements of symmetric ciphers are message integrity checks, encryption, entity authentication and, etc., while non-repudiation and key management are moreover functions provided by asymmetric ciphers. A few creators reasoned that in and programming both Asymmetric ciphers [1] are computationally all the more requesting. It is required the design is made with 1000–2000 gate equivalents (GE) in ISO/IEC standard on lightweight cryptography [2]. This paper indicates lightweight algorithms compared are executed on various hardware or software tools. The improved aftereffects of particular algorithm vary from the platform, or application fluctuates. Lightweight algorithms are for the most part utilized as a part of IoT innovation for more model security with least memory and power utilization. This paper highlights

---

A. Shah
Plot. no 1284/1, Sector 2 A, Gandhinagar 382002, Gujarat, India
e-mail: shah_ankit101@yahoo.co.in

M. Engineer (✉)
Plot no 107/1, Sector 2-a, Greenland Avenue, Gandhinagar 382002, Gujarat, India
e-mail: margiengineer@outlook.com

© Springer Nature Singapore Pte Ltd. 2019
S. Tiwari et al. (eds.), *Smart Innovations in Communication and Computational Sciences*, Advances in Intelligent Systems and Computing 851,
https://doi.org/10.1007/978-981-13-2414-7_27

283

the ideal classification of lightweight block cipher, stream ciphers, or even hash functions based on their behavior on a performed platform.

The paper involves a presentation of IoT in Sect. 2. This is trailed by the Introduction of lightweight algorithms including symmetric, asymmetric and hash functions in Sect. 3. After that survey about the lightweight algorithms is introduced in Sect. 4. Lastly, with the conclusion of work, the paper is enclosed.

2   Introduction of IoT

In 1998 by Kevin Ashton, the expression "IoT" was introduced for the first time is a future of internet and ubiquitous computing [3]. IoT is a shortening of "Internet of Things." IoT is an installed innovation in which "Things" are associated physically and are accessed through the web. "Things" here can be anything like home appliances, vehicles, machines, etc., which can speak with each-other without manual help. This procedure of association between smart gadgets is referred as "machine-to-machine" (M2M) communication [4]. Choices taken are influenced by enhanced computing innovation in a protest which aids them to interact with the outside condition or inward states. It can help to make smarter choices by enhancing productivity, use of benefits and improved process effectiveness. Because of support in complex fields like WSN, distributed computation, automatic identification, etc., in view of powerful and quick speed of Internet, Ubiquitous computing [4, 5] has now turned into a fact. The intention of IoT purposed is to certify whenever, anyplace, everything and everybody sorts of connection [6]. In spite of the fact that having a few issues like security and protection of clients, maintenance, communication, optimization, execution, and legal rights, etc., it is the quickest developing innovation ever. The features and services provided exceed the disadvantages of IoT. In this embedded technology there are few building blocks of architecture: Sensors/actuators, Internet gateways, Cloud/server framework and Big Data, lastly End users. The design of IoT appears in the figure underneath (Fig. 1).

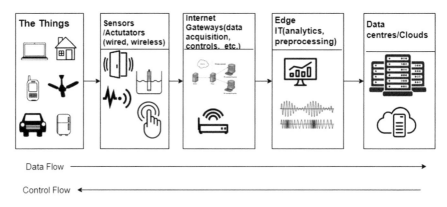

Fig. 1   Architecture of internet of things (IoT)

3 Lightweight Cryptogrpahic Algorithms and Researched Work

The primary goal of cryptography is to secure the information such that lone the sender and beneficiary can determine and work the information and no other pariah or intruder can perceive or operate it. To determine more satisfactory security with the negligible utilization of equipment and the better-improved outcome, another procedure is developed called "Lightweight Cryptography (LWC)" [6] which are easy to execute on constrained devices. Lightweight cryptographic algorithms are most reasonable for platforms like RFID (Radio Frequency Identification), WSN, FPGA (Field Programmable Gate Array), etc. There is no particular requirement to get fit in the lightweight algorithms yet, by and large, they think about more modest key size, smaller block size, littler code measure, fewer clock cycles, and so forth. The lightweight cryptographic calculation is for the most part of three kinds Block cipher, Stream cipher, or Hash function. For more complex architecture lightweight block cipher involves a keyed pseudo-random permutation which is later utilized as building blocks. The two fundamental design standards of block ciphers are Substitution-Permutation networks and Feistel cipher. For the most part, block cipher does not utilize S-box their structures or may use little S-Box (PRESENT [7]), yet some block cipher utilizes a nonlinear layer comprising in the parallel use of a few limited nonlinear capacities called "bit-sliced S-box", which are implemented by some fundamental operations like XOR, AND, and OR. While Stream ciphers produce a key stream which is XORed and can be dependent or independent, they are called synchronous or asynchronous stream ciphers separately. They essentially utilize Feedback Shift Registers (FSR) to constitute internal state easy to update. There are two kinds of FSR: Linear FSR (LFSR) (HUMMINGBIRD [8]) and Nonlinear FSR (NFSR) (Fruit v2 [9]). The shift registers having input bits as a linear function of previous states is LFSR. There are different LFSR in like manner Fibonacci, Galois, and Non-paired Galois. Similarly, NFSR is those shift registers whose input bit is a nonlinear function of its previous state. These are generally utilized for RFID or smart card applications. They involve more extensive security than LFSR against cryptanalytic attacks. Some known developments used to generate algorithms are Merkle-Demgård, Sponge, and JH-like (SPN-Hash). The most usually utilized construction is Sponge otherwise called unkeyed permutation/P-Sponge (PHOTON [10] QUARK [11] SPONGENT [12]) or random function/T-Sponge (GLUON) which was invented in 2007 [13].

4 About Lightweight Algorithm

ISO/IEC 29192, Lightweight Cryptography [14], planning to give answers for quickly developing applications that extensively utilize exceptional restricted power constrained devices, lightweight cryptography is a subcategory of cryptography as guaranteed by NIST. Devices, for example, embedded frameworks, RFID, and WSN are on the lower end of the range. The specific proposition is to utilize improvement that results in designs with better balance amidst security, execution and resource requirements for particular resource constraint environments [14]. Distinctive correlation of hardware and software implementation is troublesome as a result of contrast in measurements, measures of adequacy, and executing platform, in spite of the way that there have been a few touchtone investigations of both equipment and programming usage [15].

5 Survey of Algorithms

The listed lightweight algorithms which are compared with other algorithms and were evaluated on different platforms are described here. ECC [1] a block cipher was performed on AT94 k family FPGA and 8-bit microcontroller, was compared to lightweight symmetric cryptographic algorithms which are PRESENT, DESXL, HIGHT, CLEFIA, etc. Next is AES [6] another block cipher was implemented on different platforms like AVR and GPU, etc., while comparison was done internally and with previous work. Next is PRESENT [7] an ultra-lightweight block cipher was implemented on Mentor graphics FPGA advantage 8.1, three different architectures: Round-based, Pipelining technique, and Serialized, was implemented which output was compared with AES, SEA, and ICEBERG algorithms, and as a result round-based architecture was preferable for RFID and new arriving technologies. Hummingbird [8] is a hardware dependent ultra-lightweight block cipher was implemented on the microcontroller ATAM893-D of Atmel MARC4 and was compared to PRESENT, where it succeeds to prove better than PRESENT on target platform, but need to work on side channel attack in future. From the list, the next is PHOTON [10], a hash function was compared to KECCAK-200, KECCAK-400, and SPONGENT when implemented on Xilinx Spartan 6 FPGA, where SPONGENT implementation produce the highest throughput/area ratio and PHOTON displayed more adequate scalability in area but lowest throughput in field. DESL [16] another block cipher, is the lightweight version of DES (Data Encryption Standard) which stood out robust against many attacks of DES (types of linear cryptanalysis), was implemented on Synopsis Design Vision. It was compared to DES, DESX, DESXL, and AES which concludes that DESL is better for RFID tags by giving minimum gate equivalence comparatively. HIGHT [17], block cipher was compared internally with FPGA scalar and FPGA pipelined architecture on Verilog, Quartus, and Cyclone-II. Comparing both, scalar design requires 18% less resources and 10% less power while pipeline design has 18 times higher throughput and 60% less energy consumption [17]. TEA [18] Tiny encryption algorithm is a block cipher based on ARX (Addition/Rotation/XOR) design, a server selection algorithm having two models: Multi-level power consumption with multiple CPU's and Multi-Level Computation with multiple CPU's was compared to SEA, GEA, Round Robin, and EA. Where it comes out to be second best to obtain energy efficient server. The next is LEA

[19] Lightweight encryption algorithm, a block cipher was implemented on Verilog, Xilinx Virtex 5 series and Altera Cyclone-III series. It was compared to PRESENT, Hummingbird, Ktantan, DESL, AES, and LED (Lightweight Encryption Device). Authors proposed two design for hardware implementations: area and speed-opt design, which concludes that speed-opt version is very effective though not best among throughput/area but it is best in throughput. After that Simon [20] block cipher was implemented on ASIC application with the help of FPGA Xilinx Spartan 3, AVR ATmega 128 and MSP 430 microcontrollers, Intel Xeon E5640, and Samsung Exynos 5 dual. When it was compared to AES, PRESENT, SPECK, TWINE, and PRINCE, it was concluded that Simon and Speck are ideal for use with heterogeneous networks, they are better in implementation than AES and also very efficient to work with. Another block cipher SPECK [21] was executed by adopting the Matsuii's algorithm to find best differential and linear trails in ARX ciphers. It was implemented on Phoenix Contact ILC 350-PN controller, and WorX automation software Operating System. Different variants of speck were compared, i.e., Speck 32/48/64/96/128. In the end two new primitives were proposed MARX and SPECKEY which fulfills Markov's assumption and contain certain bounds against single-trail differential and linear cryptanalysis. The last algorithm of the table TWINE [22] variants 64/80/128-bit block size was implemented on both hardware and software. It was compared to AES, PRESENT, HIGHT, and Piccolo. It is designed to fit extremely small hardware, still it manages to give effective results on software. It is robust to many attacks but, Impossible differential cryptanalysis and Saturation cryptanalysis exploit the key schedule in TWINE-80/128 (Table 1).

ECC Elliptic Curve Cryptography, GE Gate Equivalents, H/w Hardware, S/w Software, AES Advance Encryption Standard, SPE Synergistic Processing Elements, ENC Encryption, DEC Decryption, SEA Scalable Encryption Algorithm, TEA Tiny Encryption Algorithm, LEA Link Encryption Algorithm, DESL Data Encryption Standard Lightweight, GEA GPRS Encryption Algorithm, RR Round Robin.

Table 1 Lightweight cryptographic algorithms

Sr. no.	Algorithm name	Measurement	Dependency	Application	Tools	Compared with	Comments
1.	ECC [1]	10,114 GE, 14.1 ms in $GF(2^{113})$	H/w	Pervasive computing	AT94 K microchip FPGA, $GF(2^m)$ (Galois fields)	Symmetric cryptographic algorithms	Hardware-Software co-design proposed, asymmetric requires comparatively larger chips. But performs well
2.	AES [6]	SPE: 11.7 cycles/byte (ENC), 14.1 cycles/byte (DEC), NVIDIA 8800 GTX: 0.94 C/b (ENC)	S/w	Not given	8-bit AVR microcontrollers, NVIDIA GPU, Cell Broadband Engine architecture	Hardware result comparison with already achieved results	AES-128 is successfully implemented on 3 different platform, implementation on GPU with T-table is different approach
3.	PRESENT [7]	Minimal data-path 1,000 GE, Round data-path 1561 GE, many other with different architecture	H/w	RFID	Mentor Graphics FPGA Advantage 8.1, Synopsys Design Compiler Z-2007.03-SP5, cell libraries: 350 nm MTC45000 AMIS, 250 nm SESAME-LP2 IHP, 180 nm UMCL18G212D3 UMC	AES, SEA, ICEBERG	Different architecture: Round-based data-path, Pipelined and minimal data-path, for RFID and new technology round-based is more
4.	HUMMINGBIRD [8]	2.89 ms (ENC) and 10.4 ms (DEC)	H/w	RFID tags	4-bit ATAM893-D microcontroller of Atmel MARC4	PRESENT	Security solution to active and passive RFID tags, Better than PRESENT on this platform

(continued)

Table 1 (continued)

Sr. no.	Algorithm name	Measurement	Dependency	Application	Tools	Compared with	Comments
5.	PHOTON [10]	Frequency (MHz) KECCAK-200/160/40: 144 MHz, KECCAK-400/160/80: 153 MHz, PHOTON-256/32/32: 83 MHz, SPONGENT-256/512/256: 129 MHz	H/w	RFID tags	Xilinx Spartan 6 FPGA	SPONGENT, KECCAK-200, KECCAK-400	Round function have major effect on algorithms. Throughput/area wise PHOTON comes last while SPONGENT is best. But PHOTON have sustainable scalability
6.	DESL [16]	1848 GE, 144 cycles/block	H/w	RFID	Synopsys Design Vision V-2004.06-SP2, Synopsys NanoSim, Sage-X Standard Cell Library and Cadence Silicon Ensemble 5.4	AES-128, HIGHT	Robust to many vulnerable attacks of DES, better for RFID tags, having minimal GE counts
7.	HIGHT [17]	Scalar: 18% less resources and 10% less power, Pipeline: 16% less energy and 18 times higher throughput	H/w	RFID using FPGA	Verilog_TM, Altera FPGA Quartus-II_TM, FPGA cyclone-II	FPGA Scalar and FPGA pipeline architecture	Two design of Hight: Scalar and pipeline design are implemented and compared

(continued)

Table 1 (continued)

Sr. no.	Algorithm name	Measurement	Dependency	Application	Tools	Compared with	Comments
8.	TEA [18]	Total electric energy consumption: 5100–5500 KWs, Average execution time of processes: 7.8–8.5 time unit	S/w	Energy-aware server selection algorithms in a scalable cluster	Sybase and SQL Database, multi-level power consumption (MLPC) model and the multi-level computation (MLC)model with CPU's	SEA, GEA, EA, RR	Total energy consumption of GEA is minimum, TEA is second best to find energy efficient sever
9.	LEA [19]	Xilinx Virtex 5: LEA-256 0.22 throughput/area in Area 2, Altera Cyclone-III: LEA-256 Area-1 0.15 throughput/area	S/w	RFID using FPGA's	Register Transfer Level (RTL) in Verilog, Xilinx Virtex 5 series and Altera Cyclone-III series, synthesized using Quartus-II 11.1sp2. Synopsys's Design Compiler B-2008-09.SP5. UMC 0.13 μm tech library	PRESENT, Hummingbird, Ktantan, DESL, AES, LED	Speed-opt version is very effective though not best (but in higher position) among throughput/area but it is best in only throughput output

(continued)

Table 1 (continued)

Sr. no.	Algorithm name	Measurement	Dependency	Application	Tools	Compared with	Comments
10.	Simon [20]	Low-latency encrypt-only implementations: Simon-5072 GE ans Speck-6377 GE	H/w	ASIC application	FPGA Xilinx Spartan-3, Assembly implementations on the 8-bit AVR ATmega128 and 16-bit MSP430 microcontrollers, Intel Xeon E5640, 32-bit Samsung Exynos 5 dual	AES, PRESENT, SPECK, TWINE, PRINCE	Simon and Speck ideal for use with heterogeneous networks, better in implementation than AES. Very efficient to work with
11.	SPECK [21]	PLC data v.1: Encryption time (ms): Simon-34, Speck-17; PLC data v.2: Encryption time: Simon-68, Speck-27	S/w	Not given	Phoenix Contact ILC 350-PN controller,.NET 4.2 framework, programmable according to IEC 61131 using the PC WorX	Speck and Simon 32, 48, 64, 96, 128	Implemented on PLC based on SCADA system. Two data types are used BYTE 3 and DWORD
12.	TWINE [22]	Hardware: Encryption-1,503 GE, Software: Encryption-1,011 GE	H/w	Not given	Software: ATmega163, Intel CPU with SSSE. Hardware: Synopsys DC Version D-2010.03-SP1-1	AES, PRESENT, HIGHT, Piccolo	Impossible differential cryptanalysis and Saturation cryptanalysis exploit the key schedule in TWINE-80/128

6 Conclusion

In this paper, the data about algorithms gives enough content to decide the algorithm appropriate for chose applications. The majority of the algorithms are having a dependency on hardware, so with even ultra-constrained devices, they will perform well. While few are software-dependent algorithms, however that does not mean they will not perform well on hardware, results of all specified algorithms may differ as indicated by the adjustments in hardware/software and furthermore the application. There are no settled outcomes for any algorithm whether in speed, cycles or throughput. A few algorithms are capable and secure for RFID and IoT applications. Numerous algorithms resist different attacks like: Man-in-the-center attack, Differential attacks, and key-IV attack, etc. Conclusively this paper wraps up with proficient data for choosing fitting algorithm and Hardware/Software for a particular application. Future analysts can work on different lightweight algorithms by utilizing heterogeneous platforms of hardware and software and can look at those algorithms utilizing parameters like clock cycles, speed, memory, frequency, latency, etc., to discover the proficiency and optimization of algorithms.

References

1. Eisenbarth, T., Kumar, S., Paar, C., Poschmann, A., Uhsadel, L.: A survey of Lightweight Cryptography Implementation. IEEE Design and Test of Computers, New York (2007)
2. Bansod, G., Raval, N., Pisharoty, N: Implementation of a new lightweight encryption design for embedded security. In: IEEE Transactions on Information Forensics and Security (2013)
3. Wu, M. et. al.: Research on the architecture of Internet of things. In: The Proceedings of 3rd International Conference on Advanced Computer Theory and Engineering, 20–22 Aug, Beijing, China, (2012)
4. Gaitan, N.-C., Gaitan, V.G., Ungurean, I.: A survey on the internet of things software architecture. Int. J. Adv. Comput. Sci. Appl. (IJACSA) **6**(12) (2015)
5. Singh, D., Tripathi, G., Jara, A.J.: A survey of internet-of-things: future vision, architecture, challenges and services. IEEE Word Forum on Internet of Things (WF-IoT) (2014)
6. Bos, J.W., Osvik, D.A., Stefan, D.: Fast implementations of AES on various platforms. IACR (2009)
7. Rolfes, C., Poschmann, A., Leander, G., Paar, C.: Ultra-Lightweight Implementations for Smart Devices—Security for 1000 Gate Equivalents. Springer, Germany (2008)
8. Fan, X., Hu, H., Gong, G., Smith, E.M., Engels, D.: Lightweight Implementation of Hummingbird Cryptographic Algorithm on 4-Bit Microcontrollers. IEEE (2009)
9. Ghafari, V.A., Hu, H., Chen, Y.: Fruit-v2: ultra-lightweight stream cipher with shorter internal state. Int. Assoc. Cryptol. Res (IACR) (2016)
10. Jungk, B., Lima, L.R., Hiller, M.: A Systematic Study of Lightweight Hash Functions on FPGAs. IEEE (2014)
11. Aumasson, J.-P., Henzen, L., Meierm, W., Naya-Plasencia, M.: Quark: a lightweight hash. CHES (2010)
12. Bertoni, G., Daemon, J., Peeters, M., Van Assche, G.: Sponge Functions, ECRYPT hash workshop (2007)
13. Panasenko, S., Smagin, S.: Lightweight cryptography: underlying principles and approaches. Int. J. Comput. Theory Eng. **3** (2011)

14. McKay, K.A., Bassham, L., Turan, M.S., Mouha, N.: Report on lightweight Cryptogaphy. National Institute of Standards and Technology Internal Report 8114 (2017)
15. Diehl, W., Farahmand, F., Yalla, P., Kaps, J.-P., Gaj, K.:Comparison of Hardware and Software Implementations of Selected Lightweight Block Ciphers. IEEE (2017)
16. Leander, G., Paar, C., Poschmann, A., Schramm, K.: New lightweight DES variants. Int. Assoc. Cryptol. Res. (2007)
17. Mohd1, B.J., Hayajneh, T., Khalaf, Z.A., Yousef, K.M.A.: Modeling and optimization of the lightweight HIGHT block cipher design with FPGA implementation. Security and Communication Networks. Wiley (2016)
18. Kataoka, H., Sawada, A., Duolikun, D., Enokido, T.: Energy-aware server selection algorithms in a scalable cluster. In: International Conference on Advanced Information Networking and Applications. IEEE (2016)
19. Lee, D., Kim, D.-C., Kwon, D., Kim, H.: Efficient Hardware Implementation of the Lightweight Block Encryption Algorithm LEA. Sensors (2014)
20. Beaulieu, R., Shors, D., Smith, J., Treatment-Clark, S., Weeks, B., Wingers, L.: Simon and Speck: Block Ciphers for the Internet of Things. NIST Lightweight Cryptography (2015)
21. Duka, A.V., Genge, B.: Implementation of SIMON and SPECK Lightweight Block Ciphers on Programmable Logic Controllers. IEEE (2017)
22. Suzaki, T., Minematsu, K., Morioka, S., Kobayashi, E.: TWINE: A Lightweight Block Cipher for Multiple Platforms. Springer (2012)

Fruit Disease Detection Using Rule-Based Classification

Vippon Preet Kour and Sakshi Arora

1 Introduction

Agriculture field is more than just being a feeding source in the today's world. But due to climatic and other changes over the years, crop yields and agriculture output have become prone to certain major issues which are a subject of serious concern. The world economy is going to depend very much on the agriculture as these days the production is decreasing as compared to the increase in the demand and this ratio of demand versus production is projected to be very high in the upcoming years. The various fields and the subfields associated with these agricultural units are summarized in Fig. 1.

Horticulture differs from agriculture generally in the size area of cultivation and the number of crops to be grown. It mainly focuses on the small land area and the large varieties of crops. The various professional organizations have been setup to tackle with the issues of the horticulture, e.g., Institute of Horticulture (IOH) London, American Society of Horticultural Science, Australian Society of Horticultural Science, Desh Bhagat School of Horticulture India.

V. P. Kour · S. Arora (✉)
Department of Computer Science & Engineering, SMVDU, Katra 182320,
Jammu and Kashmir, India
e-mail: sakshi@smvdu.ac.in

V. P. Kour
e-mail: preetvippon@gmail.com

© Springer Nature Singapore Pte Ltd. 2019
S. Tiwari et al. (eds.), *Smart Innovations in Communication and Computational
Sciences*, Advances in Intelligent Systems and Computing 851,
https://doi.org/10.1007/978-981-13-2414-7_28

Fig. 1 Major subfields of agriculture sector

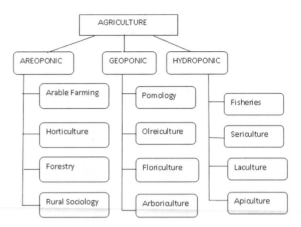

2 Literature Review

Mrunmayee et al. [1] present a diagnostic theory of Pomegranate plant diseases using the neural networks. They dealt with the concerns of plant pathology, i.e., disease detection and classification. Pranjali et al. [2] worked on the most widely grown fruit crop of India, i.e., Grape. In their study they took the grape leaves into consideration. They focused on the bacterial, fungal and viral causes of the disease, e.g., Powdery Mildew and Downy Mildew. The detection and classification of grape leaf disease was done. The concerned image is resized to size 300 * 300 and then by thresholding all green color components are extracted. K-means algorithm is used in segmentation. For feature extraction the main type of features taken into consideration are shape, color, and texture. First RGB was converted into HSV and then the image was subdivided into blocks and then mean, block variance, and skewness is computed. For classification the Linear Support Vector Machine (LSVM) is used. Revathi et al. [3] proposed an edge detection technique for classifying cotton leaf diseases. Foliar fungal disease was studied. HPCCD algorithm is proposed and is used to detect and categorize the disease. Colored image segmentation is done using RGB color models. Filtered data is then classified to identify the disease spots. Jun et al. [4] gave an automatic segmentation method to detect the defects on the citrus surface eroded by diseases using circulatory threshold segmentation. The erosion coefficient is used to quantify the erosion degree. A comparative approach of performance of circulatory threshold segmentation with respect to the Ostu segmentation method is proposed. Swati et al. [5] used support vector machine and chain code to detect and classify the apple fruit diseases. They worked on scrub apple and bitter rot like diseases. For segmentation, the morphological operations were used. For classification

Table 1 Review of different approaches used in fruit disease detection

Ref. no.	Author name	Segmentation approach	Feature extraction	Classifier
[1]	Mrunmaye et al.	K–means clustering algorithm	GLCM method	Artificial neural networks (ANN)
[2]	Paranjali et al.	K–means clustering algorithm	Color features are extracted	Support vector machine (SVM)
[4]	Jun Lu et al.	Circularity gradient method	B component of RGB color model	No classifier used
[5]	Swati et al.	Morphological techniques	Color, shape, texture and energy features are involved in feature extraction	Support vector machines (SVM) with radial kernel
[6]	Ashvani et al.	K–means clustering	SURF algorithm	Mathematical model is used
[7]	Muhammad et al.	Image gray scaling and determining	Otsu method	Membership functions are used. Fuzzy inference and defuzzification is used
[8]	Shiv Ram et al.	K–means clustering algorithm	Feature extraction based on Global color histogram (GCH), Color coherence vector (CCV)	Multi-class support vector machine (MVSM)
[9]	Manisha et al.	K–means clustering algorithm	Color, morphology and CCV	Support vector machines (SVM)

Multi-class support vector machine is employed. Ashwani et al. [6] introduced fruit disease detection technique using color, texture analysis, and artificial neural network. K-means clustering was used for segmentation and Open-CV library was used in implementation. Artificial neural network was used in pattern matching and classification. Muhammad et al. [7] implemented fuzzy logic and image processing in the orchid disease detection. The leaves were preprocessed using gray scaling, threshold segmentation and noise removal. Fuzzification, fuzzy inference, and defuzzification are used to find the results. Shiv et al. [8] detected and classified apple fruit diseases using complete binary patterns. K-means clustering method employed for segmentation, and after feature extraction the Multi-class support vector machine is used for classification. Manisha et al. [9] proposed pomegranate disease detection using image processing, a web-based tool is provided to identify the fruit disease. Feature extraction and clustering is achieved by k-means algorithm. Support vector machine is induced for classification. Further summary of the articles surveyed are given in Table 1.

Based on the above literature we have found that the mechanism for automatic disease detection would be of great use. So we proposes a fuzzy rule-based method for the detection of apple scab.

3 Apple Disease

There are different varieties of apples for example, Aceymac, Adanac, Akane, Akero, etc. Depending upon the type of apple and the climatic conditions in which it is growing, it is prone to different types of diseases and those diseases are classified as fire blight, black rot, rust, crown rot, scab, and many more. Out of the given diseases apple scab is the disease that adversely effects the growth of the apple.

Apple scab is caused by a fungus named as Venturia Inequalis, it is a fungal disease and attacks leaves as well as the fruits and is one of the globally widespread diseases of apple. The disease grows as pale black or gray brown lesions on the surface of tree bark, leaves, buds, and fruits. Lesions appear occasionally on the woody tissues of the tree. The lifecycle of the disease starts from the onset of the spring and takes about full 15 days to develop the disease. Due to the increase of leaf moisture and high temperature the secondary infection may also set in. The disease development can be predicted by the co-relation between the temperature and wetness. The lesions are mostly in the form of a cluster spread around the calyx of the fruit. Lesions are smaller in size and can range up to 1 cm in diameter.

4 Proposed Methodology for FRADD

In this study the authors have proposed Fuzzy Rule-Based Approach for Disease Detection (FRADD) for detection and classification of the most prevalent apple diseases in Kashmir valley (particularly apple scab). The basic steps of FRADD are shown in Fig. 2. In this approach, different stages are involved from image collection to the classification and these steps are depicted below. The framework of the proposed approach is shown as in Fig. 3.

For the formation of the database, the images were taken both manually and online. For manual collection of images certain angles and distance was predefined to take the picture in the presence of appropriate vision sources. The images thus collected were resized to the size of 250 * 250. Two datasets, individually of both diseased and un-diseased images were formed. These images were used for both the training as well as the testing phases.

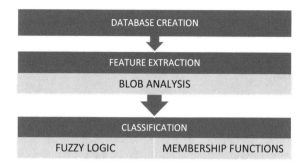

Fig. 2 The basic mechanism of the proposed approach

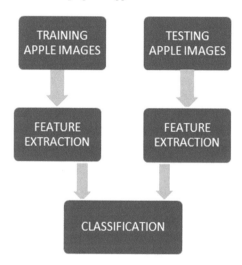

Fig. 3 Framework for the proposed work

For feature extraction, blob analysis has been used. The blob testing algorithm checks the whole image pixel by pixel from left to right. The region of interest is created by the blob analysis while checking the whole image. The point or the pixel where the disease is present is termed as blob or we can say that an image can have many regions of interest depending upon the disease distribution. In blob analysis region is any subset of pixels. There can be various blobs in a single image, depending upon the disease distribution in an image.

4.1   The Outline of Blob Analysis Method Consists of the Following Steps

4.1.1   Extraction

It is the embryonic step on the image in which image thresholding technique is implemented in order to get the region analogous to the objects being inspected. In FRADD, the image thresholding is being performed to obtain the infected region, i.e., region of interest or blob (foreground) from the un-infected region, i.e., background and is shown in Fig. 4.

4.1.2   Refinement

The given images taken into consideration for training as well as testing purpose are often associated with some sort of noise due to poor image quality or the resource from where they are taken. So in order to remove noise, the images have been refined using image refinement techniques. Blob analysis offers the advantage of noise removal in the source images that have mostly been generated by capturing using digital cameras. In blob analysis, the noise removal or the image refinement is an inbuilt step.

4.1.3   Analysis

This step aims at taking into consideration the refined regions. The refined regions are then analyzed on the basis of certain parameters, e.g., mean intensity, area, diameter, perimeter, and then the final results are computed. In FRADD, depending upon the scattering of the given disease in the image, different regions of interest or blobs are created e.g., if there are three different spots of the disease in the same image, then three blobs will be created.

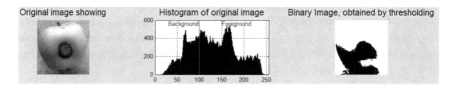

Fig. 4 Thresholding of a given image

4.1.4   Classification

For classification there are various techniques and methods followed these days, e.g., artificial neural networks, support vector machines, fuzzy logic, convolution neural networks, etc. In FRADD, the fuzzy logic is used for classification. Depending upon the values of the parameters which are used for feature extraction by using blob analysis, the fuzzy rules are generated and are thus implemented with the use of membership function.

4.1.5   Accuracy

On the basis of blob detection results and the fuzzy output, the accuracy of the FRADD is determined. The overall accuracy of the system depends upon both the training and the testing matches and un-matches. The dependency ratio of thus matched and un-matched results is thus determined.

5   Results

The FRADD is implemented by the following steps.

5.1   *Image Acquisition*

The apple images for the FRADD are taken manually by digital camera with the angle and distance values. Table 2 shows the further details for picture taking conditions. The images are then converted, i.e., resized to 250 * 250.

Table 2 Picture taking condition

Type	Condition (in.)
Height of camera	55
Distance of camera	25
Angle of camera	26
Lamp height	60
Lamp angle	75

5.2   Feature Extraction

The blob analysis is implemented for feature extraction in FRADD method. The various steps performed by blob are as under. Conversion of colored image into gray scale for extracting the background and foreground pixels, i.e., for obtaining the blobs given in Figs. 5, 6 and 7.

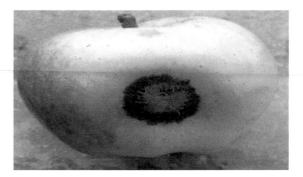

Fig. 5   Original diseased fruit image

Fig. 6   Original without disease fruit image

Fig. 7 Grayed picture

5.2.1 Image Thresholding

The thresholding of foreground and background image is performed to obtain the pixels of region of interest (blob) from the actual image. It confirms the intensity of the disease in a particular image. A histogram generated thus confirms the results given in Fig. 8.

The foreground represents the value of the blob or diseased area. The background represents the rest of the fruit or unaffected area.

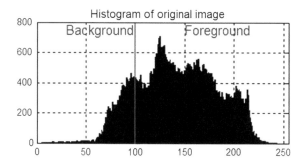

Fig. 8 Histogram showing the results of thresholding

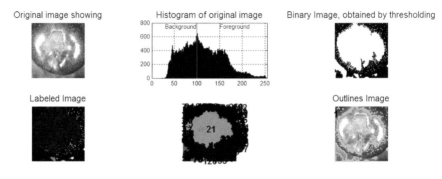

Fig. 9 Blob working in FRRAD method

5.2.2 Conversion of Labeled Gray Scale Image to Colored Image

After the detection of the blobs on the basis of the certain parameters, e.g., mean intensity, area, perimeter, diameter, the gray scale image is thus converted back into colored image shown in Fig. 9.

(i)   Mean intensity = Background pixels - Foreground pixels.
(ii) Area: since the blobs are circular, therefore circular area formulae is used.
(iii) Diameter of the can range up to 1 cm.
(iv) Perimeter: Perimeter of individual blobs is taken and it depends upon the radius of the individual blob. Then the combined perimeter value of the all blobs in a particular image is computed.

5.2.3 Classification

For classification, the fuzzy logic is implemented with the aid of trapezoidal membership function. The parameters on the basis of which feature extraction is done are used for designing fuzzy rules with the help of trapezoidal membership function.

(i)   Trapezoidal membership function: The function is defined as

$$y = trapmf(x, [a, b, c, d]) \tag{1}$$

$$(x; a, b, c, d) = \begin{cases} 0, & x \le a \\ \frac{x-a}{b-a}, & a \le x \le b \\ 1, & b \le x \le c \\ \frac{d-x}{d-c}, & c \le x \le d \\ 0, & d \le x \end{cases} \tag{2}$$

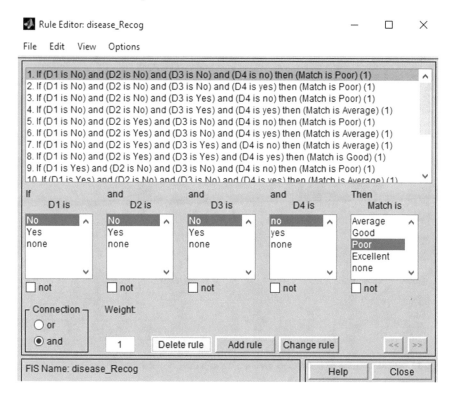

Fig. 10 Fuzzy rules

(ii) Fuzzy Logic: The detection of the disease is confirmed on the basis of the fuzzy rules shown in Fig. 10.

Here I the fuzzy rules the parameters are represented as

- D1 = mean intensity
- D2 = Area
- D3 = Perimeter
- D4 = Diameter

Four values are put forth for classification depending upon the rules and those values are:

- Poor: No matching of the parameter values at all.
- Average: Matching of the one or two parameter values.
- Good: Matching of at-least two parameter values.
- Excellent: Matching of all the parameter values, i.e., the closest by value Fig. 11.

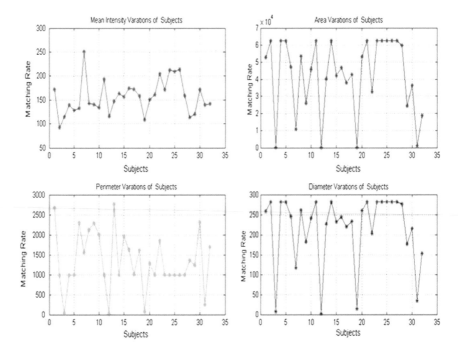

Fig. 11 Representation of parameters graphically with their respective subjects

5.2.4 Blob Analysis and Fuzzy Results

The blob analysis of both the training images with and without apple scab disease is shown in Fig. 12. The instance1 to instance 20 represent the training images, whereas the parameters with their corresponding numerical values are also shown:

The fuzzy logic of training images is shown in Fig. 13 along with their accuracy rate.

Figure 14 shows the matching rate of training images with the training images. While performing the testing steps, the threshold value is found to be 80%. So the images with value greater than 80 percent have the accurate match. The instance1 to instance20 represent the training images while as t1 to t10 represent the test images.

Image	Mean Intensity	Area	Perimeter	Diameter
Instance1	172.2563544	52837	2681.841846	259.3726235
instance2	92.064912	62500	996	282.0947913
Instance3	114.3658537	41	33.55634919	7.225151994
Instance4	139.48584	62500	996	282.0947913
Instance5	128.518576	62500	996	282.0947913
Instance6	132.6767372	47101	2299.290547	244.8894767
Instance7	251.4696619	10795	1566.198052	117.2374551
Instance8	143.4165998	53591	2125.141269	261.2167308
Instance9	140.5254905	25892	2294.375901	181.5673933
Instance10	133.7075636	45904	2007.207286	241.7577053
Instance11	193.422432	62500	996	282.0947913
Instance12	116	2	2	1.595769122
Instance13	147.2784867	40282	2767.176982	226.4699435
Instance14	163.0144	62500	996	282.0947913
Instance15	156.3341067	42241	1960.496608	231.9114305
Instance16	174.4224454	46709	1627.550432	243.8682962
Insatnce17	171.7612633	37933	1000.34733	219.7675945
Instance18	158.932616	42844	1609.366666	233.5608594
Instance19	107.4630872	149	74.97056275	13.77362305
Instance20	150.2951807	53120	1286.399062	260.0663081

Fig. 12 Blob analysis of training images

Image	Mean Intensity	Area	Perimeter	Diameter	Matching Rate	Fuzzy Output
Instance1	172.2563544	52837	2681.841846	259.3726235	63.42838442	A
instance2	92.064912	62500	996	282.0947918	58.64925801	A
Instance3	114.3658537	41	33.55634919	7.225151994	12.48764352	P
Instance4	139.48584	62500	996	282.0947918	65.09972677	A
Instance5	128.518576	62500	996	282.0947918	62.50224848	A
Instance6	132.6767372	47101	2299.290547	244.8894767	61.05072098	A
Instance7	251.4696619	10795	1566.198052	117.2374551	42.49322465	P
Instance8	143.4165998	53591	2125.141269	261.2167308	67.96463209	A
Instance9	140.5254905	25892	2294.375901	181.5673933	51.07017009	A
Instance10	133.7075636	45904	2007.207286	241.7577053	62.09674796	A
Instance11	193.422432	62500	996	282.0947918	60.79230331	A
Instance12	116	2	2	1.595769122	9.230715063	P
Instance13	147.2784867	40282	2767.176982	226.4699436	56.79425436	A
Instance14	163.0144	62500	996	282.0947918	65.09972677	A
Instance15	156.3341067	42241	1960.496608	231.9114305	61.01065387	A
Instance16	174.4224454	46709	1627.550432	243.8682962	63.85730096	A
Insatnce17	171.7612633	37933	1000.34733	219.7675946	59.99605646	A
Instance18	158.932616	42844	1609.366666	233.5608594	63.49471279	A
Instance19	107.4630872	149	74.97056275	13.77362306	16.36771707	P
Instance20	150.2951807	53120	1286.399062	260.0663081	81.90401205	E

Fig. 13 Fuzzy output of training images

Images	t1	t2	t3	t4	t5	t6	t7	t8	t9	t10
Instance1	A	A	A	A	A	A	A	A	A	A
instance2	A	p	A	A	A	A	A	A	A	p
Instance3	p	p	p	p	p	p	p	p	p	p
Instance4	G	A	A	A	A	A	G	A	A	A
Instance5	A	A	A	A	A	A	A	A	A	A
Instance6	A	A	A	A	A	A	A	A	A	A
Instance7	p	A	p	p	p	p	p	p	p	p
Instance8	A	A	A	A	A	A	A	A	A	A
Instance9	p	A	p	p	p	p	p	A	A	A
Instance10	A	A	A	A	A	A	A	A	A	A
Instance11	G	A	G	G	G	G	G	A	A	A
Instance12	p	p	p	p	p	p	p	p	p	p
Instance13	A	A	A	p	p	p	A	A	A	A
Instance14	E	A	G	A	A	A	A	A	A	A
Instance15	A	A	A	A	A	A	A	A	A	A
Instance16	A	A	A	A	A	A	A	A	A	A
Insatnce17	A	A	A	A	A	A	A	A	A	A
Instance18	A	A	A	A	A	A	A	A	A	A
Instance19	p	p	p	p	p	p	p	p	p	p
Instance20	A	A	A	A	A	A	A	A	A	A

Fig. 14 Representation of output of training images with the testing images in the linguistic form

Figure 15 shows the matching rate of training images with the testing images graphically. Here the threshold value 80 percent achieved by the images show that the system is detecting the diseases accurately. The subjects having the values equal or more than 80 percent show the best matches.

Figure 16 shows the matching rate of training images with training images numerically while as the graphical representation of these results is shown in Fig. 17, which also confirms the threshold value.

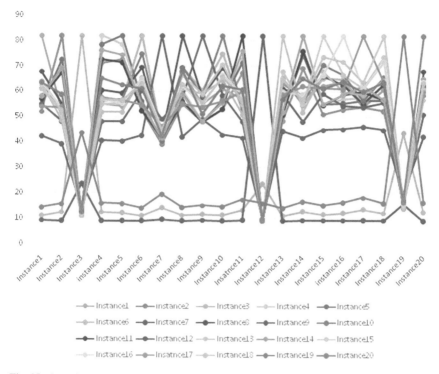

Fig. 15 Graphical representation of matching rate of training images with testing images

5.3 Accuracy Rate

The accuracy rate of the system is found to be 91 percent. Since the total test cases taken were 12 and out of 12 cases, 11 cases matched exactly.

Therefore accuracy is found by the formula

$$Accuracy = \left(\frac{Total\ number\ of\ match\ cases}{Total\ number\ of\ cases} \right) \tag{3}$$

$$Accuracy = \left(\frac{11}{12} \right) = 91.66\% \tag{4}$$

IMAGE	Instance1	Instance2	Instance3	instance4	Instance5	Instance6
Instance1	81.90401205	51.94283582	11.22711522	55.01279786	54.13099972	63.92818514
instance2	51.94283582	81.90401205	12.46759389	71.13838686	72.44660138	52.48647112
Instance3	11.22711522	12.46759389	81.90401205	12.42922521	12.25997149	11.10908213
Instance4	55.01279786	71.13838686	12.42922521	81.90401205	78.40565605	56.09051597
Instance5	54.13099972	72.44660138	12.25997149	78.40565605	81.90401205	56.09051597
Instance6	63.92818514	52.48647112	11.10908213	56.09051597	56.09051597	81.90401205
Instance7	42.34475727	39.33273174	14.33863413	40.76336429	40.38398216	42.64791998
Instance8	67.67739408	54.8744959	11.28535742	60.33893478	59.30421621	69.40828548
Instance9	54.95469411	48.03815435	11.63788624	48.03815435	48.03815435	60.28988524
Instance10	61.09442918	53.01587821	11.36108813	56.5757194	56.5757194	74.72523355
Instance11	57.01766618	67.22720909	13.24736129	72.56122177	71.23913505	52.39941106
Instance12	9.321090188	9.195111447	23.66019566	9.166474197	9.11050573	9.165730735
Instance13	63.16878781	48.49073153	11.09911518	51.76227676	51.76227676	65.23983843
Instance14	57.96680561	69.11865314	12.78141493	76.15630071	74.23370583	54.60726928
Instance15	62.261458	51.12122677	11.71891484	55.0520101	55.05132114	65.22284898
Instance16	63.92653006	52.97477761	12.3049018	56.88653357	55.43059328	64.75776845
Insatnce17	53.91820442	53.759877	13.79758074	57.64178841	56.35978603	54.63983889
Instance18	60.76679283	52.46399325	12.16502191	56.84776553	56.35515928	63.01076325
Instance19	14.33767011	15.67141879	43.74654334	16.01926014	15.75146945	14.13104794
Instance20	63.42838442	58.64925801	12.48764352	65.09972677	62.50224848	61.05072098

Fig. 16 Matching rate of training images with the training images

6 Conclusion and Future Work

In this article, we proposed a fuzzy logic based model for automatic detection of apple diseases. We have studied different approaches and their accuracy and performance level. The operations performed on the images were done very precisely and keenly keeping in mind the relativity of the disease parameters. Blob analysis was used to extract feature due to its flexibility and various inbuilt functions. When fuzzy logic is implemented on both the training as well as the testing images, the accuracy rate is found to be 91.66%. On the basis of the accuracy rate of the system, results are viable and very precise and thus the system has achieved higher performance. While studying the apple diseases under FRADD model only one disease known as the apple scab is taken into consideration. There is scope to study other apple diseases as well. The number of fruit types can also be increased thus defining the more number of parameters. By using sensors and other mechanisms like drone cameras, this work can be further extended to identify some other diseases.

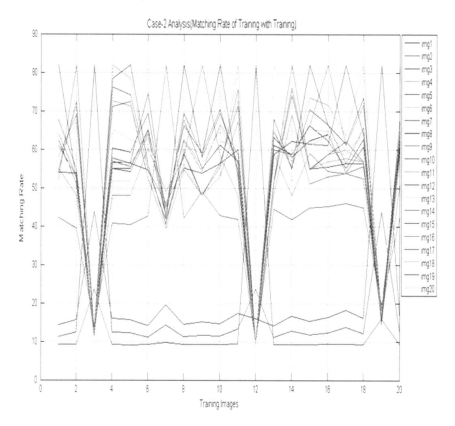

Fig. 17 Graphical representation of training images with the training images

References

1. Dhakate, M., Ingole, A.B.: Diagnosis of pomegranate plant diseases using neural network. In: Fifth National Conference on Computer Vision, Pattern Recognition, Image Processing and Graphics (NCVPRIPG), pp. 1–4, 16–19 Dec 2015
2. Padol, P.B., Yadav, A.A.: SVM classifier based grape leaf disease detection. In: Conference on Advances in Signal Processing (CASP), pp. 175–179, 9–11 June 2016
3. Revathi, P., Hemalatha, M.: Classification of Cotton Leaf Spot Diseases Using Image Processing Edge Detection Techniques. In: International Conference on Emerging Trends in Science, Engineering and Technology (INCOSET), pp. 169–173, 13–14 Dec 2012
4. Lu, J., Wu, P., Xue, J., Qiu, M., Peng, F.: Detecting defects on citrus surface based on circularity threshold segmentation. In: 12th International Conference on Fuzzy Systems and Knowledge Discovery (FSKD), pp. 1543–1547 (2015)
5. Dewliya, S., Singh, P.: Detection and classification for apple fruit diseases using support vector machine and chain code. Int. Res.J. Eng. Technol. (IRJET) **02**, 04 Aug 2015
6. Awate, A., Deshmankar, D., Amrutkar, G., Bagul, U., Sonavane, S.: Fruit disease detection using color, texture analysis and ANN. International Conference on Green Computing and Internet of Things (ICGCIoT), pp. 970–975, 8–10 Oct 2015

7. bin Mohamad Azmi, M.T., Isa, N.M.: Orchid disease detection using image processing and fuzzy logic. In: International Conference on Electrical, Electronics and System Engineering (ICEESE), pp. 37–42, 4–5 Dec 2013
8. Dubey, S.R., Jalal, A.S.: Detection and classification of apple fruit diseases using complete local binary patterns. In: Third International Conference on Computer and Communication Technology (ICCCT), pp. 346–351, 23-25 Nov 2012
9. Bhange, M., Hingoliwala, H.A.: Smart farming: pomegranate disease detection using image processing. In: Second International Symposium on Computer Vision and the Internet (Vision-Net'15), pp. 280–288, 22 Aug. 2015

Fault Tolerance Through Energy Balanced Cluster Formation (EBCF) in WSN

Hitesh Mohapatra and Amiya Kumar Rath

1 Introduction

Every invention of human intelligence more or less likely prone to failure [1, 2]. The substantial advances in technologies such as WSN, Internet of Things (IoT), Embedded System, Artificial Intelligence these all are product of human quotient. Hence, there are chances of failure in all above-said technical environments. The chances of fault occurrence may be predictable or unpredictable. There is an emergency of building an anti-fault tolerance architecture to handle such unexpected faults to maintain the smoothness of work. The present time of advance communication demands the information from human unreachable zones like volcano belt, underwater, seismic area, and so on. The WSN is a hot-cake for the researcher because of its remote and robust deployment. The very basic purpose of WSN system is to interact closely with the physical world and to perform only a limited number of dedicated functions. The robotic supervision added an extra layer of cream to this hot-cake. Basically, the WSN consists of a finite number of SNs, which are deployed in a distributed fashion over a geography. The intention behind this distributed plantation is to sense the physical environment and pass the accumulated data to the BS. The human expert at the base station is responsible for analysis and generation of meaningful report [1].

In WSN, many sensors are connected directly or indirectly to communicate the collected data with BS. The chaining of same or different type of motes further classifies the network into homogeneous and heterogeneous [3]. The capacity of SN's also varies according to the mobility of SNs. The deployment of SNs may be static or mobile that depends up on application nature [4, 5]. The several domains where WSN so far deployed are agricultural monitoring, smart city, industry automation, battlefield, space, environment, intruder detection system (IDS), weather forecasting, building architecture, navigation, healthcare monitoring, volcanic malfunction, in vehicles, commerce, etc. [1, 3].

H. Mohapatra (✉) · A. K. Rath
Veer Surendra Sai University of Technology Burla, Sambalpur, Odisha, India
e-mail: hiteshmahapatra@gmail.com

A. K. Rath
e-mail: akrath_cse@vssut.ac.in

© Springer Nature Singapore Pte Ltd. 2019
S. Tiwari et al. (eds.), *Smart Innovations in Communication and Computational Sciences*, Advances in Intelligent Systems and Computing 851,
https://doi.org/10.1007/978-981-13-2414-7_29

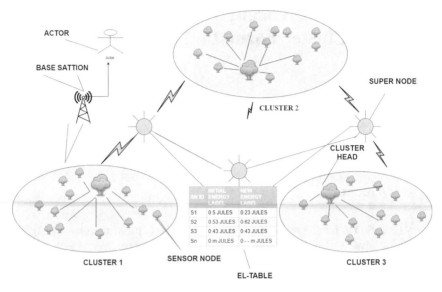

Fig. 1 WSN Architecture

The architecture of the proposed method is illustrated in Fig. 1. In our architecture we consider few SNs, CH for each cluster, Super Node (SuN), residual energy of SNs, BS, and Actor. Normally, the sensor nodes are deployed either in distributed or centralized fashion over a large geographical area by the help of drone or air plane [1]. The SNs are clubbed according to pre-fixed criteria and form clusters. Each cluster elects one CH per round by referring the residual energy label (REL) from EL-Table. CH is responsible of collecting data from sensor nodes then segregate them and forward to the sink. Sink redirects those collected data to the administrator, who is present at BS over the internet.

According to proposed WSN architecture, we deployed few SuN which introduces heterogeneity concept. The SuNs have more energy, computational capacity, and storage space in comparison to normal SNs. These SuN are placed in a quarantine manner in clusters. SuNs are liable for communicating with all SNs about there REL and update the same in a EL-Table. The EL-Table continuously updated with new REL after each round. In the proposed architecture, the consideration of mobile SNs implies the inconsistency of members with respect to clusters. Hence, the cluster will form by considering the energy level of sensor nodes in such a way that the total energy individual clusters in WSN must be equal or approximate equal. After each transaction, the mobile sensor nodes allowed to change their existing cluster and can join to a new cluster according to its remaining REL for stabilizing the strength of clusters. This proposed algorithm reduces the death of sensor nodes by reducing the imbalance in cluster strength and stops the CH election process. It increases the lifespan of the clusters which can lead to long-term communication without frequent fault handling.

This paper is segmented as follows: In Sect. 2 contains literature review. Section 3 has proposed taxonomy of fault classification. The proposed work with flow diagram is covered by Sect. 4. Section 5 has discussion and conclusion, acknowledgement, and reference respectively.

2   Literature Review

In the existing literature, many researchers have been proposed various schemes based on routing,memory,energy to overcome the fault occurrence of WSN. The EBCF algorithm is exclusively designed for hostile, unattainable and mass deployment environment. The various reasons which cause fault in the SNs are inconsistent behavior of the link, adverse environment, hardware or software failure, non-reachability of SNs, battery depletion, and noise interference [6–8]. In few cases, the BS also found faulty because of intruder attack, software or hardware failure [9, 10]. There is a probability for special nodes like SuN that it also can be faulty because of hardware or software failure, deadlock situation, and traffic congestion [11]. To recover from various faults, the paper [12] represents an analytical study. In this survey paper, various algorithms were discussed which are working for prevention, detection, identification, and isolation of faults. For better performance, several management infrastructures for WSN were discussed in [13]. The performance of a network is measured by its reliability and dependability which implies the quality of network ant its faults.The technique use for recovery and managing fault is one of the major aspects. The detection of fault in dense deployed WSN is a critical task. In paper [4], the author proposed data checkpoint method to evaluate the trust values of sensor nodes. The author considers the WSN types, i.e., static or dynamic and according to that different models were proposed for fault tolerance. This [5] work also predicates the chance of fault occurrence by evaluating the energy level of SNs.

3   Fault Tolerance in WSN

The faults in WSN can be classified into two types and further four types. The broad classification of fault can be A. Hardware fault B. Software fault. The hardware fault is again classified into two type such as 1. Energy Depletion-Based Fault and 2. Insufficient Storage Based Fault. Further, the software fault divided into 1. Computational capacity Based Fault and 2. Intruder Attack Based Fault. Figure 2 represents our taxonomy of fault classification.

Hardware Fault
In a WSN, various hardware units are embedded to accomplish the total task, i.e., from data collection to transmission [14]. The hardware units also can effected with fault by miss-handling or insufficient knowledge about operating methods. The hardware

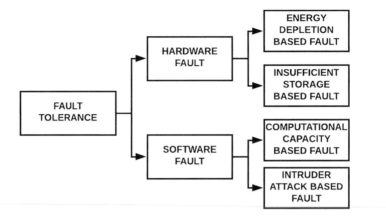

Fig. 2 Taxonomy of fault

fault can occur in these following hardware units like sensing unit, energy unit, and localization unit. The processor of SNs also can get faulty because of abnormality of data transmission. The hardware fault can be classified broadly into two types. These are:

1. Energy Depletion-Based Fault
2. Insufficient Memory Based Fault.

Software Fault

To enable the hardware units, software is indeed [15]. The software is a logical level of development. The types of fault possible with software are programmatic error, load testing failure, interface fault, overhead dealing fault, and packaging fault. The software fault can be classified broadly into two types. These are:

1. Computational Capacity Based Fault
2. Intruder Attack Based Fault.

Energy Depletion Based Fault

The miniaturization of sensing and quantification technologies allows to build out of tiny, low-power, and inexpensive sensors, actuators, and controllers. This tiny architecture constraints the SNs with low energy and low computational capacity. The mash and unattainable deployment of SNs in worse environment leads to frequent failure [16]. The approach in [12] is saving energy through solar harvesting which is again applicable to specific and limited deployed sensor nodes. However, if the number of the node is more the cost of implementing such method is an expensive one [12]. The cluster formation method which comes under unsupervised learning where the label of a node is not known and the accuracy also unknown. In this process, the nonuniform energy distribution also is a major cause of node death. In our paper, we mainly focus on uniform energy distribution whereas we are not focusing on node level rather cluster level energy uniformity.

4 Proposed Work

This section presents the working of proposed model which claims that fault tolerance can be achieved through fault prevention. In our method, we intentionally penetrate our own design fault to prevent generation of faulty report from faulty data set. The mechanism of EBCF is based on fault prevention technique [13]. In WSN, huge numbers of SNs are deployed in a specific geographic zone for close interaction [17]. The roles of motes are to sense the physical scenario and collect data accordingly. The sensor nodes are usually deployed in the remote and hostile environment hence, direct communication with BS is not suggestible because of low energy and low computational power. Traditionally in WSN, the SNs participate in cluster formation process either by election or selection method. The in-depth studies on cluster formation process conclude that cluster formation either through election or selection consumes substantial amount of energy. The core principal of EBCF is to stop this election and selection method and rescue SNs from participating in CH finalizing process. This paper proposed a method of fault prevention by assuming heterogeneous environment where sensor network is dynamic in nature. The attentiveness behind choosing of mobile SNs is to allow there free movement among clusters so that the equilibrium of total strength among clusters can be achieved [18].

The flow diagram of architecture is presented in Fig. 3. Initially, all SNs need to report their residual energy level (REL) to SuN. The SuN will maintain an EL-Table with two attribute. The first attribute is current REL and second one contains remaining residual energy level (RREL) which generates after a transmission process. Unlike the traditional cluster formation, in EBCF, the deployed SuN behaves as CH. The CH now chooses cluster members (CM) in such a way that total energy of all formed clusters must be equal or approximately equal. Initially, all SNs report their REL to SuN. The SuN forwards the same data to the BS. BS will map the CM with particular CH (SuN). The specificness maintained by BS is total energy of clusters and must be equal or approximate equal. After each iteration, the EL-Table is periodically maintained by SuN at BS with RREL.

After every round, the total energy of clusters is compared with pre-fixed threshold energy (ETH) value. If the energy label of the cluster is less than ETH (EL < ETH), then the cluster is automatically dropped from communication process until the energy regained. The regain of energy for the dropped cluster is beyond the scope of this paper. This automatic step back from process is called fault prevention method by injecting intentional fault. After every round according to REL of SNs, the BS is responsible to decide about the placing or moving of SNs from one cluster to other cluster by satisfying EBCF algorithm. The EBCF algorithm inspired from a natural environment of human dealing. For example, consider a situation where a complex task is given to solve by group of people. Let us assume that the task needs to be solved through module wise. In this case study, the group of people will be divided into number of subgroups and each subgroup is assigned with a particular module. The whole focus is on forming of cluster or team, as it implies the parallel task completion. If any particular group formed with inefficient people

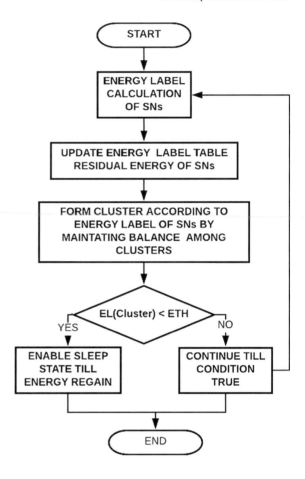

Fig. 3 Flowchart of proposed work

or weak people then they may not be able to complete their part of task which will impact on overall completion of work. If the subgroup will be formed by keeping the constraint of strength equality, then the work may complete in synchronized manner. The proposed EBCF algorithm exactly believes the same human nature, for forming energy balanced cluster. The result of this algorithm implies the equal level of energy depletion of all clusters in parallel time. This helps the administrator to take appropriate break at proper time. This proposal helps in two direction the first one is avoidance of fault and second one is prevention from collecting wrong data because of energy depletion in SNs.

Assumption scheme to peruse this work, are:

1. Each node is identified with a nonzero identity.
2. All sensor nodes are aware of super node (SuN) existence in the WSN.
3. All nodes need to report about their REL to SuN for creation and maintenance of EL-Table.
4. The initialization of function is error free.

5. Bidirectional communication exists between the SNs and SuN.

Algorithm 1 EBCF Algorithm

1: START:Search nearby SuN
      Where Min(INN)(SN, SuN).
2: APPLY:Fitness Function Algorithm for Calculating Fitness of SNs i.e REL and RREL (Energy Level)
      Fix ETH for Comparison.
3: ADVERTISE:REL of SNs to selected SuN from Step.1
4: UPDATE: Initialize EL-Table with REL value
      Then with RREL (periodically)
5: CLUSTERING:Form cluster of SNs by balancing energy among clusters.
6: COMPARE
      REL(Cluster)= ETH
      IF (REL(Cluster)< ETH)
      THAN
      Allow Clusters to Sleep
      ELSE
      REPEAT: Step: 3 (Periodically After Each Communication)
7: END

---

5 Result

The proposed protocol is implemented and the performance assessed using MATLAB R2016a in Super Computer PARAM. We executed the simulations in environment of 100 nodes in a network area of 100m × 100m. The preliminary energy of nodes is assumed equal,which implies a homogeneous network. The BS is located in the middle of the network area with coordinates (50, 50). The comparison of our protocol was performed against both homogeneous (LEACH) and heterogeneous (SEP) protocols. We simulated the network until all node dead (AND). Table 1 represents the comparative simulation result at different position w.r.t number of node dead. The first attribute of table represents the node count as first node dead (FND), 50th node represented as half node dead (HND), and 100th node dead represented as all node dead (AND). We have simulated each protocol for 2000 rounds having equal node energy, i.e., 0.5 J.

For LEACH FND at 634 number of rounds whereas HND at 802 and AND at 1149 number of rounds. In case of SEP protocol FND at 669, HND at 1003 and AND at 1476 number of rounds. In case of EBCF the FND at 1423, HND at 1541, and AND at 1741 number of rounds.

Table 1 Result table

Number of nodes	LEACH	SEP	EBCF
FND	634	669	1423
HND	802	1003	1541
AND	1149	1476	1741

6 Conclusion

The deep survey on literature concludes that prior proposed algorithms used cluster formation technique as a energy saving scheme at node level. The energy saving scheme in WSN is mainly deployed in two scenarios such as centralized and distributed fashion. According to SNs mobility, the WSN deployment again subdivided into static and dynamic type. In traditionally cluster formation technique, the major drawbacks are random election of CH and nonuniform distribution of CMs. Our proposed EBCF is unique in terms of focusing on cumulative energy of clusters rather CMs. The energy depletion of sensor nodes is directly proportional to the number of iteration carried out for a particular transmission process. Some how in these traditional algorithms, the energy depletion of SN is high which is a threat to WSN life time. In this paper, we have considered the constraints of WSN architecture and the proposed EBCF algorithmic method. This method overcomes the drawback of traditional approach by avoiding CH election and by forming energy equilibrium clusters. The energy balanced clusters survive for more number of rounds which imply the less amount of energy depletion at individual SN level. We believe that intentional fault is better than unexpected fault. EBCF is mostly suitable for human unattainable sensor deployments. The applied environment for EBCF can be underwater aqua-monitoring and its diversification. The underwater monitoring sensor cannot be provided with either through external energy resources nor by solar harvesting. In such hostile and harsh condition, EBCF algorithm is best option for energy management. This method reduces stress from individual SNs by restricting the frequent participation of CH selection. This energy balanced cluster formation enhances the overall longevity of network. This work can be extended by minimizing the energy consumption for maintaining energy label (EL) EL-Table.

Acknowledgements I would like to thank my Ph.D. Supervisor, Prof. A. K. Rath for his guidance and for providing free hand for research. I also thankful to VSSUT, TEQIP 3 for the sponsorship to attend the conference.

References

1. Munir, GA., Antoon, J.: A gordon-ross. Article No. 3. ACM Trans. **14**(1) (2015)
2. Alrajei, N., Fu, H.: A survey on fault tolerance in wireless sensor networks. SENSORCOMM09 366–371 (2014)
3. Kakamanshadi, G., Gupta, S., Singh, S.: A survey on fault tolerance techniques in wireless sensor networks. Google Sch. 168–173 (2015)
4. Mitra, S., Das, A.: Distributed fault tolerant architecture for wireless sensor network. MR3650784 94A12 (68M15) **41**(1), 47–58 (2017)
5. Park, D.S.: Fault tolerance and energy consumption scheme of a wireless sensor network. Google Sch. 1–7 (2013)
6. Younis, M., Senturk, I.F., Akkaya, K., Lee, S., Senel, F.: Topology management techniques for tolerating node failures in wireless sensor networks: a survey. Google Sch. **58**, 254–283 (2014)

7. Moreira, L., Vogt, H., Beigl, M.: A survey on fault tolerance in wireless sensor networks. Braunschweig. Google Sch. (2007)
8. Kim, S., Ko, J., Yoon, J., Lee, H.: Multiple objective metric for placing multiple base stations in wireless sensor networks. IEEE 627–631 (2007)
9. Raj, R., Ramesh, M., Kumar. S.: Fault-Tolerant clustering approaches in wireless sensor network for landslide area monitoring. Google Sch. 107–113 (2008)
10. Cardei, M., Yang, S., Wu, J.: Fault-Tolerant topology control for heterogeneous wireless sensor networks. Google Sch. 1–9 (2007)
11. Raghunathan, V., Kansal, A., Hsu, J., Friedman, J., Srivastava, M.: Design considerations for solar energy harvesting wireless embedded systems. IPSN **2005**, 457–462 (2005)
12. Mitra, S., Das, A., Mazumdar, S.: Comparative study of fault recovery techniques in wireless sensor network. IEEE, AISSMS **41**(1), 130–133 (2016)
13. Paradis, L.: Surv. Fault Manag. Wirel. Sens. Netw. **5**, 171–190 (2007). https://doi.org/10.1007/s10922-007-9062-0
14. Panigrahi, T., Panda, M., Panda, G.: Fault tolerant distributed estimation in wireless sensor networks. MR3650784 94A12 (68M15) **69**, 27–39 (2016)
15. Bin, L., Ming-Ru, D., Rong-Rong, Y., Wen-Xiao, Y.: Fault-tolerant topology in the wireless sensor networks for energy depletion and random failure. Chin. Phys. B **23**(7), 070510 (2014)
16. Das, A., Rahman, A., Basu, S.S., Chaudhuri, A.: Energy aware topology security scheme for mobile Ad Hoc network. ICCCS, ACM **41**(1), 114–118 (2011)
17. Nanda, A., Rath, A.K.: Mamdani fuzzy inference based hierarchical cost-effective routing (MFIHR) in WSNs. IEEE 397–401 (2017)
18. Kumar Rout, S., Rath, A.K., Bhagabati, C., Mohapatra, P.K.: Node localization by using fuzzy optimization technique in wireless sensor net-works. In: The Next Generation IT Summit on the Theme—Internet of Things: Connect your Worlds, pp. 176–181 (2016)

Mining Social Networks: Tollywood Reviews for Analyzing UPC by Using Big Data Framework

V. Kakulapati and S. Mahender Reddy

1 Introduction

Audiences of movies are producing online content with the movies in many online sites such as twitter and Facebook by posting their comments or reviews about films [1]. Today, potential people are habituates to check on social network review to take the decision for watching choices in addition to posting their opinion. All these are producing and investigating the movie ratings. Posting and sharing user opinions through different social websites produced large data which is called as user-produced content or simply UPC.

Telugu cinema industries also known as Tollywood has drawn attention in recent years. This drastic change is due to the reality that the directors making world-class films in the past two to three years. The recent movie like Bahubali 2 creates the world record and when the teaser is posted in social websites within 24 h millions of people watched and posted their views about the teaser. Not only Telugu speaking people but also many non-telugu speakers started watching Telugu film. So, collecting and reviewing Telugu movies reviews will be an appreciating work in the field of Telugu language processing. Web plays a censorious role in gathering the reviews.

The gap removed between the movie producer and audiences by presenting understandable opinion of viewer choices. In twirl, the prerequisite of news contents can be adapted in view of making, circulation, display, and allied advertising aids. Expecting viewer's choices and attitude, not only movie makers endorse audience contents more successfully. The comments of movie audiences give better assistance in decision on watching a particular film.

V. Kakulapati (✉) · S. Mahender Reddy
Sreenidhi Institute of Science and Technology, Yamnampet, Ghatkesar,
Hyderabad 501301, Telangana, India
e-mail: vldms@yahoo.com

S. Mahender Reddy
e-mail: mahendersheri@gmail.com

© Springer Nature Singapore Pte Ltd. 2019
S. Tiwari et al. (eds.), *Smart Innovations in Communication and Computational Sciences*, Advances in Intelligent Systems and Computing 851,
https://doi.org/10.1007/978-981-13-2414-7_30

User-Produced Content: One of the popular online site where users share their views is Twitter which leads to an expanding to increase in the user-generated content. Users are posted and shared information in the form of tweets. Through web user-related information is available in diverse forms are as follows

1. **Weblogs**—This is generating by the collection of user sessions, log lists, web links, and time of session, etc. Web log contains a log or diary of information, specific topics, or opinions.
2. **Online News**—It is catering to daily occurrences, actions, and other information from corner to corner of the world with a fast approaching in different languages.
3. **Conversation Groups**—Web pages relating to each user, manufactured goods, services, business, which providing information between the consumers/users with the suppliers/companies. Online users can share their opinion through different methods such as furnish feedback, acquiring information, posing the query, conveying messages, etc.
4. **Online Reviews**—Most popular industries like e-commerce and the entertainment industry are the largest booming industries across the web. Through the web, consumers/users share their opinions and feelings about the items and services which help to making optimal choices and decisions.

Facebook: Users on Facebook are around 850 million. Each day, there is exponential growth of uploading and people are viewing. These social networks are store, admittances, and user-generated data is also analyzed.

Twitter: Another popular social network is Twitter has more than 465 million user accounts and above 175 million tweets are posted for every day.

Google+: In Google, more than 90 million consumers and 675,000 consumers are added every day.

All these statistics give an idea about the rate of the increasing e web. By using this huge data generated with social networks, which provides massive business prospects to handle this data carefully and accurately.

Our objective of work is proposed to deal with challenges of Big Data related with storage of data and computational requirements [2, 3]. The major contributions offered in this work are as follows:

1. Twitter generated reviews on Telugu movies are collected for huge data processing on the Hadoop environment. These review data set contents are processing on the HDFS (Hadoop distributed file system);
2. The collected twitter review data facilitates in proficient and synchronized manner which is obtained by multiple-task queue.
3. Diversity of film details (i.e., actor, producer, director, writer, and story elements) are generated from heterogeneous features to characterize unprocessed data related to film reviews and audience opinions'.
4. To increase the flexibility and efficiency of Big Data applications, an enhanced Apriori algorithm based on MapReduce is utilized.

Our proposed framework can analyze enlighten movie features like actor, story plot and location, etc. Additionally to modify or confining stories for particular markets, locales, and target audience.

2 Background and Related Work

Here, a concise review of a study about UPC-based analysis. First, analyzing opinion analysis in movie reviews based on data collected from audience comments. Subsequently, confer the prediction system based on UPC for Tollywood rating analysis. Additionally, UPC analysis by recommendation systems with presented processing frameworks and capability for mining UPC is offered.

Creating feature vector [3] by using every character presence, the number of times each character presents, negation word as features. In this work, they also utilize unigram and bigram approaches effectively to formulate attribute vector in opinion analysis. Naive Bayes [4] strongly demonstrates dependent attributes for a certain problem. In Bayesian algorithm [5] model, various proficient approaches are utilized for choosing attribute, the calculation of weight and categorization. In other model, [6] designed a two-step analysis method for generating classification of tweets by using usual opinion analysis. In step one, tweets are categorized into subjective and objective and in the second step, individual tweets are classified as optimistic and pessimistic tweets. Another method [7] as accent based expression grouping and with this method regulates noisy tweets.

In Telugu language, some words have the same accent with dissimilar significances. For eliminating this type of variance, there is a method where words having the same accent are grouped and allocated regular tokens. Wu and Ren [8] proposed the weight probability to study the opinion tweets. If we found @username in the tweet, it manipulates action and facilitates to weight probability. Automatic tweets [9] expanded a method for opinion analysis by generating twitter data sets. While creating feature vector, they can utilize emoticons as a feature. They utilized a Naïve Bayesian classifier to analyze the opinion analysis. Researchers are studying to discover the user reviews regarding films, news bulletin, etc. from the tweets. In [10], the authors described the information from further open access data like movie databases.

2.1 Sentiment Analysis

In the present era, there is a huge popularity of online data, micro-blogging and website analytics which produce a large extent of information. Users are sharing their thoughts and opinions through the Internet. The web is connecting with each other via the Internet through the blog post, online discussion groups, and so on. The movie audience verifies the movie reviews or ratings of the movies proceeding to watch that movie. In south Indian films, these reviews are mainly based on heroism. The scope of data is difficult for an ordinary people to investigate with an assist of the naive technique [11].

The main concept of opinion analysis is the recognition and categorization of emotions or sentiments of every tweet. The sentiment analysis is divided into two types: 1. Aspect-based sentiment analysis. 2. Objectivity-based opinion analysis. In our approach, the opinion related to movie reviews falls in the category of characteristic-based opinion analysis. Movie reviews opinion analysis does the study of the tweets which are associated with opinions like hit, super hit, blockbuster, etc.

Opinions are measured as the expression of person feelings and sentiments. Analyzing and predicting the unknown data stored in the content by using different computational methods. The unseen data offers valuable approaching about people's targets, experience, and likeliness. Opinion mining categorizes the text into subjective and objective nature. First one is that the content contains opinion contented whereas the second one point towards that the tweet is exclusive of feeling content.

For example

- This movie by Bala Krishna and Nayantara is superb which is called subjective.

(This sentence has an opinion which is about the movie and the writer's feelings about same "superb" and then it is called subjective)

- This movie stars Bala Krishna and Trisha which is called objective.

(This sentence contains data rather than an attitude or an analysis of some being and therefore its intention)

Based on the user's opinion expressed, the subjective can be categories into three types.

1. **Positive**—I like Balayya movies.
2. **Negative**—the movie is poorly for actor fans.
3. **Neutral**—I am feeling hungry. (This is personal and does not contain any positive or negative, so it is neutral.)

2.2 Prediction System

A set of data [12, 13], which we can able to know the entries and data popularity. In this data need to predict the hidden data in the set, as precisely as possible [14]. The previous study demonstrates that the general movie user rating could be successfully predicted based on some characteristics such as budget, directors, and cast [15]. Another work observe the potential of predicting movie ratings by utilizing the machine learning and regression, such as random forests and support vector machines [16].

2.3 Recommender System

Here, systems collect data about the choice of users on different items (e.g., movies, shopping) by implicitly or explicitly [17–22]. The implicit method observed the user's behavior, for the instance, movie watching, procured the review, etc. Gathering the movie rating posted by users and then recommends movies to the like-minded public with similar experiences and curiosity in history [23].

Box office revenue is previously calculated based on the prediction of movie success in the form of reviews and ratings posting on different social networks, as such predictions would facilitate to success popularity in the preproduction periods of movie making. Movie prediction is also useful for better recommendations for prerelease movies, or as an assisting factor in a user-focused recommendation system. Such systems might found within services for the media streaming, or other similar services aimed more specifically in the direction of media discovery and recommendation.

3 Framework

With this framework have more benefits which improve the audience count.

In contrast with predictable methods, Telugu movie reviews facilitate more predictable data analysis (Figs. 1 and 2);

1. Telugu movie reviews allow to implement multiple things to analyze or explain features in different perspectives.

Fig. 1 Identifying the frequent movie attributes from twitter data

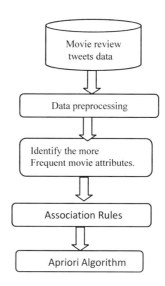

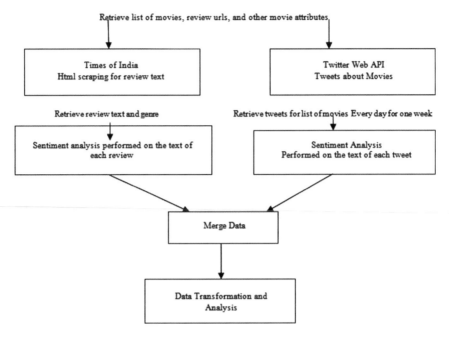

Fig. 2 Analysis of review dataset collecting from different resources

2. Telugu movie reviews produce logical results using association methods which are easily understandable and quick decision making.

3.1 Movie Database

Telugu film industry is one of the biggest hubs of the great films in India. Tollywood also produces movies more than any other industry in India. Over the past few decades, Telugu film industry has given us some of the most magnificent movies that India has ever seen. Moreover, Telugu film industry got reorganization by the Guinness World Record for its more movies production ability in the worldwide. If you are looking for best Telugu movies to watch over the weekend, we are here to help you. One more astonishing fact is that Telugu film industry produces more movies than Bollywood.

We collect movie information, such as the title, actors, director, release date, ratings, etc., and form Twitter data by utilizing Twitter API. User reviews contain many irrelevant to read between the lines, collecting for logical tweets that may include ratings prepared for mining. For this purpose, we are collecting tweets IMDb, times of India and BMS, etc. which includes more than 20,000 tweets and accumulated as

Table 1 Association attributes of movie, review and user

Movie attribute	Comment attribute	User attribute
Directors	Posted Time	User ID
Movie ID	Comment ID	Registration Time
Writers	Movie ID	Location
Actors	User ID	Number of followers
Genres	Rating	
Year	Comment	

Table 2 Collecting data from movie review

Movie	Cast	Genres	Reviewer	Rating
Jailavakusa	NTR	drama,action	IMDb	3.3
Arjunreddy	Vijaydevarakonda	drama,action	TOI	4
MCA	Nani	comedy	123telugu	3
MCA	Nani	comedy	BMS	3.5
Spyder	Mahesh babu	crime.thriller	IMDb	3.6
Spyder	Mahesh babu	crime.thriller	TOI	2.5
Dhruva	Ramcharan	thriller,action	IMDb	4.25
Bahubali2	Prabhas	drama,fantasy	BMS	4.5
Bahubali2	Prabhas	drama,fantasy	Hindustan Times	3

a data set. In this dataset, we analyzed 1200 tweets (600 positive tweets, 600 negative tweets, 600 neutral tweets) (Table 1).

3.2 Preprocessing of Tweets

All tweets in our data set are changed to similar case. So that all words of each tweet are similar case. After that removing the URLS and substituted with text. In tweets, we replaced "@username" and punctuation containing in the tweets and additional white spaces, #hashtag (Table 2).

3.3 User Sentiment Analysis

Movie users Opinion mining including optimistic or pessimistic towards the movie can be indirect from movie review for the significance of their emotion. This emotion depends on a reviewer's mood. For instance, super hit, Blockbuster, fun, family entertainer, etc. while writing the review. The users are utilizing pre-annotated corpus

Table 3 Movie reviews sentiment analysis

Negative	Positive	Sentiment
132	348	216

Fig. 3 Figure for Nani movie review association rules

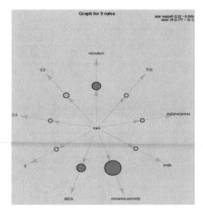

of most common words, Telugu and generating unstructured information which will establish the polarity of review and ensure for contradictions.

In our dataset, we classified these opinions into positive and negative (Table 3). For example, Review: Touch Chesi Chudu—Routine Touch.

4 Experiment Evaluation

In this phase, tweets from the dataset are able to analyze the movie ratings to a particular movie. Based on these we can develop single actor popularity. We construct association rules for the single actor and then analyzing the entire dataset for different actors and their movie reviews.

4.1 Generating the Frequent Review

In our evaluation, only strong association rules are generated. Frequent reviews satisfy the minimum support, association rules satisfy the confidence threshold (Fig. 3 and Table 4).

An improved Apriori algorithm is applying for identifying frequent reviews from the data set. Enhanced Apriori is applying to enumerate movie attributes, such as the rating of movies (Ri) and popularity (Pi), respectively (Fig. 4 and Table 5).

Table 4 Actor and movie association rule with support and confidence

S. No.	Actor association		Support	Confidence	Count
[1]	{ninnukori}	{nani}	0.04950495	1	5
[2]	{MCA}	{nani}	0.04950495	1	5
[3]	{romance, comedy}	{nani}	0.09900990	1	10
[4]	{BMS, MCA}	{nani}	0.00990099	1	1
[5]	{3.5, BMS}	{nani}	0.00990099	1	1
[6]	{BMS, romance, comedy}	{nani}	0.00990099	1	1

Fig. 4 Association rules with support and confidence for entire dataset

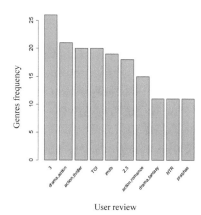

Table 5 Association rules and finding support and confidence for entire dataset

S. No.	lns	rns	Support	Confidence	Lift	Count
[1]	{CAST}	{GENRE}	0.00990099	1	101	1
[2]	{GENRE}	{CAST}	0.00990099	1	101	1
[3]	{CAST}	{MOVIE}	0.00990099	1	101	1
[4]	{MOVIE}	{CAST}	0.00990099	1	101	1
[5]	{CAST}	{RATING}	0.00990099	1	101	1
[6]	{RATING}	{CAST}	0.00990099	1	101	1

```
Categorize the more frequent movie attributes.

Read D = {Ij , Tj}

For every transaction in D if the
transaction Ti incorporates Ij then
Tij=1

else

Tij =0

end if

end for

end

Generate the set of frequent 1 movie
attribute L1 For every feature Ij of
Matrix

If sum Ij >=MS then L1 = Ij (where MS= min_sup)

 Else

Remove Ij from matrix.

End if

k:=2; //k represents the pass number//

While (Lk-1 ≠ ø) do

 begin

 for every k movie attribute calculate SC(Sup_ count)

Support count for k movie attribute s (Ii,Ij……….,Ik

SC(Ii,Ij,Ik )= RC1x(T1i.T1j .T1k)+……+ RCnx(Tni.Tnj..Tnk.)

 if SC >= MS then

 Lk := All candidates in Ck with minimum
support; end if

For every movie attribute in LK remove infrequent Movie attribute
feature from dataset

Reorganize dataset

 end for

 k := k + 1;

end for

end

end
```

5   Movie Recommender System

Based on user reviews posted in the social network, it will give popularity for a particular movie. Movie reviews make a conventional taste profile, and then recommend movies for users to watch. User likes to watch worth seeing before spending their earnings on a particular movie, online reviews gives number of options to understand whether movie is worth seeing before watch. The recommendation of particular movies depends on movie ratings to users provide relevant information to the like-minded people. These reviews and ratings increase the popularity of the movie.

6   Conclusions

In this work, we discuss a user reviews and their views in Twitter offers potential and proficient way for the developing of huge corpora that can provide useful information to movie makers. Analysis on social networks producing user sharing and posting data is still in its pettiness and therefore there is more research in this field must continue. In this work, we analyzed the small content of social networks with the application of big data. For future research directions and improvements several possibilities are exist.

7   Future Enhancement

Further on it might also be relevant to evaluate whether or not the prediction performance and Twitter reviews are generalizable over multiple datasets with multiple topics. Applying different machine learning techniques for both larger and smaller datasets in order to drawn more general conclusions, as the current experiment setup only incorporated a single though well-established dataset. Likewise, it might also be of interest to evaluate whether or not the same is applicable to box office predictions, to further widen the scope of the study.

References

1. Yang, J., Yecies, B.: Mining Chinese social media UGC: a big data framework for analyzing Douban movie reviews. J. Big Data (2016) https://doi.org/10.1186/s40537-015-0037-9
2. Zang, W., Zhang, P., Zhou, C., Guo, L.: Comparative study between incremental and ensemble learning on data streams: case study. J. Big Data (2014)
3. Liu, X., Wang, X., Matwin, S., Nathalie, J.: Meta-mapreduce for scalable data mining. J. Big Data **2**(1), 14 (2015)

4. Domingos, P., Pazzani, M.: On the optimality of the simple bayesian classifier under zero-one loss. Mach. Learn. **29**(2–3), 103130 (1997)
5. Niu, Z., Yin, Z., Kong, X.: Sentiment classification for microblog by machine learning. In: 2012 Fourth International Conference on Computational and Information Sciences (ICCIS), vol. 286289, pp. 286–289. IEEE (2012)
6. Barbosa, L., Feng, J.: Robust sentiment detection on twitter from biased and noisy data. In: 23rd International Conference on Computational Linguistics: Posters, vol. 3644 (2010)
7. Celikyilmaz, A., Hakkani-Tur, D., Feng, J.: Probabilistic Model-based sentiment analysis of twitter messages. In: Spoken Language Technology Workshop (SLT), 2010 IEEE, vol. 7984. IEEE (2010)
8. Wu, Y., Ren, F.: Learning sentimental influence in twitter. In: 2011 International Conference on Future Computer Sciences and Application (ICFCSA), vol. 119122. IEEE (2011)
9. Pak, A., Paroubek, P.: Twitter as a corpus for sentiment analysis and opinion mining. In: Proceedings of LREC, vol. 2010 (2010)
10. Peddinti, V., Chintalapoodi, P., Kiran, V.M.: Domain adaptation in sentiment analysis of twitter. In: Analyzing Microtext Workshop. AAAI (2011)
11. Amolik, A., et al.: Twitter sentiment analysis of movie reviews using machine learning techniques. Int. J. Eng. Technol (IJET), e-ISSN: 0975-4024
12. GroupLens is a Research Lab in the Department of Computer Science and Engineering at the University Minnesota
13. http://grouplens.org/datasets/movielens/latest/
14. Paterek, A.: Predicting movie ratings and recommender systems—a monograph. http://arek-paterek.com/book
15. Asad, K.I., Ahmed, T. Rahman, M.S.: Movie popularity classification based on inherent movie attributes using C4.5, PART and correlation coefficient. In: International Conference on Informatics, Electronics and Vision (ICIEV), (2012)
16. Persson, K.: Predicting movie ratings a comparative study on random forests and support vector machines. http://www.diva-portal.org/…/FULLTEXT01.pdf
17. Ruan, Y., et al.: An Integrated Recommender Algorithm for Rating Prediction. arXiv:199/abs/1608.02021.pdf
18. Lu, J., Wu, D., Mao, M., Wang, W., Zhang, G.: Recommender system application developments: a survey. Decis. Support Syst. **74**, 12–32 (2015)
19. Bobadilla, J., Ortega, F., Hernando, A., Gutiérrez, A.: Recommender systems survey. Knowl. Based Syst. **46**, 109–32 (2013)
20. Chen, L., Chen, G., Wang, F.: Recommender systems based on user reviews: the state of the art. User Model. User-Adap. Inter. **25**, 99–154 (2015)
21. Ja, K., Riedl, J.: Recommender systems: from algorithms to user experience. User Model. User-Adap. Inter. **22**, 101–23 (2012)
22. Katarya, R., Verma, O.P.: Recent developments in affective recommender systems. Phys. Stat. Mech. Appl. **461**, 182–190 (2016)
23. Katarya, R., et al.: An effective collaborative movie recommender system with cuckoo search. Egypt. Inf. J. (2016)

Cloud-Based E-Learning: Using Cloud Computing Platform for an Effective E-Learning

Shams Tabrez Siddiqui, Shadab Alam, Zaki Ahmad Khan and Ashok Gupta

1 Introduction

In the recent years, cloud computing emerged as an advanced technology that accelerated the innovations for computing industry. Cloud computing changed the way; the applications are developed and accessed on Internet. This technology provides the services through Internet. Cloud computing is a computing model of networks of network, and the main task is to corroborate users to use the hardware and software resources on the demand basis by paying money as per the usage. Clouds can be joining up of the resources either physical or virtualized on a centralized propagated data centers. A cloud consists of various types of workloads like batch style backend jobs and client facing applications [1]. The maintenance of IT infrastructure and management becomes less due to devise of cloud computing with the convenience of economy and scalability; it assists enterprise without any oppression. Cloud environment requires less software and hardware infrastructure [2].

E-learning is an Internet-based learning system. Due to the emergence and benefits of cloud computing, the use of this platform in education especially for e-learning mode gains attention by many software developers and vendors. E-learning has a significant impact on teaching and learning environment. E-learning embraces numer-

S. T. Siddiqui (✉) · S. Alam · A. Gupta
Department of Computer Science, Jazan University, Jizan, Saudi Arabia
e-mail: stabrezsiddiqui@gmail.com

S. Alam
e-mail: s4shadab@gmail.com

A. Gupta
e-mail: kgupta.ashok@gmail.com

Z. A. Khan
Nanjing Agricultural University, Nanjing-Jiangsu, China
e-mail: khanzaki05@gmail.com

© Springer Nature Singapore Pte Ltd. 2019
S. Tiwari et al. (eds.), *Smart Innovations in Communication and Computational Sciences*, Advances in Intelligent Systems and Computing 851,
https://doi.org/10.1007/978-981-13-2414-7_31

ous media activities that deliver text, images, audio, and animation to the student, getting notes and classes by the instructors or teachers. E-learning system usually requires many hardware and software resources to establish system with healthy investments. Education and learning are very crucial part of day in today's life. No person is able to survive a better life properly without education [2, 3]. There are several formats and techniques for getting knowledge. The most providential and precious modus operandi for education is e-learning. Generally, it entrusts to deliberate the use of networked information and communication technology (ICT) in teaching, training, and learning [4, 5]. E-learning components comprise many formats, online learners community, and content experts and developers. So far pacing the e-learning system to enhance the latest multimedia and communication technology, we need cloud computing, which seems to be the best solution. Now several questions arise in the vein of cloud computing [6]. Cloud computing is an excellent alternative for the academic organizations and institutions which are having less budget for hosting and operating their online learning systems.

2 Cloud Computing

Cloud computing can be defined as "*a new style of computing in which dynamically scalable and often virtualized resources are provided as a service over the Internet.*". The cloud can be seen as a collection of large groups of computers interconnected to each other [7]. These interconnected computers can be personal computers or network servers inside public or private organizations. While using cloud computing platform, you do not need over expense on hardware or software for building or maintaining IT infrastructure for individuals [4]. The resources of an IT system offered as services to the users through Internet [8]. To utilize the computing resources like application software, you have to admit the facility of third-party organization, via the Internet. However, cloud-based learning systems require high speed and reliable Internet access. Assorted service providers, like Google, Amazon, Microsoft, Yahoo, etc., are there to support educational systems [9, 10]. Currently, from the cloud service providers, the users of cloud computing expect the following types of services;

Anything-as-a-Service (XAAS): Collective term refers to the delivery of anything anywhere as a service.
Infrastructure-as-a-Service (IaaS): IaaS is at the foundational level of cloud computing and it provides computing infrastructure in the vein of operating systems, networks, and virtual machines, e.g., Google Compute Engine, Amazon EC2, and Rackspace.
Platform-as-a-Service (PaaS): In the cloud computing, PaaS is an established model for running applications without any annoy of maintaining the hardware as well as software infrastructure at the user side, e.g., Google App Engine, Windows Azure, and Force.com.

Software-as-a-Service (SaaS): These application software facilities are provided to the users on demand. Users need not worry about its installation, setup, and running of the application because service provider will take care of it, e.g., Google Apps, WhatsApp, and Microsoft Office 365.

Storage-as-a-Service (SaaS): In cloud computing, most of the cloud service providers include storage so that applications can store structured and nonstructured data in their cloud, e.g., Mozy, Google Drive, and Dropbox.

Network-as-a-Service (NaaS): In the cloud computing, this delivers the network services virtually over the Internet to the users or organizations on the basis of pay-per-use or on a monthly subscription.

Depending upon the requirements, the users can choose one or more services. E-learning services with cloud-based platform minimize the setup cost, and easier to maintain as well as it offers innumerable benefits to the end users in terms of compatibility [11, 12]. But the issues surrounding the security of the cloud are unclear.

3 Technological Challenges in Cloud Computing

Cloud computing has appeared to be an extremely compelling worldview as indicated by its features, such as on-demand self-service where the clients can easily facilitate without any interaction; a deliberate and measured service permitting to pay-per-use to the organizations; from heterogeneous networks, the clients can facilitate with the pooling of resources that can serve multiple clients; unlimited capabilities for the clients and having a elasticity nature [13–15].

Security and Privacy: Security from malicious, spyware, and from other potential threats to privacy.

Availability: Even with the use of redundant systems, availability means a guarantee for regular service (24 × 7).

Confidence: Audits and certification of security must be provided to increase the confidence of the users or clients.

Fault tolerance and recovery: Fault tolerance enables a system to continue operating properly even after the failure of its some components.

Scalability: Under changing demands of the users, providing an additional resource to them with an intelligent management.

Energy efficiency: Using microprocessors with a lower energy consumption to reduce the electric charge and make adaptable for use.

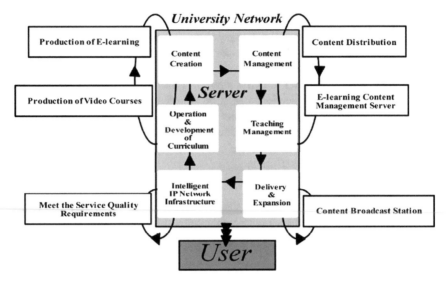

Fig. 1 Framework of e-learning [26]

4 Methods for Providing Cloud-Based E-Learning System

E-learning is cognitive science principles and excellent solution for learners having effective multimedia learning using electronic educational technology. E-learning is a fast, efficient procedure for learning and cost-effective as per the user's ease as well as organization's profitability. Figure 1 shows the framework of e-learning, whereas Fig. 2 shows how cloud-based e-learning works. E-learning has plentiful direct as well as positive effects on the learners [16, 17]:

- Learner's ratio is increased at national as well as global level.
- It simply assists the learners in learning specific program or subject.
- It improved the long-term retention of data and information.
- Traveling cost is not required.

E-learning demand is increasing day by day, and it is not going to vanish in coming years. Some e-learning trends of 2017 are discussed below [17].

4.1 E-Learning Trends

Mobile Learning Enhancement: In current years, due to the massive increase in the smartphone users, the quality of mobile learning is improving gradually. Every day, the users of smartphones are increasing and they want to facilitate themselves. The users of smartphone do not want to use laptops or desktop PC, to access various applications. As a result, mobile e-learning trend is increasing more rapidly. Mobile

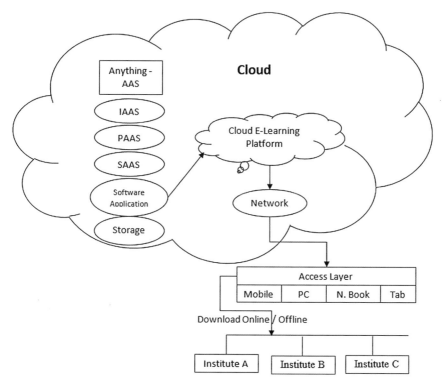

Fig. 2 Cloud-based e-learning

e-learning is attracting more and more users and can improve the knowledge of the learners anywhere anytime. Due to this, the organizations are paying a more attention toward this learning tool.

YouTube/Video-Based Learning: Nowadays, learning based on videos is one of the compelling and effective methods of learning. In YouTube or video-based learning, the academicians, researchers, and practitioners are sharing their recorded video lectures. The videos help students and employees to improve their understandings, training, and knowledge of specific course.

Forum-Based Learning: **Another proficient technique is forum**-based learning, in which the learners exchange their questions, ideas, and problems on a common platform. In response to questions, different answers or ideas are shared by an expert that helps to improve the skills of the learner and boost his confidence.

Social E-Learning: Social e-learning is a modern tool for effective e-learning and the latest trend of 2017–2018, for the learners. The learner is having a facility to participate and share his views and experiences, to discover and explore suitable learning. Learners find this tool quite rich and dependable for better understanding and to improve efficiency.

4.2 Benefits of Cloud Computing for E-Learning Solutions

As we know that many educational institutions do not have resources and infrastructure that are needed to run or install e-learning solution. Some of the versions of base applications of BlackBoard and MOODLE are cloud-oriented for e-Learning software [18]. Different educational organizations are widely using e-learning at several levels for continuous education, academic courses, company training, etc. Implementing e-learning systems with cloud computing has numerous advantages over other e-learning systems [19].

- Low cost.
- Instant software updates.
- Cloud-based e-learning platforms are safe.
- Increases employee productivity and performance.
- Allows for seamless collaboration in the distributed workforce.
- Improved document format compatibility.
- Internal IT support is not required.
- Improves employee retention.
- Cloud-based e-learning platforms are dependable.
- Benefits for students' point of view.

    (a) Take online course.
    (b) Send homework or assignments, projects,
    (c) Send feedback, and
    (d) Take exams.

- Benefits for teachers or trainers:

    (a) Deal with content management,
    (b) Communicate with students (forums),
    (c) Prepare tests,
    (d) Evaluate tests,
    (e) Assess students homework or assignments, projects, and
    (f) Send feedback.

5   Comparing Cloud-Based Authoring Tools

E-learning has exploded in recent years and the platforms of e-learning mainly used for interactivity of human with computers. A vast number of e-learning tools and platforms are there in the market vying for your attention. The advents of cloud-based authorizing tools are relatively effective and emerging trend in e-learning [20]. This emerging trend provides innovative ideas and thinking for multiscreen course editing and proofreading. While using any one of the feature-rich online learning platforms, anyone from anywhere having access to the internet and compatible device like desktop, laptop, or smartphone can create virtually any type of training program. Before using or sharing any of the existing tools from the market, it is better to choose and compare the cloud-based e-learning tool with the top existing tools of cloud-based learning. Some of the existing tools have been explored with its pros and cons.

5.1   Easy Generator

Easy generator is a complete cloud-based learning software that has a simple tool to create required or in-demand courses, without having coding expertise. On easy generator, define objectives and encourage the instructors to measure the learning success through a strong instructive framework. It does not provide management features to the students but they can see their performance and result by their name and email before ending the course [20, 21].

Pros

- User interface scripting and programing is not needed for creating an interactive course. It is extremely user-friendly.
- Ability to add more co-authors as per the requirement of the courses. Mobile-friendly responsive courses.
- For learning and training, this next generation authoring tool is an excellent example.
- Easily customizable questions and responses for correct and incorrect feedback.
- It is easy to commence without any experience or instructional or design expertise.
- Facilitate with comprehensive sharing options like website embedding, SCORM, HTML downloads, and private link sharing.

Cons

- Limited interactive elements; however, it is not tested but the "Academy" plan facilitates you to build and add templates.
- It becomes pricey when multiple authors entail to access the platform.

Price

- A free trial of easy generator is available with a freemium model of two rudimentary types.
- Cost is $39 per author per month but for advance feature $59 per month and for educational institutions 50% off.

5.2 Lectora Online

Recently, Lectora online cloud-based authoring tool released version two with new features and its interface is easy to get hang of [20, 22].

Pros

- 30 days free trial is available.
- Ability to create learning management system compatible assignments.
- Ability to add Go Animate professional animated videos directly into the Online Media Library of Lectora.
- Easily download the resources without subscription like SCORM, HTML, and Tin Can.

Cons

- HTML is not mobile friendly and responsive.
- Privacy not provided.
- Each time when the course is published error checking is required.
- Especially for beginners most interactive elements can result in errors.

Price

- Lectora online price is $159 per month and $1,908 per year.

5.3 eCoach

eCoach is a new arrival authoring tools to the market. eCoach is a multi-device authoring tools which authorize anyone to create new online courses. Creating online courses easily and quickly using media as well as resources from around the web [20, 23].

Pros

- A free 30-day trial of is available.
- There are variety of templates, easy to work with.
- Using a course code to share resources via a private URL.
- Easy to learn user interface.

- Completely mobile-friendly learning resources.
- Great looking and consistent format for learners to work with.

Cons

- Templates are form-based, each time preview before adding new content.
- No SCORM exporting or HTML downloads.
- Some question templates are hard to find.

Price

- For teachers, the cost is $29.95 per month and for other users $49.95 per month.

5.4 Ruzuku

Ruzuku is an authoring tool but it is used more as a learning repository. Students' data cannot be tracked. It also allows to build a course by uploading information, files, activities, and then sharing with the learners [20, 24].

Pros

- A free 14-day trial is available.
- Simplest to use. Build courses can be easily sold.
- Ability to track student participation and performance.
- It simply allows to create a signup page including sign-in pages providing the feature of custom enrollment of the course.

Cons

- Courses are not engaging or interactive.
- Having limited activity options. Limited styling options with no previews.
- During testing, page loading time is not good.
- Blended learning, short courses, seminars, or workshops where the focus of the training and learning is not purely online.

Price

- For 25 enrollments $49 per month and for unlimited access with full features about $997 per year.

5.5 iSpring Learn LMS

iSpring Learn LMS cloud-based learning management tool. It is easy-peasy to use, and it provides the full control over learning solution generally to the organizations which arrange training and learning. Easily you can check your effectiveness of the training while monitoring and generating a detailed report. It is completely scalable and course authoring tool that powers the complete e-learning process including learners' evaluation progress. The LMS facilitate the learners to communicate with the trainer and give feedback for the instant communication [25].

Pros

- A 30-day trial with full functions, with no limitations and customization.
- Generate a powerful set of a detailed report of the activities and performance of the learner. Provisions of third-party SCORM courses.
- Advanced gratified security apparatuses.
- The mobile apps are available for Android tablets and iPad.
- The mobile app permits learners to save the obligatory courses as per this need and complete offline easily without any time boundation.
- An unlimited number of both public and private views.
- There is no limit for the courses, the trainer creates, and upload as many as he wants.

Cons

- iSpring is PC-compatible desktop authoring tool.
- If there are many courses, you cannot switch from tiles view, you need to scroll a lot.
- Third-party SCORM quizzes save overall score; it does not save detailed answers of individuals.

Price

- The Starter Plan is $97.00 and the Pro Plan is $197.00 per month for up to 50 and 250 users.
- For 1,000 users, Enterprise Plan is of $579.00 per month.

6 Conclusion

This paper discusses top five e-learning tools that significantly going to escalate, as the number of learners is going to increase in the coming years. The proportion of e-learners is growing at rapid pace with each passing day. The institutes and organizations are exploring and emphasizing more on e-learning systems to enhance the learning skills of employees/students. Different organizations and institutes are promoting the cloud-based e-learning for the authentic and reliable learning experience. From the growing e-learning solutions, all are taking benefit and advantages in several ways and exploring the knowledge apart from learning. With the help of the cloud computing for e-learning, all the essential and valuable data uploaded by numerous institutions/organization will be easily shared by the users of that network or cloud. Cloud-based e-learning systems are must faster, cost-effective, and efficient than traditional e-learning systems. Users do not need high configuration hardware for cloud-based e-learning; he does not have to install any software on the system only he needs an Internet connection to access or share the data and hardware. None of the cloud-based tools is perfect.

In this paper, we discussed few of the existing tools with their pros and cons, in which some platforms are too complex. Easy generator is great substitutions for the learners looking for a great course with a minimum of fuss. Ruzuku is simple authoring tool that allows embedding or link sharing package. In future work, we will perform an evaluation of existing cloud-based e-learning tools and proposed the cloud-based architecture for e-learning that overcomes the shortcomings of existing cloud-based learning tools.

References

1. Padmaja, K., Seshadri, R.: A review on cloud computing technologies and security issues. Indian J. Sci. Technol. **9**(45) (2016)
2. Alshwaier, A., Youssef, A., Emam, A.: A new trend for e-learning in ksa using educational clouds. Adv. Comput. Int. J. (ACIJ) **3**(1) (2012)
3. Pireva K., Kefalas P., Dranidis D., Hatziapostolou T., Cowling A.: Cloud e-learning: a new challenge for multi-agent systems. In: 8th International KES Conference on Agent and Multi-Agent Systems: Technologies and Applications, vol. 296. Springer, Cham (2014)
4. Jyoti, U.B., Ahmed, M.: E-learning using cloud computing. Int. J. Sci. Mod. Eng. (IJISME) **1**(2) (2013)
5. El Mhouti, A., Erradi, M., Nasseh, A.: Using cloud computing services in e-learning process: Benefits and challenges. Educ. Inf. Technol. **23**(2), 893–909 (2018)
6. Roveicy, M.R.R., Bidgoli, A.M.: Migrating from conventional e-learning to cloud-based e-learning: a case study of armangarayan co. In: Computer Science On-line Conference, pp. 62–67. Springer, Cham (2017)
7. Sampson, D.G., Isaias, P., Ifenthaler, D., Spector, J.M. (Eds).: Ubiquitous and Mobile Learning in the Digital Age. Springer Science and Business Media (2012)
8. Sosinksy, B.: Cloud Computing Bible. Wiley (2011)
9. Bibi, G., Ahmed, I.S.: A Comprehensive survey on e-learning system in cloud computing environment. Eng. Sci. Technol. Int. Res. J. **1**(1) (2017)
10. Alam, A., Ansari, A.M., Barga, M.M., Albhaishi, A.: Is mobile cloud computing efficient for e-learning?. In: Proceedings of the Second International Conference on Computer and Communication Technologies, pp. 109–123. Springer, New Delhi (2016)
11. Babu, S.R., Kulkarni, K.G., Sekaran, K.C.: A generic agent based cloud computing architecture for e-learning. In: ICT and Critical Infrastructure: Proceedings of the 48th Annual Convention of Computer Society of India, vol. I, pp. 523–533. Springer, Cham (2014)
12. Ahmad, M.O., Khan, R.Z.: The cloud computing: a systematic review. Int. J. Innovative Res. Comput. Commun. Eng. (IJIRCCE) **3**(5) (2015)
13. Fernandez, A., Peralta, D., Herrera, F., Benitez, J.M.: An overview of e-learning in cloud computing. In: Workshop on Learning Technology for Education in Cloud (LTEC'12), pp. 35–46. Springer, Berlin, Heidelberg (2012)
14. Hwang, K., Dongarra, J., Fox, G.C.: Distributed and Cloud Computing: From Parallel Processing to the Internet of Things. Morgan Kaufmann (2013)
15. Shuaib, M., Samad, A. Siddiqui, S.T.: Multi-layer security analysis of hybrid cloud. In: 6th International Conference on System Modeling and Advancement in Research Trends, SMART-2017 (2017). IEEE, Moradabad, pp. 526–531
16. Paul, M., Das, A.: SLA based e-learning service provisioning in cloud. Information Systems Design and Intelligent Applications, pp. 49–57. Springer, New Delhi (2016)

17. https://solutiondots.com/blog/top-4-e-learning-trends-2017.html. Accessed 03 Mar 2018
18. Subramanian, P., Zainuddin, N., Alatawi, S., Javabdeh, T., Hussin, A.: A study of Comparison between moodle and blackboard based on case studies for better LMS. J. Inf. Syst. Res. Innov. (2014) ISSN 2289–1358
19. Alavi, E., Mohan, M.C.: An e-learning system architecture based on new business paradigm using cloud computing. Int. J. Eng. Sci. Res. Technol. **2**(10), 2990–2999 (2013)
20. https://elearningindustry.com/7-cloud-based-authoring-tools-compared. Accessed 24 Mar 2018
21. https://www.easygenerator.com/. Accessed 25 Feb 2018
22. http://lectoraonline.com/app/BN0057/product.html. Accessed 06 Mar 2018
23. https://ecoach.com/. Accessed 25 Feb 2018
24. https://www.ruzuku.com/?utm_campaign=elearningindustry.com&utm_source=%2F7-cloud-based-authoring-tools-compared&utm_medium=link. Accessed 14 Mar 2018
25. https://www.ispringsolutions.com/ispring-learn/pricing.html. Accessed 20 Mar 2018
26. Riahi, G.: E-learning systems based on cloud computing: a review. Procedia Comput. Sci. **62**, 352–359 (2015)

Deadline-Aware Scheduling for Scientific Workflows in IaaS Cloud

Mainak Adhikari and Tarachand Amgoth

1 Introduction

Cloud computing is a buzzword in a distributed environment and has gained popularity due to the dynamic nature and the efficient resource utilization [1, 2]. The cloud providers provide various types of services over a network such as IaaS, PaaS, SaaS, DaaS, IdaaS, and so on. Among them, the IaaS cloud offers a large scale of computing resources to execute the user submitted tasks in the form of the virtual machine (VM) instances and virtual local area networks which established the logical connections between them [3, 4]. An individual IaaS provider consists of a large number of cloud data centers that host thousands of servers with a different set of VM instances. The cloud providers offer flexibility and elasticity to the user for varying their requirements for different computing resources [5]. This gives a confidence to the customers that the system provides a better QoS in the form of service-level agreement [6]. One of the objectives of the IaaS providers is to increase their revenues by providing QoS to the users and minimizing the SLA violations. More users can be accommodated if the system resources are perfectly utilized [7, 8].

The cloud data center receives a diverse set of applications as a form of a workflow from the users. A workflow consists of a set of interrelated tasks which is represented as a form of a direct acyclic graph. The workflows are commonly used to represent various scientific problems in the field of bioinformatics, physics, and astronomy [9]. Each workflow requires a diverse set of resources to execute the tasks in order to meet various QoS constraints in a distributed environment. There are two main stages to execute a workflow application such as resource provisioning and

M. Adhikari · T. Amgoth (✉)
Department of Computer Science & Engineering, Indian Institute
of Technology (Indian School of Mines), Dhanbad, India
e-mail: tarachand.ism@gmail.com

M. Adhikari
e-mail: mainak.ism@gmail.com

© Springer Nature Singapore Pte Ltd. 2019
S. Tiwari et al. (eds.), *Smart Innovations in Communication and Computational Sciences*, Advances in Intelligent Systems and Computing 851,
https://doi.org/10.1007/978-981-13-2414-7_32

scheduling. In resource provisioning phase, several resources are selected to run the tasks efficiently and deployed them on the server. In scheduling phase, an optimized schedule of the tasks is formed to minimize the makespan and schedule length of a workflow. Previous workflow scheduling algorithms are mostly focused on the scheduling phase while ignoring the resource provisioning phase of the workflow. However, with the emergence of cloud in a distributed environment, the particular challenges of workflow scheduling need to be verified and solved.

Workflow scheduling is one of the significant research issues in IaaS cloud. One of the important issues in workflow scheduling is to minimize the makespan of the workflow and meets the QoS constraint. Another important issue in the cloud environment is to select the suitable VM instances for each task of the workflow to minimize the overall makespan. The main aim of the work is to tackle the resource provisioning and task scheduling in an optimized way based on some dynamic approach. On the other hand, improper resource management in cloud data center may increase the makespan of the workflow and failed to meet the deadline.

In this paper, we develop a dynamic deadline-aware workflow scheduling strategy in IaaS cloud, namely, DAWS. The algorithm divides into two phases, resource provisioning and task scheduling. In the first phase, the algorithm devises an efficient strategy to assign a sub-deadline to each of the tasks of a workflow and selects the best-fit VM instances for each task to minimize the total execution time of the workflow. On the other hand, the task scheduling phase finds an optimal schedule of the tasks based on a rank-based method and deploys the tasks to the selected VM instances on the servers. This may reduce the makespan and schedule length of the workflow while meeting the QoS constraint of the workflow. The proposed algorithm is evaluated with the several state-of-the-art algorithms, such as IC-PCP [15], JIT [16], and PDC [19] workflow scheduling using various performance metrics. The main contributions of this work are discussed below:

(1) We extensively reviewed some of the existing workflow scheduling strategies along with their advantages and drawbacks.
(2) We develop an efficient strategy to assign a deadline for each task of a workflow and set the best-fit VM instances which can minimize the makespan of the application.
(3) We develop an effective scheduling strategy to find the optimal schedule of the tasks which may minimize the makespan of the workflow.
(4) Finally, the performance of the proposed DAWS algorithm is compared over the existing workflow scheduling strategies using various performance metrics.

The rest of the parts of the paper are organized as follows. The review results of the existing workflow scheduling strategies are discussed in Sect. 2. The data center model and the problem statement of the work are discussed in Sect. 3. The proposed works and the pseudocode are developed in Sect. 4. Performance analyses of the algorithm are discussed in Sect. 5, and the conclusion and future scope are discussed in Sect. 6.

2 Related Work

Nowadays, workflow scheduling is a hot research topic in a cloud environment. Various static and dynamic workflow scheduling strategies are widely studied in a cloud environment. Here, we extensibly review some of the important workflow scheduling strategies [10–19] published in recent times. They are as follows.

Malawski et al. proposed various types of static and dynamic workflow scheduling algorithms in cloud environment [10]. The algorithm deployed the tasks to the leased VM instances on the cloud data center and met various QoS constraints. However, the algorithm considered only homogeneous VM instance while ignoring the dynamicity of the clouds. Byun et al. developed a dynamic workflow scheduling strategy in cloud environment [11]. The algorithm estimated the number of resources for executing an application and deployed the tasks to those resources on a leased basis. The algorithm took the advantage of the flexibility of the resources on the server but failed to minimize the makespan of the workflow. Ghafarian et al. developed a partition-based workflow scheduling strategy in the cloud [12]. The algorithm partitioned the workflow into sub-workflows which reduced the dependencies among the tasks and found an optimal schedule of the tasks for minimizing the makespan. However, the algorithm failed to reduce the total execution cost of the workflow. Chen et al. developed a workflow scheduling strategy to reduce the schedule length of the tasks [13]. The algorithm produced an optimal schedule of the tasks and deployed them to the VM instances on the server. This may optimize the schedule length of the workflow but failed to minimize the execution cost and met the deadline for the workflow. Casas et al. developed a scheduling strategy for increasing the resource utilization on the cloud data center [14]. The algorithm splits the workflow into multiple subparts and found an optimal schedule for the tasks. The algorithm reduced the execution time and monitory cost of the workflow but failed to minimize the makespan while meeting the deadline. Abrishami et al. developed a scheduling strategy using partial critical path approach which minimized the makespan and the cost of the workflow [15]. This algorithm initiated an enormous number of same types of VM instance for executing the tasks but failed to utilize the resources efficiently on the server.

Sahni et al. developed a dynamic scheduling strategy—JIT to minimize the cost of the workflow and met the deadline [16]. The algorithm combined the pipeline tasks to a single task to reduce the communication cost of the workflow. The algorithm found an optimal schedule of the tasks but failed to utilize the resources efficiently on the servers. Yuan et al. developed a workflow scheduling strategy called deadline early tree algorithm [17]. The algorithm partitioned the tasks of the workflow into two categories. All tasks belonging to the critical path were deployed to the resources using dynamic programming while the other followed the tasks of the critical path. The algorithm met the deadline but failed to minimize the cost and the resource utilization on the servers. Bittencourt et al. developed a scheduling strategy for optimizing the cost of hybrid cloud environment [18]. Initially, the algorithm scheduled the tasks to the resources of the local data center; if the resources failed to meet the deadline then leased the additional resources from the remote data center of the cloud.

The algorithm minimized the makespan but failed to meet the deadline. Arabnejad et al. proposed a dynamic deadline-aware scheduling strategy, proportional deadline constrained, to optimize the makespan while meeting the QoS constraint [19]. The algorithm found an optimal schedule of the tasks but failed to utilize the resources efficiently on the servers.

From the review of the related works, it is clear that all the abovementioned algorithms schedule the tasks to the resources in a static or dynamic manner without concerning about the resource provisioning of the workflow. This may increase the makespan of the workflow without meeting the QoS constraint. To meet the above-mentioned challenges, we developed a dynamic deadline-aware workflow scheduling strategy in IaaS cloud. The algorithm calculates the sub-deadline of each task and deploys the tasks to the best-fit VM instances with an optimal schedule on the server. This may minimize the makespan of the workflow while meeting the deadline.

3 Preliminaries

Here, we first discuss the workflow model followed by the problem formulation of the proposed work.

3.1 Workflow Model

A workflow is represented by a DAG $w(T, E)$, where T is a set of interrelated tasks which are represented as $T = \{T_1, T_2, T_3, \ldots, T_k\}$ and E is the set of edges which shows the dependencies among the tasks. Each edge, $E_{ij}(T_i, T_j)$ represents a dependency constraint which indicates that the task T_j will start its execution only when all the predecessor tasks T_i completed their execution on the server. A join task and a non-join task are defined as the task of having single and multiple predecessor tasks simultaneously. Each edge consists of a weight, called transmission cost (TC_{ij}), between tasks T_i and T_j. The transmission cost represents the amount of time required to transmit the output data of tasks T_i from one server S_1 to the tasks T_j, and executes on another server S_2 only when the tasks are executed on two different servers in cloud data center. In a deadline-aware workflow scheduling strategy, each workflow has a deadline D_w and the workflow should complete the execution of the tasks within the deadline D_w. The tasks belonging to the same level of a workflow can be executed in parallel order on the VM instances. Please note that we consider all links in the inter-process networks are contention free and the tasks of a workflow can compute and communicate simultaneously.

Table 1 Four representative VM instances on Amazon EC2

Instance Type	CPU	Memory (GB)
m4.large	2 EC2 Units	8
m4.xlarge	4 EC2 Units	16
m4.2xlarge	8 EC2 Units	32
m4.4xlarge	16 EC2 Units	64

3.2 Problem Statement

The IaaS cloud providers receive the applications as a form of a workflow from the users and deploy the tasks to the best-fit VM instances on the servers. Let consider the cloud data center consists of n number of heterogeneous servers that are represented as $S = \{s_1, s_2, s_3, \ldots, s_n\}$ and each S hosts m numbers of VM instances according to the capacity that is represented as $VI = \{v_1, v_2, v_3, \ldots, v_m\}$. The major objective of our work is to select the suitable VM instances for the tasks based on the sub-deadline such that the makespan is minimized while meeting the deadline. Here, we consider a schedule with a set of four tuples, $SC = \{VT, M, TET, MS\}$, where VT represents the set of VM instances, task to VM instance mapping, total execution time, and makespan. The cloud data center consists of various types of VM instances, and each VM instance requires a set of resources to execute the tasks. VM instances are classified based on the resources consumed by them. As an example, Table 1 consists of four types of VM instances of Amazon EC2.

In task-VM mapping, each task of the workflow needs to be deployed to the best-fit VM instances on the server that minimized the execution time of the task. The execution time of a task depends on the size of the task and the resource capacity of the assigned VM instance which is defined as follows:

$$ET_{ki} = \frac{SI_k}{C_i}, \tag{1}$$

where SI_k and C_i represent the size of the task k (represents as million instructions) and the CPU capacity of VM instance i (represents as million instructions per second). The total execution time (TET_w) of the wth workflow is defined as the summation of the execution time of the tasks present in that workflow, which is formulated as

$$TET_w = \sum_{x=1}^{k} ET_{xi}. \tag{2}$$

The major objective of the proposed DAWS algorithm is to select the suitable VM instances for the tasks which will minimize the total execution time and makespan while meeting the deadline of the workflow. Finally, the mathematical formulation of the objective of DAWS algorithm is formulated as follows:

$$\text{Minimize } TET_w \text{ and } \textbf{Minimize } MS_w$$
$$\textbf{Subject to}$$
$$MS_w <= D_w$$

4 Deadline-Aware Workflow Scheduling (DAWS) Algorithm

The DAWS algorithm divides into two phases, namely, resource provisioning and task scheduling phases. They are discussed as follows.

4.1 Resource Provisioning

In this phase, the DAWS algorithm assigns a sub-deadline to each task, selects the best-fit VM instances to execute the tasks with minimum time, and meets the sub-deadline. The selection of the VM instances for the tasks depends on the sub-deadlines. Consider a workflow w with a set of tasks $T = (T_1, T_2, T_3, \ldots T_k)$ with their corresponding sizes $(SI_1, SI_2, SI_3, \ldots, SI_k)$ is waiting to execute. Here, we consider that the server can deploy m heterogeneous VM instances, i.e., $v_1, v_2, v_3, \ldots, v_m$, where the resource capacities of the VM instances are $C_{v_1} > C_{v_2} > C_{v_3} >, \ldots, > C_{vm}$. The algorithm estimates the expected time to compute (ETC) for each task over every VM instances. The ETC matrix of the workflow w over various VM instances is shown as follows:

ETC =		v_1	v_2	v_3	$\cdots$	v_m
	T_1	ETC_{11}	ETC_{12}	ETC_{13}	$\cdots$	ETC_{1m}
	T_2	ETC_{21}	ETC_{22}	ETC_{23}	$\cdots$	ETC_{2m}
	T_3	ETC_{31}	ETC_{32}	ETC_{33}	$\cdots$	ETC_{3m}
	$\vdots$	$\vdots$	$\vdots$	$\vdots$	$\vdots$	$\vdots$
	T_k	ETC_{k1}	ETC_{k2}	ETC_{k3}	$\cdots$	ETC_{km}

The sub-deadlines of the tasks depend on the deadline of the workflow (D_w) and the average execution time (AET_k) of the tasks over various VM instances. The average execution time (AET_k) of the task k over m distinct VM instances is defined as follows:

$$AET_k = \sum_{i=1}^{m} ET_{ki}. \tag{3}$$

The sub-deadline of a task is calculated as follows:

$$D_{kl} = D_w; when\, l = Leave\, node$$
$$D_k = Min(D_{kl-1} - (TC_{kl-1} + AET_k)); when\, l = Non - leave\, node. \quad (4)$$

The sub-deadlines of the tasks are calculated in a bottom-up fashion. The sub-deadlines of all leaf node tasks of a workflow are equal to the deadline for the selected workflow. On the other hand, the sub-deadlines of the non-leaf nodes depend on the sub-deadlines of the immediate successor tasks and the average execution time of the current task, which is formulated in Eq. (4). Next, the algorithm calculates the required minimum resource capacity VM instance for task k to complete the execution within the sub-deadline (D_k) based on the following equation:

$$C_k = \frac{SI_k}{D_k}. \quad (5)$$

Finally, the algorithm selects the best-fit VM instances for the tasks according to the required resource capacity of the VM instances and the availability of the resources on the servers. This may minimize the total execution time of the tasks in the cloud data center which is calculated using Eq. (1).

4.2 Task Scheduling

This phase assigns the priority of each task to be set with a rank value (r_k) which is based on the forward and backward pass computations of the workflow. The optimal schedule of the task list is generated by arranging the task in decreasing order according to their rank value. The algorithm finds the earliest start time (EST_k) and latest finish time (LFT_k) of each task for forwarding pass computation of the workflow. The earliest start time (EST_k) of task k is calculated using the following equations:

$$EST_{kl} = 0; when\, l = 0$$
$$EST_{kl} = Max(ET_{k-1l-1} + TC_{k-1l-1}); when\, l \neq 0. \quad (6)$$

Let CT is the completion time of the leaf node of the workflow and CT_k represents the completion time of a non-leaf node task k. Then, the latest finish time (LFT_k) of task k is calculated using the following equations:

$$LFT_{kl} = CT; when\, l = leaf\, Node$$
$$LFT_{kl} = Min(CT_{kl+1} - (ET_{k+1l+1} + TC_{kl})); when\, l = Non\, leaf\, Node. \quad (7)$$

The rank of the tasks is calculated based on the following equation:

$$r_k = (EST_{kl} + LFT_{kl}). \quad (8)$$

DAWS Algorithm:	
Input: Tasks of the workflows	
Output: Type of VM instances	
Resource provisioning:	
1:	**Begin**
2:	**For each** task T_k of workflow w
3:	**For each** type of VM instance
4:	Calculate the ETC matrix
5:	**End for**
6:	**End for**
7:	**For each** task T_k of workflow w
8:	Calculate the average execution time for each tasks using Eq. (3)
9:	Calculate the sub-deadline for each tasks using Eq. (4)
10:	**End for**
11:	**For each** task T_k of workflow w
12:	**For each** type of VM instance
13:	Calculate the minimum capacity VM instance each task of a workflow
14:	**End for**
15:	**End for**
Task Scheduling:	
16:	**For each** task T_k of workflow w
17:	**For each** type of VM instance
18:	Calculate the EST and EFT using Eq. (6) and Eq. (7) respectively.
19:	Calculate the rank of the task using Eq. (8)
20:	**End for**
21:	Generate an optimal schedule of the tasks using rank value
22:	Assign the tasks to the selected VM instances
23:	**End for**
24:	**End**

Fig. 1 The pseudocode of DAWS algorithm

The tasks are arranged in ascending order according to the rank value of the tasks and generate an optimal schedule for deploying the tasks to the VM instances. Finally, the tasks are deployed to the selected VM instances on the server. The pseudocode of the DAWS algorithm is shown in Fig. 1.

5   Performance Analysis

The proposed DAWS algorithm is evaluated over four well-known scientific work-flows, namely, Montage, Epigenomics, LIGO, and CyberShake which are described by Juve et al. [9]. The pictorial presentations of the workflows are shown in Fig. 2. Each of the workflows has a different structure based on their properties. The Montage workflow represents the astronomy application to create custom mosaics of the sky using the help of some images. The Epigenomics workflows are generated from the field of bioinformatics. LIGO workflows are applied in the field of Physics with a plan to detect gravitational waves. Finally, CyberShake workflow is characterized by the earthquake hazards using the help of synthetic seismograms. For experimental purpose, we consider that each of the workflows consists of 1500 number of tasks with

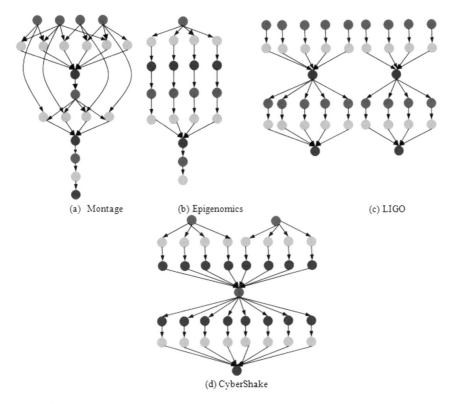

(a) Montage   (b) Epigenomics   (c) LIGO

(d) CyberShake

Fig. 2 Scientific workflows

different sizes. The workflows are generated using the workflow generator [20]. In our experiment, we use several heterogeneous VM instances with different resource usages which are shown in Table 1.

5.1 Comparison Results

Here, the proposed DAWS algorithm is compared with the various state-of-the-art algorithms such as IC-PCP [15], JIT [16], and PDC [19] over various performance metrics such as makespan evaluation, scheduled length ratio (SLR), throughput, reliability, and resource utilization.

Makespan Evaluation: Makespan of the workflow is the completion time of the leaf node tasks in that application. The value obtained for the makespan of each scientific workflow is shown in Fig. 3. The existing algorithms assigned the tasks to the VM instances without knowing the capacity of the resources which may increase the execution time as well as the makespan of the workflow. On the other hand,

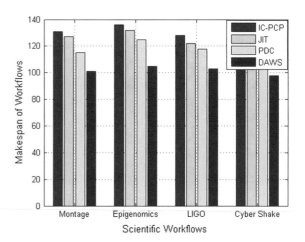

Fig. 3 Makespan evaluation

DAWS algorithm deploys the tasks to the best-fit VM instances which may execute the tasks as earliest as compared with the existing algorithms which may minimize the makespan of the workflow. The makespan of DAWS algorithm is better than IC-PCP algorithm by 20%, the JIT algorithm by 16%, and the PDC algorithm by 11%.

SLR Evaluation: The SLR of each task is calculated based on the ratio between the completion time of the task to the makespan of the workflow. The workflow scheduling algorithm that gives the higher SLR of a workflow is the best algorithm with respect to the performance. Average SLR values of the tasks are used for our experiment to show the prominence of our algorithm. The SLR value of a task is reflected while changing the makespan of the workflow. The value obtained for the average SLR of each scientific workflow is shown in Fig. 4. The existing algorithms deployed the tasks to the VM instances without generating a perfect schedule of the tasks. However, the DAWS algorithm finds an optimal task schedule based on a rank-based method that increased the average SLR value compared with the existing ones. The average SLR value of DAWS algorithm is better than IC-PCP algorithm by 18%, the JIT algorithm by 15%, and the PDC algorithm by 10%.

Throughput Evaluation: Throughput of a server defines as the number of tasks completed their execution within a time interval. The graphical representations of the experimental results are shown in Fig. 5. The throughput of the server depends on the execution time of the tasks. The existing approaches did not find the best-fit VM instances for the tasks to minimize the execution time. However, the DAWS algorithm selects the best-fit VM instances for the tasks based on their sub-deadlines which may reduce the execution time and increase the throughput of the workflow. The throughput of DAWS algorithm is better than IC-PCP algorithm by 17%, the JIT algorithm by 13%, and the PDC algorithm by 10%.

Reliability: The reliability of a scheduling algorithm depends on the number of tasks that should complete their execution within the deadline. To analyze the algorithm in terms of reliability, we plotted the percentage of the deadline for various

Fig. 4 Average SLR

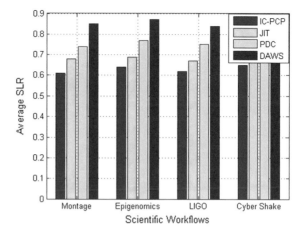

Fig. 5 Throughput evaluation

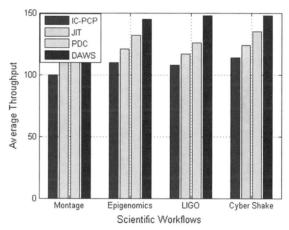

types of tasks set. The percentage represents the number of tasks of a workflow that meet the deadline. The graphical representations of the experimental results are shown in Fig. 6. The DAWS algorithm calculates the sub-deadline of the tasks of the workflow and deploys the tasks to the best-fit VM instances according to the requirements of the tasks. However, the existing approaches failed to meet the deadline of the tasks in all cases of the workflow. This may prove that the DAWS algorithm is more reliable than the existing approaches. The reliability of DAWS algorithm is better than IC-PCP algorithm by 22%, the JIT algorithm by 17%, and the PDC algorithm by 14%.

Resource Utilization: The resource utilization of the servers indicates the amount of time the resources are rescheduled to execute the tasks in cloud data center. Rescheduling of the resources may reduce the number of resources and the number of servers to execute the requested tasks. Higher utilization of the resources may execute a number of tasks in parallel on cloud data center. The existing scheduling

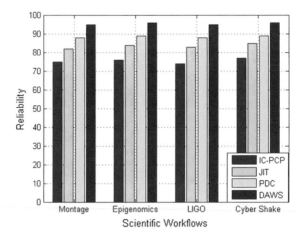

Fig. 6 Reliability of scheduling algorithms

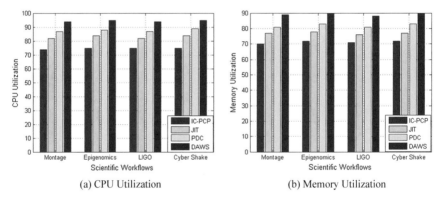

(a) CPU Utilization   (b) Memory Utilization

Fig. 7 Resource utilization

approaches tried to minimize the execution time of the tasks but failed to utilize the resources efficiently. On the other hand, the DAWS algorithm selects the best-fit VM instances for each task of the workflow according to the requirements of the tasks on the servers. This may increase the utilization of the resources on the servers. In our experimental purpose, we are considering two types of resources of each VM instance—CPU and memory—and analyze the algorithm in terms of resource utilization; we plotted the percentage graph over various types of tasks set which are shown in Fig. 7. The resource utilization of DAWS algorithm is better than IC-PCP algorithm by 17%, the JIT algorithm by 14%, and the PDC algorithm by 11%.

The proposed DAWS algorithm calculates the sub-deadline of the tasks based on the actual deadline of the workflow and deploys the tasks to the suitable VM instances on the server which will reduce the total execution time and makespan of the workflow while meeting the deadline. On the other hand, the algorithm also finds an optimal schedule of the tasks to minimize the makespan of the workflow. This may improve the reliability and resource utilization of the servers.

6 Conclusion

In this work, we have designed a dynamic deadline-aware workflow scheduling strategy for IaaS cloud. The algorithm divides into two phases, resource provisioning and task scheduling. First, DAWS algorithm calculates the sub-deadline of each task of the workflow and selects the best-fit VM instances for the tasks according to the requirements of the tasks and the availability of the server. In task scheduling phase, DAWS algorithm finds an optimal schedule for the tasks based on a rank-based policy and deploys the tasks to the selected VM instances. This may minimize the total execution time and the makespan of the workflow while meeting the deadline. The simulation results show the superior performance of the proposed algorithm over the existing state-of-the-art algorithms, such as IC-PCP, JIT, and PDC algorithm using four well-known scientific workflows. We will extend DAWS to load balancing of the servers with proper resource utilization based on some optimization techniques in the future research.

References

1. Singh, A., Juneja, D., Malhotra, M.: A novel agent-based autonomous and service composition framework for cost optimization of resource provisioning in cloud computing. J. King Saud Univ. Comput. Inf. Sci. **29**, 19–28 (2017)
2. Mell, P., Grance, T.: The NIST definition of cloud computing—recommendations of the National Institute of Standards and Technology (*Special Publication 800-145*). NIST, Gaithersburg (2011)
3. Buyya, R., Broberg, J., Goscinski, A.M. (eds.): Cloud Computing: Principles and Paradigms, vol. 87. Wiley Publication (2010)
4. Suresh, S., Sakthivel, S.: A novel performance constrained power management framework for cloud computing using an adaptive node scaling approach. Comput. Electr. Eng. **50**, 30–44 (2017)
5. Adhikari, M., Amgoth, T.: Heuristic-based load-balancing algorithm for IaaS cloud. Future Gener. Comput. Syst. **81**, 156–165 (2018)
6. Adhikari, M., Koley, S.: Cloud computing: a multi-workflow scheduling algorithm with dynamic reusability. Arab. J. Sci. Eng. **43**, 645–660 (2018)
7. Banerjee, S., Adhikari, M., Kar, S., Biswas, U.: Development and Analysis of a New Cloudlet Allocation Strategy for QoS Improvement in Cloud. Arab. J. Sci. Eng. **40**, 1409–1425 (2014)
8. Garg, S.K., Toosi, A.N., Gopalaiyengar, S.K., Buyya, R.: SLA-based virtual machine management for heterogeneous workloads in a cloud data center. J. Netw. Comput. Appl. **45**, 108–120 (2014)
9. Juve, G., Chervenak, A., Deelman, E., Bharathi, S., Mehta, G., Vahi, K.: Characterizing and profiling scientific workflows. Future Gener. Comput. Syst. **29**, 682–692 (2012)
10. Malawski, M., Juve, G., Deelman, E., Nabrzyski, J.: Cost-and deadline-constrained provisioning for scientific workflow ensembles in IaaS clouds. In: Proceeding of International Confer-

ence High-Performance Computing, Networking, Storage and Analysis (SC), vol. 22, pp. 1–6 (2012)

11. Byun, E.K., Kee, Y.S., Kim, J.S., Maeng, S.: Cost optimized provisioning of elastic resources for application workflows. Future Gener. Comput. Syst. **27**, 1011–1026 (2011)

12. Ghafarian, T., Javadi, B.: Cloud-aware data-intensive workflow scheduling on volunteer computing systems. Future Gener. Comput. Syst. **51**, 87–97 (2015)

13. Chen, W., Xie, G., Li, R., Bai, Y., Fan, C., Li, K.: Efficient task scheduling for budget constrained parallel applications on heterogeneous cloud computing systems. Future Gener. Comput. Syst. **74**, 1–11 (2017)

14. Casas, I., Taheri, J., Ranjan, R., Wang, L., Zomaya, A.Y.: A balanced scheduler with data reuse and replication for scientific workflows in cloud computing systems. Future Gener. Comput. Syst. **74**, 168–178 (2017)

15. Abrishami, S., Naghibzadeh, M.: Deadline-constrained workflow scheduling in software as a service Cloud. Scientia Iranica Trans. D Comput. Sci. Eng. Electr. Eng. **19**, 680–689 (2011)

16. Sahni, J., Vidyarthi, D.: A cost-effective deadline-constrained dynamic scheduling algorithm for scientific workflows in a cloud environment. IEEE Trans. Cloud Comput. **1**, 99–112 (2015)

17. Yuan, Y., Li, X., Wang, Q., Zhu, X.: Deadline division-based heuristic for cost optimization in workflow scheduling. J. Inform. Sci. **179**, 2562–2575 (2009)

18. Bittencourt, L., Madeira, E.: HCOC: a cost optimization algorithm for workflow scheduling in hybrid clouds. J. Internet Serv. Appl. **2**, 207–227 (2011)

19. Arabnejad, V., Bubendorfer, K., Ng, B.: Scheduling deadline constrained scientific workflows on dynamically provisioned cloud resources (2017). http://dx.doi.org/10.1016/j.future.2017.01.002

20. da Silva, R.F., Chen, W., Juve, G., Vahi, K., Deelman, E.: Community resources for enabling research in distributed scientific workflows. In: Proceedings of the IEEE International Conference on E-Science, (e-Science), vol. 1, pp. 177–184. IEEE (2014)

Table Detection and Metadata Extraction in Document Images

Anand Gupta, Devendra Tiwari, Tarasha Khurana and Sagorika Das

1 Introduction

Tables are widely used in scientific documents, newspaper articles, historical records, forms and financial scripts because they provide a concise yet meaningful way to list interrelated data. Recreation of tabular data from printed documents to digital format is an important task for increasing the knowledge base for efficient implementation of data science techniques.

Consider the following scenario:

A disease X is prevalent in city Y for the past 50 years. Using data science techniques, data in digital format can help to find trends such as when, where and whom disease X can impact the most. This analysis can help in taking preventive measures efficiently. However, records of only past 5 years are stored digitally. In most of the cases, older records such as date, place, name of person, etc. are preserved in the form of printed documents and organized in tabular structures such as ledgers or record books. For the advancement of this medical analysis, digitalization of this printed tabular data is crucial.

Relational integrity of data is important in trend analysis, for example, considering the above scenario if data is placed incorrectly in a different cell during regeneration of a digital copy, the trend analysis may not be accurate. In such cases, metadata

A. Gupta (✉) · T. Khurana · S. Das
Netaji Subhas Institute of Technology, New Delhi, India
e-mail: omaranand@nsitonline.in

T. Khurana
e-mail: tarashak.co@nsit.net.in

S. Das
e-mail: sagorikad.co@nsit.net.in

D. Tiwari (✉)
College of Engineering and Technology, Bikaner, India
e-mail: dev.gl.tiwari@gmail.com

© Springer Nature Singapore Pte Ltd. 2019
S. Tiwari et al. (eds.), *Smart Innovations in Communication and Computational Sciences*, Advances in Intelligent Systems and Computing 851,
https://doi.org/10.1007/978-981-13-2414-7_33

(a)   **(b)**

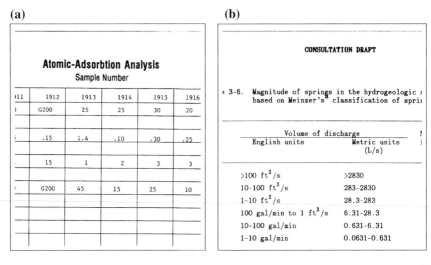

Fig. 1 Examples of tables of various formats from the UNLV dataset [1]. **a** Presents a table with the complete skeleton having both vertical and horizontal lines. **b** Presents a table with a partial skeleton, consisting of only the horizontal rule lines

plays a crucial role in maintaining the relational integrity of the table. To extract the required metadata, efficient detection of table regions is a pre-requisite.

Usually, a typical table detection system focuses on two important aspects:

1. the physical location, size and aspect ratio of table in a raw document image.
2. the logical properties of the table itself, such as the number of rows it contains, the number of columns, placement of the horizontally merged cells, placement of the vertically merged cells and text alignments used in each cell.

Both kinds of information, if present, aid in perfect recreation of tables in content-creating software from raw document images. In this paper, while detecting the table regions in document images, metadata is also extracted, i.e. logical properties of the table, and organize this information in a data structure that contains number of rows and columns and a complete list of cell-wise contents in the table detected. Another salient feature of this paper is that unlike the approaches in literature which use the table skeleton explicitly by checking the intersection points of vertical and horizontal lines, this dependency of the proposed pipeline on the table skeleton such as in Fig. 1a is removed and thus it enables to process those tables also which are without a visible skeletal structure such as those in Fig. 1b. Horizontal and vertical rule lines that construct a table are referred to as table skeleton. Existing research published in this problem domain is surveyed and discussed in Sect. 1.1.

1.1 Related Work

A number of strategies have been developed in order to detect table regions in document images. Comparative surveys and evaluation frameworks such as the one by Klampfl et al. [2] and Zanibbi et al. [3] highlight well the challenges faced by table detection techniques. Several approaches like [4, 5] exploit the presence of line separators or skeletons of tables, while others like [6–8] work for tables even without a visible skeletal structure. Skeleton-based approaches in [4, 5] depend on the intersection of line separators for their table detection approaches, while the separator-independent approaches generally exploit the larger gap between columns. Authors of [5] detect horizontal and vertical lines present in document images and train a classifier to detect tables based on the properties of the detected lines. In the work of Ghanmi and Belaid [9], tables from handwritten documents are extracted using conditional random fields. In Gatos et al. [4] approach, the authors detect horizontal and vertical lines and progressively identify all possible types of line intersection. Table reconstruction is then achieved by drawing the corresponding horizontal and vertical lines that connect all line intersection pairs.

Mandal et al. [8] identified tables on the basis of substantially larger gaps between columns and rows. The approach works even in the absence of any horizontal or vertical rulings. In the work of Bansal et al. [6], they use a context-dependent learning-based framework to recognize tables and assign labels to different document elements such as table header, table trailer, table cell and non-table regions. Similarly, in the work of Fang et al. [10], an approach is proposed to detect table headers from an already extracted table. They identify a set of features to segregate headers from tabular data and build a classifier to detect table headers. In the work of Kieninger [7], a bottom-up approach is proposed for table analysis. A system called T-Recs is developed which works for ASCII file. The system takes word bounding box information as input, and these word boxes are clustered with a bottom-up approach into distinct regions which are designated as tables if they satisfy certain heuristic rules.

Approaches which have been discussed above address a subset of problems from a larger problem domain that consists of correcting skew in document images, detection of table regions and extraction of their metadata. All aforementioned problems have been studied and it has been found that table processing can be enhanced further by combining these tasks in a single pipeline. The challenges which are identified in the literature survey are summarized in Sect. 1.2.

1.2 Challenges

The drawbacks of the approaches mentioned above have imposed the challenges to develop a system that gives improved and enhanced information about the tables in document images. Following are the challenges:

Skeleton Independence Some approaches like [4, 5] are dependent on-line separator information of tables in order to detect them and fail when the table skeleton is not visible. Therefore, the challenge here is to detect all possible layouts of table regions without the help of line separator information from document images.

Table Metadata Most of the approaches discussed [6, 10] extract peripheral table information like table heading, column headings, subheadings, table captions and footnotes. However, the real challenge is the extraction of table metadata, number of columns and number of rows. In the process of table recreation, the most important information needed is the table metadata. The headings, subheadings and captions become secondary in that case.

Content Extraction Another challenge is the extraction of cell contents of tables and its representation in a structured manner which is beneficial during digitalization. However, this task has not been incorporated in any of the discussed approaches.

1.3 Contribution

The contributions of the paper keeping in mind the challenges discussed in Sect. 1.2 are as follows:

1. The approach derived in this paper works well in situations where table skeleton is visible or invisible and is able to process full page tables very well. The dependence on table rule lines is removed by horizontal and vertical line removal in the preprocessing stage.
2. The proposed approach is able to compute the number of columns and number of rows in the detected table region and thus addresses the importance of table metadata. This step in the process has direct significance in regeneration of tables.
3. The approach is augmented with an additional stage to separate out and arrange recognized text of table cells into a data structure.

1.4 Organization

The approach proposed in this paper is discussed in Sect. 2. Dataset and intermediate results of the proposed approach are addressed in Sect. 3. Section 4 provides the result analysis of the experiments, and the approach in this paper is concluded in Sect. 5.

2 Proposed Approach

In this paper, an integrated approach is proposed to detect table regions in raw document images and simultaneously extract their structural and textual contents. By

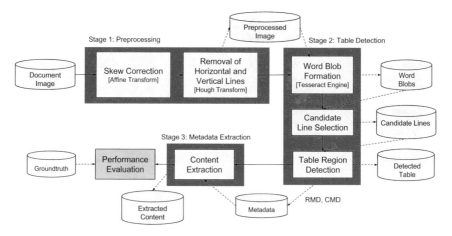

Fig. 2 Proposed approach

integrated approach, it is meant that the entire process flow is a single program divided into three major stages as shown in Fig. 2. These stages are

1. Preprocessing stage,
2. Table detection stage and
3. Metadata extraction stage.

2.1 Preprocessing Stage

A raw document image, shown in Fig. 3a, is first passed through the preprocessing stage. It is initially skew corrected by applying affine transformations. Then, all horizontal and vertical lines from images are removed using Hough transform in order to send a cleaner input image to Tesseract [11]. After the preprocessing steps, the image looks as shown in Fig. 3b. Since even the table borders, cell borders, row rulings and column rulings are removed in this step, therefore, skeletal tables and non-skeletal tables are treated homogeneously in the following stages. Hence, the proposed approach successfully works on both these kinds of tables.

2.2 Table Detection Stage

This stage is further divided into three substages: Word blob formation, candidate line selection and table region detection.

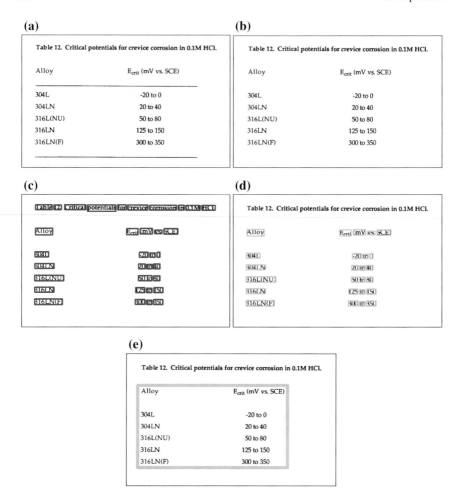

Fig. 3 Intermediate results of proposed approach

2.2.1 Word Blob Formation

In this substage, the alphanumeric characters are grouped together into word blobs by first sending the preprocessed image to the Tesseract OCR Engine, which gives an hOCR output of each word blob with its upper left bounding box coordinates and lower right bounding box coordinates in pixels. These coordinates are accompanied by their respective textual contents. All detected word blobs for a document image are shown with blue bounding boxes in Fig. 3c.

Here, Tesseract is chosen because of its remarkably accurate optical character recognition ability in a number of languages and more importantly, its intermediate output format which helps to process the image further based on the well-organized information provided.

2.2.2 Candidate Line Selection

Once the bounding box coordinates of all word blobs are obtained, they are grouped into lines and are subsequently sent for horizontal threshold computation by Otsu's method [12]. It is computed as

$$\sigma_w^2(t) = \omega_0(t) * \sigma_{\omega_0}^2(t) + \omega_1(t) * \sigma_{\omega_1}^2(t) \tag{1}$$

where t is the computed threshold at probabilities ω_0 and ω_1 of the two classes 0 and 1 and σ^2 is the variance such that the given equation gives a minimum variance for w, σ_w^2. Even though the traditional Otsu's method is used in image processing to binarize (cluster) an image by computing a threshold that minimizes the intra-class variance, the same technique is deployed here using inter-word gaps between all word blobs instead of pixel intensities. After the computation of t, which is the horizontal threshold, all lines that contain pair of word blobs with word gap greater than the horizontal threshold are termed as the candidate textlines.

A single pass of selection of candidate textlines poses two issues: (a) it may contain extraneous lines and (b) it may miss potential candidate lines. To address both the issues, a second pass is conducted in which

1. Those lines are ruled out in which not more than 2 word gaps are found to be greater than the threshold. This is done to ensure that the smallest tables (which may contain a minimum of two columns) remain unaffected.
2. Those lines are included which are present between two detected candidate textlines which failed to be categorized as one because of no cell entry or singular cell entries in that row.

After this pass, candidate textlines are obtained as shown with green bounding boxes around word blobs in Fig. 3d.

2.2.3 Table Region Detection

Once the candidate textlines are found, the minimum upper left coordinate and maximum lower right coordinate from all the bounding boxes are found and a detected table region is obtained. Figure 3e shows this detected table region. At this point, column metadata (CMD) has already been obtained and the computation of interline gap yields the row metadata (RMD). This is done by computing the interline distance between every pair of adjacent candidate textlines. Distance between textlines is calculated using coordinates of the word blobs provided by Tesseract. The fact that the distance between two adjacent candidate textlines belonging to different cells is notably larger than distance between adjacent candidate textlines belonging to the same cell is utilized. After computing the interline distances, vertical threshold is evaluated using Otsu's method, which is used to identify RMD. CMD refers to the number of columns in the table, and RMD refers to the number of rows.

(a)   **(b)**

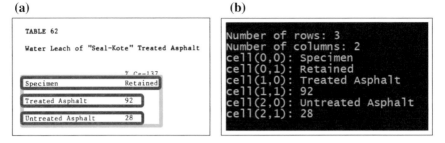

Fig. 4 Metadata extraction

2.3 Metadata Extraction Stage

After detecting the table region in the document image, the extraction of tabular contents can also be done. The words detected by Tesseract, having bounding box completely within the detected region, become a part of the table. To arrange these contents cell-wise, blobs are sorted first in the x-direction and then in the y-direction. In case of multiline data in a single cell, CMD and RMD help to arrange these blobs in a cell-wise matrix. For a detected table region in Fig. 4a, the corresponding cell-wise contents are stored in a 0-indexed matrix and all metadata is gathered from Tesseract hOCR output as in Fig. 4b.

3 Experimental Setup

To implement the above approach, Python 2.7.12 with OpenCV 3.1.0 has been used. The dataset used for the evaluation of the derived approach is described below.

UNLV Dataset For the table detection and structural/textual content extraction task, the UNLV dataset annotations as provided by Shahab et al. [1] are used. According to the official documentation of the UNLV dataset, a total of 2889 document images with tables of different aspect ratios and sizes are scanned. These images are available in two resolutions: 200 DPI and 300 DPI and three different formats: bitonal, grey and fax. A total of about 427 images in the bitonal format at 300 DPI resolution have table zones marked and since the original dataset does not contain the structural annotation of the tables, the table structure ground truths of these 427 images as provided by Shahab et al. [1] are used. They provide this ground truth annotation in XML format with the following key information: row boundaries, column boundaries and each cell's bounding boxes (which automatically includes information about row-spanning and column-spanning cells).

Table 1 Results of the approach against the UNLV dataset

Metrics	Values
Total images	427
Successful detection	354
Detection accuracy	82.90%
Average IoU	71.33%
Row detection error	0.2929
Column detection error	0.3325

4 Result and Discussion

In the past, table detection algorithms have been successfully tested on the popular UNLV dataset as it contains rich annotations of a variety of tables in scanned document images. Hence, UNLV dataset is chosen to evaluate the proposed approach too. Since general object detection accuracy is calculated using the Intersection over Union (IoU) metric, the same has been used here. IoU is described below.

IoU is an evaluation metric that measures the extent of successful detection by calculating the ratio of area of overlap to area of union of the ground truth bounding box and detected bounding box:

$$\frac{Area(Groundtruth) \cap Area(Detection)}{Area(Groundtruth) \cup Area(Detection)} \tag{2}$$

All IoU values fall between 0 and 1. To compute the final detection accuracy, the standard threshold of 0.5 is taken. IoUs above 0.5 are regarded as successful detection, and the rest are rejected as failure cases. For row and column detection, regular error metric is used which is calculated using the ratio of missed rows/columns to the actual number of rows and columns in the detected table. Evaluation results of the approach are given in Table 1 and are discussed below.

From a total of 427 annotated images in the UNLV dataset, 354 images have IoU greater than 0.5 with the proposed approach. Therefore, the accuracy is calculated to be 82.9%. Average IoU is the mean of IoUs of all successful and unsuccessful cases and is always less than the detection accuracy. It is calculated to be 71.33%. Since the average IoU is close to the detection accuracy, it shows that the unsuccessful cases are only because of partial table detection. This concludes that the proposed approach performs well overall. For the performance of metadata extraction, row detection error and column detection error are computed which are 0.2929 and 0.3325, respectively. These figures show that the rows are detected better than the columns.

Results of the proposed pipeline on the UNLV dataset are shown in Fig. 5. The outer green bounding box represents detected table region and detected rows have smaller red bounding boxes. Groups of text are successfully recognized as table rows in almost every case. In Fig. 5a and c, a part of a full page detected table is shown with

(a)
(b)
(c)

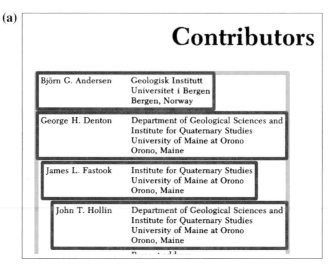

Fig. 5 Results after applying proposed approach on the UNLV image dataset

first four and first six rows, respectively. It can be seen that even multiline rows are detected efficiently with the help of vertical threshold used in the proposed approach. In Fig. 5b, entire detected table is shown along with division into separate rows.

The subsequent step of content extraction uses Tesseract text recognition outputs for each word bounding box, and words that belong to the same cell are concatenated into a string. These cell contents are then arranged into a 2D matrix according to the number of rows and columns found in the metadata stage. A sample has been shown in Fig. 4.

5 Conclusion

This paper focuses on providing a novel and integrated approach for the problem of table detection and extraction of its metadata and contents in document images with varying layouts. As mentioned earlier, the fact that a candidate text line has substantially larger inter-word distances between table columns as compared to words in a non-candidate textline is used for table region detection.

Metadata obtained from this detection stage enables to extract the cell-wise contents of the table and arrange it in a data structure. The proposed approach is experimentally verified and an accuracy of 82.9% is achieved for table detection.

Results obtained show that the proposed approach successfully detects various types and layouts of tables: with or without a skeleton, single-column tables, multi-column tables and full page tables. However, cases where document images contain graphical components having line diagrams, graphs and grid plots are prone to incorrect detection. Some failure cases have been analysed in Fig. 6. In Fig. 6a, detection

(a)

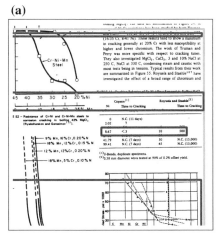

(b)

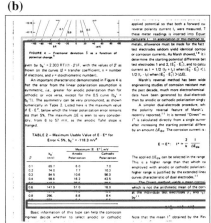

Fig. 6 Failure cases

is hampered due to excessive table-like figures, while in Fig. 6b the presence of mathematical equations is suspected to alter the ideal value of Otsu's threshold. Future work in this area can overcome the above failure cases by embedding suitable page segmentation techniques in the preprocessing stage of the proposed approach.

References

1. Shahab, A., Shafait, F., Kieninger, T., Dengel, A.: An open approach towards the benchmarking of table structure recognition systems. In: Proceedings of the 9th IAPR International Workshop on Document Analysis Systems, pp. 113–120. ACM (2010)
2. Klampfl, S., Jack, K., Kern, R.: A comparison of two unsupervised table recognition methods from digital scientific articles. D-Lib Mag. **20**(11), 7 (2014)
3. Zanibbi, R., Blostein, D., Cordy, J.R.: A survey of table recognition. Doc. Anal. Recognit. **7**(1), 1–16 (2004)
4. Gatos, B., Danatsas, D., Pratikakis, I., Perantonis, S.J.: Automatic table detection in document images. In: Singh, S., Singh, M., Apte, C., Perner, P. (eds.) Pattern Recognition and Data Mining, pp. 609–618. Springer, Berlin, Heidelberg (2005)
5. Kasar, T., Barlas, P., Adam, S., Chatelain, C., Paquet, T.: Learning to detect tables in scanned document images using line information. In: 2013 12th International Conference on Document Analysis and Recognition (ICDAR), pp. 1185–1189. IEEE (2013)
6. Bansal, A., Harit, G., Roy, S.D.: Table extraction from document images using fixed point model. In: Proceedings of the 2014 Indian Conference on Computer Vision Graphics and Image Processing, p. 67. ACM (2014)
7. Kieninger, T.G.: Table structure recognition based on robust block segmentation. In: Document Recognition V, vol. 3305, pp. 22–33. International Society for Optics and Photonics (1998)
8. Mandal, S., Chowdhury, S., Das, A.K., Chanda, B.: A simple and effective table detection system from document images. Int. J. Doc. Anal. Recognit. (IJDAR) **8**(2–3), 172–182 (2006)
9. Ghanmi, N., Belaid, A.: Table detection in handwritten chemistry documents using conditional random fields. In: 2014 14th International Conference on Frontiers in Handwriting Recognition, pp. 146–151 (2014). https://doi.org/10.1109/ICFHR.2014.32
10. Fang, J., Mitra, P., Tang, Z., Giles, C.L.: Table header detection and classification. In: Twenty-Sixth AAAI Conference on Artificial Intelligence (2012)
11. Smith, R.: An overview of the Tesseract OCR engine. In: Ninth International Conference on Document Analysis and Recognition (ICDAR 2007) (2007)
12. Otsu, N.: A threshold selection method from gray-level histograms. IEEE Trans. Syst. Man Cybern. **9**(1), 62–66 (1979)

Interactive Mobile Application to Determine and Enhance User's Skills in Their Respective Field of Interest

Akshay Talke, Rohit Kr. Singh, Sanyam Raj, Virendra Patil, Ameya Jawalgekar and Ajitkumar Shitole

1 Introduction

Current Indian Education System promotes rat race. So our legislature should make innovatory training its first need and attempt to address issues. Generally students choose career path which do not match their interest due to lack of knowledge of different career options or advice from peers. Since individuals tend to lose interest in a particular field because of lack of proper knowledge and guidance, proposed system will develop user's interest by providing them with required information at suitable time [1].

According to Indian Census 2011, adolescent and youth population of India was nearly equal to population of 18 Western Asian countries [2]. It is predicted by Census Board of India that by the year 2030, India would have the highest adolescent and youth population in the world [2]. As human resource is the major factor in the overall development of the country, we should cater the need of skilled individuals which

A. Talke · R. Kr. Singh · S. Raj · V. Patil (✉) · A. Jawalgekar · A. Shitole
International Institute of Information Technology, Pune, India
e-mail: viren21096@gmail.com

A. Talke
e-mail: akshaytalke@gmail.com

R. Kr. Singh
e-mail: rohit.12.ks@gmail.com

S. Raj
e-mail: sanyamraj22@gmail.com

A. Jawalgekar
e-mail: jawalgekar007@gmail.com

A. Shitole
e-mail: ajitkumars@isquareit.edu.in

© Springer Nature Singapore Pte Ltd. 2019
S. Tiwari et al. (eds.), *Smart Innovations in Communication and Computational Sciences*, Advances in Intelligent Systems and Computing 851,
https://doi.org/10.1007/978-981-13-2414-7_34

can be achieved by focussing on the Education System. Hence, our target audience for the proposed system is the teenagers.

Our system aims to Suggest, Assist and Motivate Students in their respective field of interest. According to user's interest the proposed system would try to nurture them in their field of interest, which could be anything including academics, so that the individual can pursue his/her Interest and could consider it as a career. As one's field of interest can vary drastically, we need to consider each and every domain which can provide a career path. If we categorize various career options under general umbrellas, we get Academics, Arts and Sports as major domains. Therefore, we have considered three domains namely Academics, Sports, and Arts which can be further extended in near future.

2   Existing Methodologies

There are different solutions available to the issue which is being presented.

- Some of the esteemed institutes designate counselors/guides for mentoring students throughout their understudies [3].
- There are different online destinations and distinct mobile applications accessible for career guidance [4].
- Different aptitude tests are accessible for students to decide their interest in light of their inclination.

The limitations of existing solutions are as follows:

- In most of the educational systems, students have to choose a career path at the age of around 14–16. Although their meanings of interests are still in progress from one representation to another. Ultimately, individuals remain in dilemma and they may have to follow a career path which they have no prior idea of.
- Vast majority of the establishments are precluded from securing the counseling facility as it incurs high additional cost which cannot be afforded by every institution.
- The Counselors being humans, can be biased depending on their clients and their opinion may vary due to their past experiences [5]. The manual examination of each individual is not possible due to constraints such as time, variety of career paths and increasing number of students.
- The issue related with online sites and mobile applications is that it requires the intrigue or career choice to be known by the individual in advance.
- One test is not sufficiently adequate to decide the intrigue or career path of an individual.

3   Background and Related Work

This issue has been partially addressed earlier and few systems have been proposed for the same. A system was proposed in which four steps were followed namely (a) knowledge of user's profile in our case: the examination of students' vocational interests and competencies, (b) Personalization of user's environment, (c) Personalized recommendations and (d) Evaluation of students' feedback [6]. But this system only deals with Academic-oriented courses. Another System was proposed which focuses on finding interest of an individual based on one's technical skills through social media [7]. The data in this system was gathered and processed through different API's. In addition, a survey was done to analyzed on what factors do high school students tend to choose their career [7]. This system again was constrained to single domain, i.e., academics and lacked real-time counseling or building user profiles.

4   Proposed Design and Architecture

The system proposed in [7] considers only textual data from Facebook as input. To overcome all the conceivable disadvantages of the current framework and related works, we have proposed an interactive mobile application to upgrade user's expertise in his/her respective field of interest with help of Real-time Counseling.

In order to have more accuracy and confidence in prediction, we are considering textual data, images, location data, likes and follows from Facebook, Twitter, and Instagram. Additionally, we are also considering the results of the tests based on Gardner's Theory of Multiple Intelligences. Our system also tracks user's activity on our application which also contributes in determining the field of interest. We are not limiting our system just to determine the field of interest, rather it provides with continuous feeds in form of text(s), images and videos. This helps in guiding, motivating and assisting users.

4.1   Design

Our Framework comprises of a front-end mobile application and a back-end server. The essential element of our framework being real-time tracking, for now we have considered three data sources viz. "location data", "tests (based on Gardner's Theory of Multiple Intelligences)" and "social media" like Facebook, Twitter, Instagram for the collection of user data. The mobile application comprises of a profile for every user to create a personalized environment [8]. The three major features of our application for each profile are "Guidance", "Suggestion", and "Motivation". The data collected is then classified into three domains mentioned earlier using Facebook

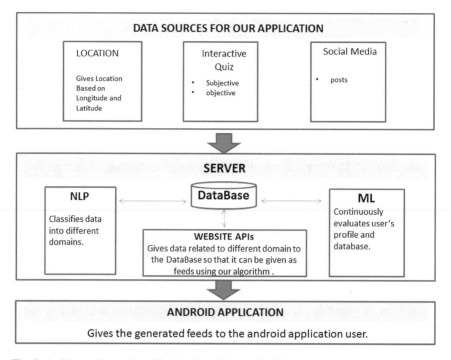

Fig. 1 Architectural overview of back-end and front-end of the system

Graph API, Google Vision API and DatumBox API. Based on this classification, feeds are generated according to the user's interests on their profiles (Fig. 1).

4.2   Algorithm

Variables Used:

'n': Total number of Domains
'm': Total number of Sources
's': Sources
'd': Domains

Weightage Matrix 'X':

$$
X = \begin{bmatrix}
X(s_1 d_1) & X(s_1 d_2) & \cdots & X(s_1 d_n) \\
X(s_2 d_1) & X(s_2 d_2) & \cdots & X(s_2 d_n) \\
\vdots & \vdots & & \vdots \\
X(s_m d_1) & X(s_m d_2) & \cdots & X(s_m d_n)
\end{bmatrix}
\tag{1}
$$

In Eq. 1, $X(s_i d_j)$ stands for Weightage given for ith Source and jth Domain. Number of rows in Matrix 'X' refers to number of Sources while number of columns refers to number of Domains.

Profile Matrix 'P':

$$P = \begin{bmatrix} P(s_1 d_1) & P(s_1 d_2) & \cdots & P(s_1 d_n) \\ P(s_2 d_1) & P(s_2 d_2) & \cdots & P(s_2 d_n) \\ \vdots & \vdots & & \vdots \\ P(s_m d_1) & P(s_m d_2) & \cdots & P(s_m d_n) \end{bmatrix} \qquad (2)$$

In Eq. 2, $P(s_i d_j)$ stands for profile interest of each user in ith Source and jth Domain.

Td_i: Total Interest in Domain 'i'

$$Td_i = \sum_{k=1}^{m} X(s_k d_i) \cdot P(s_k d_i) \qquad (3)$$

Td_i %: Percentage of Interest in Domain 'i'

$$Percentage \, Td_i = \frac{\sum_{k=1}^{m} X(s_k d_i) \cdot P(s_k d_i)}{\sum_{j=1}^{n} \left(\sum_{k=1}^{m} X(s_k d_j) \cdot P(s_k d_j) \right)} \qquad (4)$$

Algorithmic Steps.

START
Step 1: Read data for all sources
Step 2: Manipulate values in matrix P
Step 3: Compute $(X) \cdot (P)$ [Dot Product]
Step 4: Compute $TD_i(\%)$ [for all i $= 1$:n] (Using Eq. 4)
Step 5: Round off each $TD_i(\%)$ to nearest Tens
Step 6: Generate feeds in the same percentage
END

5 Experimental Survey and Results

To test our proposed system, we conducted an experiment.

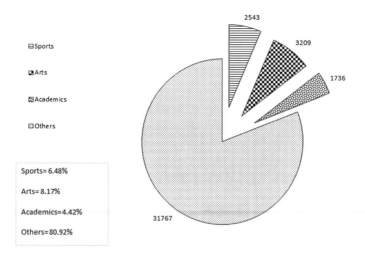

Fig. 2 Percentage of posts classified

5.1 Input

- Social Media platform acts as one of the data sources for our system, Facebook being the most popular Social Media platform we chose to survey 100 random Facebook profiles (If unsatisfied with the accuracy of the system, we may increase number of profiles under evaluation up to 1000 in future).

5.2 Processing

- We manually examined all the posts on the Facebook timeline of the input user profiles and classified the posts into three domains mentioned earlier (Academics, Sports, Arts). The posts which did not lie in the three mentioned domains were classified into the domain "Others".
- The following figure shows statistics of overall Posts classified into different domains (Fig. 2).
- The number of classified posts in each domain helped us in calculating Weightage Matrix 'X' (for Source Facebook).
- Using Weightage Matrix (X) and Profile Matrix (P), the users' interests were calculated by TD_i (Refer Eq. 3).
- The deduced interests were conveyed to the users and their feedback was taken.

Fig. 3 Graphical representation of result obtained of a user

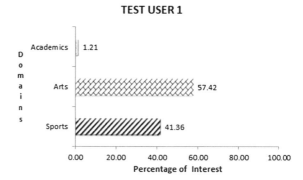

5.3 Output

- 70% agreed to overall statistical prediction (based on algorithm). However, they feel minor changes were required to match their real-life interests (Fig. 3).
- 30% completely agreed to the deduced results.

6 Future Scope and Conclusion

Presently the system has three domains as mentioned previously. The users who will utilize this framework are students within the age group of twelve to eighteen. This system works for selection of career path according to user's interest.

In the future, we aim at

- Elaborating above-mentioned general domains into specific categories and adding different domains other than the three specified earlier, so that each and every area of interest is taken care of.
- The system would also be extended to users of other age groups. For users of higher age groups this would go about as an application which would help them pursue their hobby.
- This application would act as an instructor and a companion in everyday life in coming future.

We have successfully discovered area of interest of the users and aim at guiding them to pursue their career in that field with high accuracy, unbiased behavior and more affordability. As discussed earlier, the current solutions have limitations like high cost, biased counseling, less viability of tests and prior knowledge of interest, which have been overcome effectively by our system. Our proposed system would gather data from various data sources, process it using our proposed algorithm and provide real-time counseling by suggesting, assisting, and motivating the users. If any of the data sources fail, it would affect the quality of the gathered information which in turn would affect the overall system, however this can be countered by collecting more data from other data sources. Our experimental survey has demonstrated the efficiency of our algorithm. Through this application, we intend to channelize the energy used by the students from other usages of smart phones towards making and moulding their future, consequently contributing to the overall development of the nation.

References

1. Why You Should Share to Social Media in the Afternoon + More of the Latest Social Media Research. https://blog.bufferapp.com/new-social-media-research
2. http://www.censusindia.gov.in/2011-Documents/PPT_World_Population/Size_Growth_and_Composition_of_Adolescent_and_Youth_Population_in_India.ppt
3. Wen, J., Shih, W.-L.: Information aptitude scale development for vocational high school in Taiwan. In: Proceedings of the 3rd IEEE International Conference on Advanced Technologies (2003)
4. https://www.shiksha.com/
5. Tiffany Iskander, E., Gore, P., Bergerson, A.A., Furse, C.: Gender disparity in engineering: results and analysis from school counsellors survey and national vignette. In: 2013 IEEE Antennas and Propagation Society International Symposium (APSURSI) (2013)
6. Alimam, M.A., Seghiouer, H., El Yusufi, Y.: Building profiles based on ontology for career recommendation in E-Ieaming context. In: International Conference on Multimedia Computing and Systems (ICMCS) (2014)
7. Pisalayon, N., Sae-Lim, J., Rojanasit, N., Chiravirakul, P.: Identifying personal skills and possible fields of study based on personal interests on social media content. In: 6th ICT International Student Project Conference (ICT-ISPC) (2017)
8. Ginon, B., Jean-Daubias, S.: Consideration of knowledges, skills and preferences for multi-aspect customizing activities on learners' profiles. In: Revue STICEF, vol. 19 (2012)

Back-Propagated Neural Network on MapReduce Frameworks: A Survey

Jenish Dhanani, Rupa Mehta, Dipti Rana and Bharat Tidke

1 Introduction

An advancement of Internet technology has resulted in generation of vast data every second. Approximately, 2.5 terabytes of data is generated by various sources each day [13], such as web services (i.e., E-commerce services, financial services, healthcare service, search engines, etc.), Internet Of Things (IOT) (i.e., Sensors, mobile devices, etc.), and various social networks (i.e., YouTube, Facebook, Twitter, etc.) [4, 8, 14, 30]. It has resulted in the evolution of big data which is defined by Gartner, an international research agency [3], as "*high-volume, high-velocity and/or high-variety information assets that demand cost-effective, innovative forms of information processing that enable enhanced insight, decision-making, and process automation*". Neural Network (NN) is commonly applied to various machine learning, deep learning algorithms, and big data analytics [8, 23, 27]. It has been successfully applied in many other domains like bio-informatics [7], health care [15], remote sensing [17], finances [24], e-commerce [31], etc.

Backpropagation (BP) algorithm is broadly used for training in neural network. It is computationally expensive for big data applications, as it performs huge number of iterations and vast mathematical calculations [16, 19]. It requires significant amount of time and memory for the big data. High dimensionality of big data demands exponential growth of connections in neural network [28], results in poor performance of BPNN model. BPNN model for big data needs the distributed and parallel framework like MapReduce. MapReduce framework exploits the cluster of commodity

J. Dhanani (✉) · R. Mehta · D. Rana · B. Tidke
S. V. National Institute of Technology, Surat, India
e-mail: jenishdhanani26@gmail.com

R. Mehta
e-mail: rgm@coed.svnit.ac.in

D. Rana
e-mail: dpr@coed.svnit.ac.in

B. Tidke
e-mail: batidke@gmail.com

© Springer Nature Singapore Pte Ltd. 2019
S. Tiwari et al. (eds.), *Smart Innovations in Communication and Computational Sciences*, Advances in Intelligent Systems and Computing 851,
https://doi.org/10.1007/978-981-13-2414-7_35

hardware to process the smaller chunks of large volume of data [1, 10]. Mapper and Reducer are the major tasks in MapReduce framework. An input data from distributed file system is processed by Mapper based on user-defined logic. The data is resulted in form of *<Key, Value>* pairs by Mappers which subsequently called by Reducers. The resultant data is utilized by subsequent user-defined logic over an input.

BPNN utilizes the MapReduce framework to address poor performance issue for large volume of data [9]. There are state-of-the-art surveys that discuss the variants of MapReduce. Li et al. [18] focused on many challenges of MapReduce frameworks including multiple access of distributed file system during iterative process and complexity in processing of stream data. Various extensions of MapReduce frameworks are proposed by the same group of researchers like Phoenix [25], Phoenix++ [29], and Twister [12]. BPNN is most widely used machine learning technique especially in prediction modeling [7, 15, 17, 24, 31]. This paper specifically focuses on BPNN models, built in the MapReduce environment, considering large volume of testing data [21], large volume of training data [5, 9, 19–22], and huge neural network with large number of neurons [21, 26]. MapReduce-based BPNN approaches exploit single or multiple MapReduce jobs to build the complete model.

Rest of the paper is organized as follows: In Sect. 2, various MapReduce-Based BPNN (MRBPNN) approaches such as single-pass and multi-pass MRBPNN are described. In Sect. 3, a comparative study of MRBPNN is presented. The research is summarized with challenges in Sect. 4.

2 MapReduce-Based BPNN (MRBPNN)

MapReduce is widely used solution to handle the large volume of data. The complex computation in iterative task of BPNN for huge data has been one of the major challenges for researchers. Chu et al. [9] proposed BPNN with distributed way of MapReduce framework. This section presents various MRBPNN approaches found in literature which are categorized as follows:

- Multi-pass MRBPNN (MPMRBPNN) [5, 9, 19–22, 26].
- Single-Pass MRBPNN (SPMRBPNN) [21].

MPMRBPNN requires several passes through the MapReduce job to build the complete BPNN model. An individual job updates the parameters (i.e., weights and bias). SPMRBPNN builds the local BPNN models from individual chunk of data. Further, these generated models are merged together to build complete BPNN model.

2.1 Multi-pass MRBPNN (MPMRBPNN)

MPMRBPNN requires multiple MapReduce jobs to build the complete BPNN model as shown in Fig. 1. Single MapReduce job consists of stages, namely, *Mapper*,

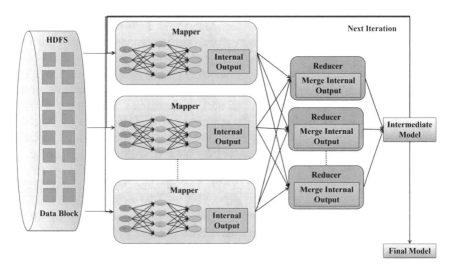

Fig. 1 Multi-pass MRBPNN (MPMRBPNN)

Sort/Shuffling, and *Reducer*. Mapper processes an individual chunk of data inputted from distributed file system that produce the intermediate result. It is stored in the form of *<Key, Value>* pairs on local file system. Subsequently, sorting and shuffling are performed based on *Key* with intention to distribute *<Key, Value>* pairs among Reducers. Once all the Mappers finish their jobs, Reducers fetch *<Key, Value>* pairs and subsequently merge all *<Key, Value>* pairs together follows the user's logic.

Chu et al. [9] proposed MapReduce-based binary BPNN (MRNN) with three-layered structure by utilizing batch gradient on single- and multi-core system. The training data is divided into chunks where individual chunk processed by individual Mapper. Each Mapper calculates the partial gradient for weights of all instances. Partial gradient updates the weights by back-propagating errors in the network. Once all Mappers finish their jobs, Reducer performs batch gradient descent by summing all partial gradients to update weights. Such MapReduce job is repeatedly performed until desired BPNN model is built or terminating condition is reached.

Liu et al. [22] designed MapReduce-based architecture to parallelize BPNN (MBNN) in cloud environment for large-scale mobile data. Similar to MRNN, each Mapper computes updated value for all weights of each training instance. It results in *<Key, Value>* pairs where *Key* is weight and *Value* is updated weight for every training instance. Consequently, Reducer updates weights by averaging all the *Value* of identical *Key*. This process is repetitively performed until desired BPNN model is built or terminating condition is reached.

MRNN and MBNN can effectively handle large volume of data by utilizing the MapReduce. The large number of *<Key, Value>* pairs generated results in large network I/O between the Mappers and Reducers. To reduce the large network I/O, Binhan et al. [5] proposed MapReduce-based Training in Mapper Neural Network (MR-TMNN). In the MR-TMNN, data is divided into several chunks. Each Mapper builds local BPNN from one chunk of data. Mapper results in <*Key, Value*> pairs of local BPNN model. MR-TMNN generates the intermediate <*Key, Value*> pairs for each local BPNN model, while MRNN and MBNN generate the intermediate <*Key, Value*> pairs for each instance. Reducer updates all weights by weighted average. The same process is performed until the expected model is built or terminating condition is reached. This approach achieves less network overhead due to less number of <*Key, Value*> pairs generated by Mappers but at the cost of degraded accuracy. To build well-trained BPNN, this process is executed iteratively. It is observed that MRNN, MBNN, and MR-TMNN are not suited for huge network consisting large number of neurons.

Liu et al. [21] specifically built the MRBPNN architecture (MRBPNN_3) for large network. In MapReduce environment, entire BPNN model is distributed in multiple MapReduce jobs. MRBPNN_3 utilizes MapReduce job at every layer of BPNN (excluding input layer) which performs the sequential training of instances. For each stage, output of Reducer will be input to next stage. This process is applicable to all the layers except the output layer. For each instance, Mapper results in <*Key, Value*> pairs of instances where *Key* is the index of Reducer and *Value* contains the output of neuron, weights, threshold, and target of instance. Subsequently, Reducer outputs <*Key, Value*> pairs where *Key* is the index of Mapper and *Value* contains the output of neuron, weights, threshold, and target of instance. This process is performed until last layer is reached where Reducer performs an additional functionality of backpropagation and weight updation.

Similar to previous approach, Ren et al. [26] proposed Fine-Grained Parallel MapReduce-Based Backpropagation (FP-MRBP) approach for large neural network. It utilizes the Layer-wise Parallelism and Layer-wise Integration (LPLI) strategy where each layer (excluding input layer) is distributed in parallel structure consisting of multiple neurons. Mapper calculates the activation values of neurons for the structure. Successively, Reducer integrates activation values of all the structures. This process is repeatedly performed until last layer is reached which performs backpropagation to update weights.

Liu et al. [19] further modified their work to improve the precision by utilizing the Cascading model in evaluation phase for Parallelizing BPNN (CPBPNN). Training instances are grouped where each group consists of instances with identical label. Each group is chunked into bootstrapped samples [11] which are input to an individual Mapper. Each Mapper builds the BPNN model having a strong capability to identify the particular class of instances due to training of identical labeled instances. Cascading is adopted in evaluation phase where testing data divided into chunks. Individual chunk evaluated in cascading manner by the individual group's BPNN models. Specific group's Mappers evaluate instances, which results in <*Key, Value*> pairs. Subsequently, Reducers apply the majority voting. Accordingly, correctly identified instances are filtered out from testing instances and rest will be evaluated by the next specific group's Mappers. This process is repeatedly performed on all groups in cascading manner. However, there is a need to identify the correct evaluation order of group's BPNN models to avoid scenarios where earlier classifier may incorrectly classify. These instances are filtered out for actual class model.

2.2   Single-Pass MRBPNN (SPMRBPNN)

MPMRBPNN process is computationally very expensive due to high start-up cost of MapReduce job. This limitation is resolved by the SPMRBPNN that completes all stages (i.e., Mapper, Sort/Shuffle, and Reduce) of MapReduce in one pass only as depicted in Fig. 2. Mapper builds the local BPNN model on individual data chunks which are stored on local file system. Reducer merges all local BPNN models to generate final BPNN model. SPMRBPNN also resolves the issue of huge network I/O between the Mapper and Reducer by sending *<Key, Value>* pairs of local model rather than individual instances.

Liu et al. [21] proposed two different SPMRBPNN architectures for different scenarios as follows:

- Large volume of testing data (MRBPNN_1) and
- Large volume of training data with bootstrapping and majority voting (MRBPNN_2).

MRBPNN_1 is specifically designed for large volume of testing data where it is stored in distributed file system by dividing it into several chunks. However, all Mappers build identical BPNN models which are trained by complete training data. Each Mapper evaluates individual testing chunk which results in *<Key, Value>* pairs of testing instances where *Key* is instance and *Value* is predicted output of instance. Reducer stores the final output on distributed file system by clubbing all *<Key, Value>* pairs collected over Mappers. MRBPNN_1 is computationally expensive for large volume training data as all Mappers build identical BPNN model which is overcome in MRBPNN_2. It is designed to handle large volume of training data. MRBPNN_2 utilizes the balance bootstrapping and majority voting technique to improve the performance. Each Mapper builds BPNN models on the individual chunk of training

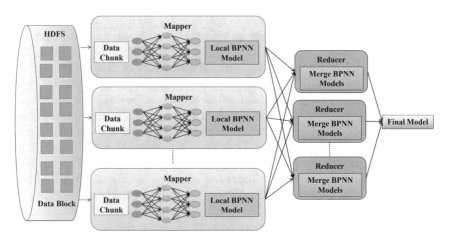

Fig. 2 Single-pass MRBPNN (SPMRBPNN)

data which is stored in distributed file system. However, chunking is performed by the balance bootstrapping method. Once Mapper finishes training process, it starts evaluating all testing instances. Reducer does similar to MRBPNN_1's Reducer. Lie et al. [20] implemented and compared previous approach in different distributed frameworks such as Hadoop [1], Haloop [6], and Spark [2]. It is observed that MRBPNN_2 improves the accuracy at additional cost for preprocessing of bootstrapping and majority voting.

3   Comparative Study of MRBPNN

This section presents the comparative study of various MRBPNN approaches proposed by the research community, based on several parameters. Table 1 compares different MRBPNN approaches with respect to Mapper computations, Reducer computations, granularity of Mapper's output, number of MapReduce jobs required to build the BPNN model, and Mapper's *<Key, Value>* pairs. In Table 1, K indicates number of iterations required to build the BPNN model, L indicates number of layers in neural network, G indicates number of test data chunks, C indicates the number of classes in training data, and *MR job* indicates the MapReduce job. The key element of any MapReduce architecture is the design of Mapper and Reducer, as it directly affects the performance. It should be neither overloaded nor under loaded to achieve maximum performance. Mapper produces output at different granularities such as neurons, instances, and models level, and it is directly impacted on the number of *<Key, Value>* pairs. Once all Mappers finish their jobs, Reducers pull *<Key, Value>* pairs through network. Therefore, size and number of *<Key, Value>* pairs generated by Mappers should be minimal to decrease the network I/O. Likewise, multiple MapReduce jobs also affect time to build the BPNN model. Multiple MapReduce job's start-up time is high due to huge network I/O, as it requires to store the output which will be utilized in next MapReduce job. Additionally, it also requires orchestrating in MapReduce jobs [6].

Various frameworks such as MRNN [9], MBNN [22], MR-TMNN [5], FP-MRBP [26], MRBPNN_3 [21], and CPBPNN [19] are proposed with multiple MapReduce jobs. The Mapper produces intermediate output by processing each instance of training data, and Reducer merges the intermediate output. As shown in Table 1, MRNN and MBNN require one MapReduce job for each iteration where computational load is less due to less number of neurons. Large network consists of large number of neurons that required huge computation. Thus, these approaches are not suited for large network. Large network can be handled by the FP-MRBP and MRBPNN_3 which build the MapReduce job for every single layer excluding input layer, e.g., FP-MRBP and MRBPNN_3 which require $K(L-1)$ MapReduce jobs. However, CPBPNN proposed a cascading-based model to improve the precision where (CG) MapReduce jobs required. Multiple MapReduce jobs have been escaped in research of Liu et al. [20, 21] by utilizing single MapReduce job to build BPNN. In MRBPNN_1, MRBPNN_2, and MR-TMNN, Mapper has huge computational burden as local

Table 1 Comparative study of MRBPNN approaches

Approach	Mapper computations	Reducer computations	Mapper's granularity	No. of MR jobs	Mapper	Output	Remark
					Key	Value	
MRNN	Partial gradient of weights	Batch gradient by summing all partial gradients	Instance	K	*Key*	*Value*	Generate <*Key, Value*> pairs for weights of each instance
MBNN	Update value of weights	Average *values* having same *Key*	Instance	K	Weight	Updated value	Generate <*Key, Value*>pairs for weights of each instance
MRBPNN_3	Output of neurons	Forward output to next layer; the output layer updates weights	Neuron	$K(L-1)$	Index of Mapper	Neurons output,weights, threshold and target	Built for the large BPNN
FP-MRBP	Activation value, error and updates weights	Integrates the activation values error and updates weights	Neuron	$K(L-1)$	Current layer	Activation value, error and updated weights	Built for the large BPNN
CPBPNN	BPNN model on similar class bootstrapped sample, also predict the output of testing instances	Majority voting	Instance	CG	Index of Reducer	Predicted output	To improve the precision
MR-TMNN	Local BPNN model	Build BPNN model by weighted average	Model	K	Weight and threshold	Weight value_number Threshold value_number	Speed up due to less number of <*Key, Value*> pairs

(continued)

Table 1 (continued)

Approach	Mapper computations	Reducer computations	Mapper's granularity	No. of MR jobs	Mapper		Output		Remark
					Key		Value		
MRBPNN_1	BPNN model, also predict the output of testing instances	Classify the testing instances	Instance	1	Instance		Predicted output		Built for large testing data
MRBPNN_2	BPNN model on bootstrapped sample, also predict the output of testing instances	Majority voting	Instance	1	Instance		Predicted output		Built for large training data

BPNN model is built from the chunk of data, e.g., MRBPNN_1 builds the model from all training data, MRBPNN_2 builds the local BPNN from bootstrapped samples, and MR-TMNN builds the local BPNN from chunk of training data. On the other hand, Reducers are less loaded as compared to Mappers. Mapper of different approaches produces the output at different granularities, e.g., Mapper of FP-MRBP and MRBPNN_3 produces output for neurons. Moreover, Mapper of MRNN and MBNN produces *<Key, Value>* pairs for each weight at every instance. However, Mapper of MR-TMNN produces *<Key, Value>* pairs at the end of training for each weight which speeds up the training. On the other hand, classification accuracy might degrade as BPNN model is built from less training data. MRBPNN_2 adopted bootstrapping and majority voting to improve the accuracy. The further improvement in precision is achieved in CPBPNN by adopting the cascading for evaluation of testing data.

4 Summary and Challenges

BPNN is one of the famous and extensively used algorithms in machine learning and deep learning. The flooding of digital data has become a major challenge for many conventional machine learning applications and demands for the processes to handle the big data. Research infers that the performance of BPNN is poor for large volume of data as it is computation intensive. In contrast, MapReduce is tremendously well-working framework for large volume of data. BPNN can utilize the MapReduce, an effective solution to handle big data in the distributed and parallel environment. This paper presented various BPNN algorithms on MapReduce framework along with the comparative study. It is observed that majority approaches required multiple MapReduce jobs to build well-trained models such as MRNN, MBNN, MR-TMNN, FP-MRBP, MRBPNN_3, and CPBPNN and rest required one MapReduce job such as MRBPNN_1 and MRBPNN_2. Many approaches having high burden on the Mappers such as MR-TMNN, MRBPNN_1, and MRBPNN_2. In contrast to them, many approaches have large network I/O between the Mapper and Reducer such as MRNN and MBNN. From the extensive literature survey, many challenges which seek attention from the research community are as follows:

- **Iterative algorithms**: In traditional iterative algorithms, outcome of each iteration is utilized by next iteration as input. In case of MapReduce, such output needs to be stored in distributed file system which incurs the huge communication I/O. Additionally, MapReduce needs extra efforts (i.e., bootstrapping and majority voting) for iterative models (i.e., BPNN) to achieve benchmark performance. Thus, there is a need of model with single MapReduce job with less efforts and communication I/O.

- **Real-time system**: MapReduce is the best suited for batch data processing applications. It is high latency framework as it includes various time-consuming phases like map, sort/shuffle, and reduce. Thus, native MapReduce is not well suited for real-time applications. There is a need of additional mechanism that enhances the performance of MapReduce which can efficiently process data in real-time environment.

References

1. Apache hadoop. https://hadoop.apache.org/. Accessed 08 Dec 2017
2. Apache spark. https://spark.apache.org/. Accessed 20 Dec 2017
3. Big data. https://www.gartner.com/it-glossary/big-data. Accessed 01 Jan 2018
4. Alekhya, G.S.S.L., Lydia, E.L., Challa, N.: Big data analytics: a survey. Int. J. Appl. Innov. Eng. Manag. (IJAIEM) **5**(10), 090–106 (2016)
5. Binhan, Z., Wang, W., Zhang, X.: Training backpropagation neural network in MapReduce. In: Proceedings of International Conference on Computer, Communications and Information Technology (CCIT 2014), pp. 22–25. Atlantis-Press (2014)
6. Bu, Y., Howe, B., Balazinska, M., Ernst, M.D.: Haloop: efficient iterative data processing on large clusters. Proc. VLDB Endow. **3**(1–2), 285–296 (2010)
7. Chen, K., Kurgan, L.A.: Neural networks in bioinformatics. Handbook of Natural Computing, pp. 565–583. Springer, Berlin, Heidelberg (2012)
8. Chen, M., Mao, S., Liu, Y.: Big data: a survey. Mob. Netw. Appl. **19**(2), 171–209 (2014)
9. Chu, C.T., Kim, S.K., Lin, Y.A., Yu, Y., Bradski, G., Ng, A.Y., Olukotun, K.: Map-reduce for machine learning on multicore. In: Proceedings of NIPS, vol. 6, pp. 281–288. Vancouver, BC (2006)
10. Dean, J., Ghemawat, S.: Mapreduce: simplified data processing on large clusters. Commun. ACM **51**(1), 107–113 (2008)
11. Efron, B., Tibshirani, R.J.: An Introduction to the Bootstrap. CRC press (1994)
12. Ekanayake, J., Li, H., Zhang, B., Gunarathne, T., Bae, S.H., Qiu, J., Fox, G.: Twister: a runtime for iterative MapReduce. In: Proceedings of the 19th ACM International Symposium on High Performance Distributed Computing, pp. 810–818. ACM (2010)
13. Every day big data statistics. http://www.vcloudnews.com/every-day-big-data-statistics-2-5-quintillion-bytes-of-data-created-daily/. Accessed 01 Feb 2018
14. Gantz, J., Reinsel, D.: Extracting value from chaos. IDC iview **1142**(2011), 1–12 (2011)
15. Ghavami, P., Kapur, K.: Prognostics & artificial neural network applications in patient healthcare. In: 2011 IEEE Conference on Prognostics and Health Management (PHM), pp. 1–7. IEEE (2011)
16. Hecht-Nielsen, R.: Theory of the backpropagation neural network. Neural Networks for Perception. Elsevier, pp. 65–93 (1992)
17. Jiang, J., Zhang, J., Yang, G., Zhang, D., Zhang, L.: Application of back propagation neural network in the classification of high resolution remote sensing image: take remote sensing image of Beijing for instance. In: 2010 18th International Conference on Geoinformatics, pp. 1–6. IEEE (2010)
18. Li, R., Hu, H., Li, H., Wu, Y., Yang, J.: Mapreduce parallel programming model: a state-of-the-art survey. Int. J. Parallel Program. **44**(4), 832–866 (2016)
19. Liu, Y., Jing, W., Xu, L.: Parallelizing backpropagation neural network using MapReduce and cascading model. Comput. Intell. Neurosci. **2016** (2016)
20. Liu, Y., Xu, L., Li, M.: The parallelization of back propagation neural network in MapReduce and Spark. Int. J. Parallel Program. **45**(4), 760–779 (2017)

21. Liu, Y., Yang, J., Huang, Y., Xu, L., Li, S., Qi, M.: MapReduce based parallel neural networks in enabling large scale machine learning. Comput. Intell. Neurosci. **2015**, 1–13 (2015)
22. Liu, Z., Li, H., Miao, G.: MapReduce-based backpropagation neural network over large scale mobile data. In: 2010 Sixth International Conference on Natural Computation (ICNC), vol. 4, pp. 1726–1730. IEEE (2010)
23. Nasrabadi, N.M.: Pattern recognition and machine learning. J. Electron. Imaging **16**(4), 049901 (2007)
24. Patel, S.D., Quadros, D., Patil, V., Saxena, H.: Stock prediction using neural networks. Int. J. Eng. Manag. Res. (IJEMR) **7**(2), 490–493 (2017)
25. Ranger, C., Raghuraman, R., Penmetsa, A., Bradski, G., Kozyrakis, C.: Evaluating MapReduce for multi-core and multiprocessor systems. In: IEEE 13th International Symposium on High Performance Computer Architecture, HPCA 2007, pp. 13–24. IEEE (2007)
26. Ren, G., Hua, Q., Deng, P., Yang, C.: FP-MRBP: fine-grained parallel MapReduce back propagation algorithm. In: International Conference on Artificial Neural Networks, pp. 680–687. Springer (2017)
27. Schmidhuber, J.: Deep learning in neural networks: an overview. Neural Netw. **61**, 85–117 (2015)
28. Sharma, C.: Big data analytics using neural networks. Master's thesis, San Jose State University, San Jose, CA 95192 (2014)
29. Talbot, J., Yoo, R.M., Kozyrakis, C.: Phoenix++: modular MapReduce for shared-memory systems. In: Proceedings of the Second International Workshop on MapReduce and its Applications, pp. 9–16. ACM (2011)
30. Tidke, B., Mehta, R., Dhanani, J.: A comprehensive survey and open challenges of mining bigdata. In: International Conference on Information and Communication Technology for Intelligent Systems, pp. 441–448. Springer (2017)
31. Zhao, K., Wang, C.: Sales forecast in e-commerce using convolutional neural network (2017). CoRR arXiv:1708.07946

TWEESENT: A Web Application on Sentiment Analysis

Sweta Swain and K. R. Seeja

1 Introduction

With the growth in information technology, there is an increase in social media usage such as Twitter, LinkedIn, and Facebook where people communicate or share information with each other to maintain personal or professional relationship across the globe. Twitter is used as an analytical tool for mining social media data due to its real-time accessibility feature and easy terms of service for detailed analysis. Twitter acquires a large portion of population which reflects the opinion or response of the country as whole. Thus, the analysis of twitter data to detect public sentiment towards an event represent the response of the population as whole which serves as a powerful survey methodology. Sentiment detection from twitter data is a challenging task because of ill format of text, syntactical error and writing style of users. Twitter also has limitations that it does not ask for demographic information like age and gender while registration, which is key parameters for fine-grained analysis and to study the dynamic changes between the emotions and user attributes. In this paper, a web based application called "TweeSent" is presented that can predict the polarity and emotion from the input twitter data.

S. Swain · K. R. Seeja (✉)
Department of Computer Science and Engineering, Indira Gandhi Delhi Technical University for Women, New Delhi, Delhi, India
e-mail: krseeja@gmail.com; seeja@igdtuw.ac.in

S. Swain
e-mail: swetaswain86@gmail.com

© Springer Nature Singapore Pte Ltd. 2019
S. Tiwari et al. (eds.), *Smart Innovations in Communication and Computational Sciences*, Advances in Intelligent Systems and Computing 851,
https://doi.org/10.1007/978-981-13-2414-7_36

393

2 Literature Review

Sentiment analysis deals with the mood or opinion detection within a text of target topic or event using natural language processing, text analysis, and computational linguistics. Lots of research work [1, 2] has been done on sentiment analysis focusing on classifying the text into positive or negative category as compared to the emotion. Aman et al. [3] used knowledge based approach to classify the sentence into six emotions (joy, happiness, sadness, angry, surprise, anticipation and fear) and non-emotion. Bollen et al. [4] identified tension, depression, anger, vigor, fatigue and confusion in the text of tweet other than these six emotions. Also, Lexicon based approaches are found to perform [5, 6] better than machine learning approaches to classify the tweets. Social media user's opinions, thoughts, beliefs and emotion vary with the demographics attributes like gender, age, occupation, and location. Inferring demographic attributes of twitter user is of great interest among researchers due to the unavailability of demographic information on twitter's account. The first study related to demographic attributes of twitter user was done by Mislove et al. [7], they focuses on geography and gender. Later, Rao et al. [8] and Schwatz et al. [9] used facebook status updates to detect gender and age of user using content analysis. Age estimation is challenging task, as people grows older they express general views towards any topic, this creates ambiguity for analysts to make proper classification and only 3.34% of user mention their age related information [10] in the description field. Sloan et al. [11] discussed how to infer demographic information like gender, age, location using the tweets associated metadata. Another approach is to use the content of the tweet to categorize gender [12] and location [13] of user by searching specific keywords. Firstname [14] is mostly used for predicting age of twitter users, i.e., the first name matches with the most popular name over the time period present in US social security administration data. Nyugen et al. [15] found that with the prediction of users age group, they can also estimate the exact age and life stages using the text of tweet. Siswanto et al. [16] uses the age of twitter user estimated first to estimate the job attributes using the text of tweets. There are few work related to sentiment classification of tweet based on attributes. Bi et al. [17] classified sentiment across gender, location, time, and occupation using news data. In this research, an application considering the demographic attributes of user which is difficult to infer from user tweets is developed for analyzing the user sentiments.

3 Proposed Methodology

In this paper an emotion/polarity detection method is proposed to classify the sentiment of twitter users in terms of emotions (joy, happiness, sadness, angry, trust, surprise, anticipation, and fear) and polarity (positive and negative) using lexicon based approach. In order to perform fine-grained sentiment analysis of tweets, the sentiment based tweets are classified across demographic attributes of users such as

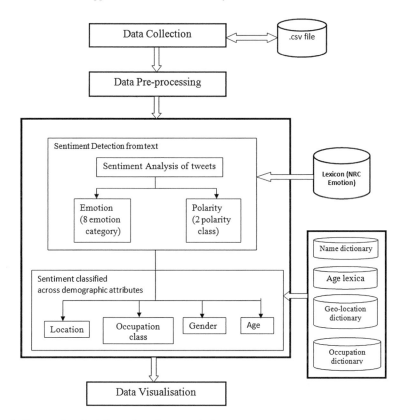

Fig. 1 Proposed methodology of sentiment analysis

age, gender, location, and occupation class. In the proposed method shown in Fig. 1. the tweets are extracted from twitter with associated metadata field. Subsequently preprocessing mechanism is applied on the tweets to remove errors and irregular languages and to make it suitable for sentiment classification. The sentiment based tweets are then visualized graphically in the data visualization process, in order to infer the sentiment state of the twitter users easily.

4   Implementation

The main aim of developing TweeSent is to study the response of public towards an event and automatically classify the tweets into different emotion categories across demographic attributes of users like location, age, occupation, and gender. In order to design front and back end of TweeSent web application, shiny dashboard package of R is used.

4.1 Data Collection and Preprocessing

The dataset is prepared by extracting the tweets from twitter using demonetization and demonetisation as search term and it consist of 5249 tweets. In our work, we are interested in the fullname, text, location, created time, and description field of tweets to analyze the sentiment of tweets across demographic attributes of users. The raw twitter data undergoes various preprocessing steps like Non-english tweets removal, Punctuation, tab and white space removal, RT entities and user handles removal, Emoticons Removal, Link Removal, Numbers Removal, Duplicate letters removal, Stop words Removal, Removal of suffixes from words and Lowercase conversion. The preprocessed structured data consist of 4602 tweets after removing the non-English tweets. The tweets dataset generally consists of retweets which is retweeted by another twitter user. In this work, the retweets are not removed and treated as original tweets as they validate and amplify the message tweeted.

4.2 Emotion/Polarity Analysis of Tweets

In order to identify the sentiment of the tweets, a lexicon-based approach is applied on tweets using a lexicon called "NRC emotion lexicon" [18]. Emotion lexicon consists of list of words and their associated emotion in each emotion category, to identify emotion from text. According to lexicon based approach, each word of the tweets is matched against the lexicon and assigned a score based on the words matched as per word sense association score. After score assignment, identify and remove the tweets having no emotional content (with zero association score) or belongs to more than one emotion/polarity category from the tweet collection. The resultant dataset has 1460 tweets having emotion context and the rest 3142 tweets are discarded for emotion detection. Similarly the resultant dataset has 2799 tweets having polarity context and the rest 1803 tweets are discarded for polarity detection. Next, the tweets are categorized into that emotion/polarity category which has the maximum association score. The tweets are classified into eight basic emotions (joy, sadness, disgust, anticipation, trust, fear, surprise and anger) and two polarity categories (positive and negative).

4.3 Location-Based Analysis of Tweets

Location is an important parameter and should be considered before implementing any policy, rule, decision and initiative by government. The emotion/polarity of tweets varies with the location due to circumstances, resources, regional belief, and environment factors. In order to extract location from the location field, the data need to be preprocessed and matched with the location in Geo-location dictionary. The matched location is assigned with corresponding state in the Geo-location dictionary and the sentiment classified tweets are again classified based on the location of the tweet author. In this work, it is found that approximately 70% of tweets have reported their location and around 68% of tweets are classified with associated location.

4.4   Gender Based Analysis of Tweets

The gender of the twitter users can be identified by comparing the full name with the lists of female and male names obtained from Indian baby names called as name dictionary. The name field associated with every tweet needs to be preprocessed to extract the fullname, which is matched with the name dictionary and assigned corresponding gender class. It is found that 99.5% tweets having associated name field and majority of the users (around 71.8%) are male users. But women are found more expressive than the men.

4.5   Occupation Class Based Analysis of Tweets

The occupation related term is extracted first from the description field of the tweets and then assigned the corresponding occupational class using job lexica. In our dataset, 23.10% description field is empty while 19% of tweets having occupation related information among the tweets with description field. The tweets are classified into 91 occupation class categories.

4.6   Age Group Based Analysis of Tweets

Age can be used to analyze the group of population actively participating in the initiative and to find people of all age groups involved in the discussion/debate. For estimating age range, analyze the vocabulary of user description field and search for particular words like mother, father, retired, grandfather, student, boy, girl and so on using age lexica [9]. The tweets are classified based on the word matched in terms of age. In this work, the sentiment based tweets are classified into three age groups (i.e., 16–23 years, 24–29 years and 30–67 years).

5   Result and Discussion

Sentiment analysis of tweets in terms of emotion and polarity is shown in Fig. 2. It is found that more people responded positively towards demonetisation and have

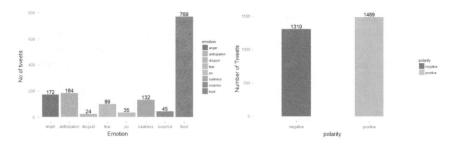

Fig. 2 Sentiment analysis of tweets based on emotion/polarity

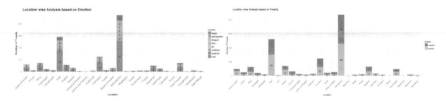

Fig. 3 Location based analysis of tweets in terms of emotion/polarity

trust over the initiative taken by government. But, there is less difference between negative and positive response, thus people are showing anger towards the initiative. At the same time people anticipate that it may leads to the betterment of the country.

The demographics and social background affects the thought and feelings of the user which results in different sentiment of the users for the same event or topic. So, sentiment analysis of tweet across demographic attributes of users helps to learn the sentiment variation among users and helps in determining the impact of the initiative on the country. The location-based analysis of tweets is shown in Fig. 3. According to the analysis Maharashtra and Delhi respond with maximum number of sentiment tweets. Other states like Andhra Pradesh, Karnataka, Bihar, and Uttar Pradesh also responded and actively participated in debate/discussion related to the event. Thus, the location analysis of the tweets helps to analyze the cash crisis and efficient implementation of demonetization in all states of India.

In India, male population is more than the female and also in terms of internet users and literacy rate. But, females are more expressive and actively participate in social media discussions. The result shown in Fig. 4 indicates that both male and female shows trust but males have expectation from this initiative.

Demonetization not only affects the economy of India but also people lives. People from every social class got affected. The result indicates that engineering professionals responded with highest number of tweets who mostly work in IT firm or any company in India as shown in Fig. 5 and have trust on the government.

The main aim of Demonetization is making India a cashless society. The use of digital medium for transaction is popular among youth of nation. From the result, we can infer that people among 24–29 years (48.4%) shared their emotion more than

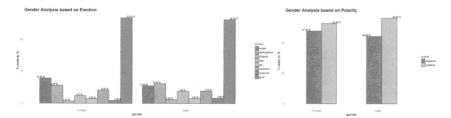

Fig. 4 Gender based analysis of tweets in terms of emotion/polarity

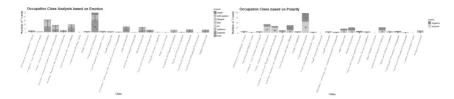

Fig. 5 Occupation based analysis of tweets in terms of emotion/polarity

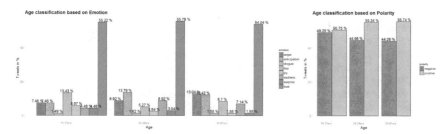

Fig. 6 Classification of emotion/polarity based tweets across age group

other age group. 55.78% have trust and 13.79% have anticipation feelings towards demonetisation as shown in Fig. 6.

6   Conclusion

In this research TweeSent, a web application has been developed to analyze the sentiment of twitter users. TweeSent automatically classify the tweets into emotion and polarity class, to determine sentiment state of the user across demographic attributes of tweeters like location, gender, age, and occupation class. Sentiment analysis is helpful for government to evaluate the performance of their policy and initiative from people's perception instead of traditional surveys. The performance of TweeSent is evaluated with the tweets related to demonetization in India. It is found that the outcome of the analysis provided by TweeSent has strong correlation with the statistics and report provided by government.

References

1. Jurek, A., Mulvenna, M.D., Bi, Y.: Improved lexicon-based sentiment analysis for social media analytics. Secur. Inform. **4**(1), 9 (2015)
2. Andreevskaia, A., Bergler, S., Urseanu, M.: All blogs are not made equal: exploring genre differences in sentiment tagging of blogs. In: ICWSM, Mar 2007
3. Aman, S., Szpakowicz, S.: Identifying expressions of emotion in text. In: Text, Speech and Dialogue, pp. 196–205. Springer, Berlin, Heidelberg (2007)
4. Bollen, J., Mao, H., Pepe, A.: Modeling public mood and emotion: Twitter sentiment and socio-economic phenomena. In: ICWSM, vol. 11, pp. 450–453 (2011)
5. Kundi, F.M., Khan, A., Ahmad, S., Asghar, M.Z.: Lexicon-based sentiment analysis in the social web. J. Basic Appl. Sci. Res. **4**(6), 238–248 (2014)
6. Westling, A., Brynielsson, J., Gustavi, T.: Mining the web for sympathy: the pussy riot case. In: 2014 IEEE Joint Intelligence and Security Informatics Conference (JISIC), pp. 123–128. IEEE, Sept 2014
7. Mislove, A., Lehmann, S., Ahn, Y.Y., Onnela, J.P., Rosenquist, J.N.: Understanding the demographics of Twitter users. In: 5th ICWSM, vol. 11, (2011)
8. Rao, D., Yarowsky, D., Shreevats, A., Gupta, M.: Classifying latent user attributes in twitter. In: Proceedings of the 2nd International Workshop on Search and Mining User-Generated Contents, pp. 37–44. ACM, Oct 2010
9. Schwartz, H.A., Eichstaedt, J.C., Kern, M.L., Dziurzynski, L., Ramones, S.M.: Personality, Gender, and Age in the Language of Social Media: The Open-Vocabulary Approach. PLoS ONE **8**(9), e73791 (2013). https://doi.org/10.1371/journal.pone.0073791. PMID: 24086296
10. Ludu, P.S.: Inferring latent attributes of an Indian Twitter user using celebrities and class influencers. In: Proceedings of the 1st ACM Workshop on Social Media World Sensors, pp. 9–15. ACM, Sept 2015
11. Sloan, L., Morgan, J., Burnap, P., Williams, M.: Who tweets? Deriving the demographic characteristics of age, occupation and social class from Twitter user meta-data. PLoS One **10**(3), e0115545 (2015)
12. Bamman, D., Eisenstein, J., Schnoebelen, T.: Gender identity and lexical variation in social media. J. Socioling. **18**(2), 135–160 (2014)
13. Mandel, B., Culotta, A., Boulahanis, J., Stark, D., Lewis, B., Rodrigue, J.: A demographic analysis of online sentiment during hurricane irene. In: Proceedings of the Second Workshop on Language in Social Media, pp. 27–36. Association for Computational Linguistics, June 2012
14. Oktay, H., Firat, A., Ertem, Z.: Demographic Breakdown of Twitter Users: An Analysis Based on Names. Academy of Science and Engineering (ASE) (2014)
15. Nguyen, D.P., Gravel, R., Trieschnigg, R.B., Meder, T.: "How old do you think I am?" A study of language and age in Twitter. In: Proceedings of the Seventh International AAAI Conference on Weblogs and Social Media, ICWSM 2013, 08–10 July 2013, Cambridge, MA, USA, pp. 439-448. AAAI Press. ISBN 978-1-57735-610-3
16. Siswanto, E., Khodra, M.L.: Predicting latent attributes of Twitter user by employing lexical features. In: 2013 International Conference on Information Technology and Electrical Engineering (ICITEE), pp. 176–180. IEEE, Oct 2013
17. Bi, Y., Li, M., Leow, D., Huang, R.: Political Sentiment Visualization Data Analysis and Visualization using Voxgov US Federal Government Media Releases (2014)
18. Mohammad, S.M., Turney, P.D.: NRC emotion lexicon. NRC Technical Report (2013)

Personalized Secured API for Application Developer

R. Maheswari, S. Sheeba Rani, P. Sharmila and S. Rajarao

1 Introduction

If it a simple messaging app or banking app, the end user will never try to put their privacy at risk for many reasons, after all it is a personal device. But most of the malware apps looks like legit apps and often asks for all kind of permissions while in stalling and by which it puts the device into huge risk. As soon as mobile data is turned on, about 82% of mobile apps collect and track the current and last location. An astounding 57% of apps even track when a phone is being used. While most of the time this tracking is harmless, your device may be at risk when one app starts over collecting information compared to similar apps in the same category [1]. There are enough security measures to take care of threats. But what is secure today can be broken tomorrow or be broken by "something" or "somehow". Cybercriminals are always two steps ahead of the present security mechanisms available [2]. It is always better to create something that will serve our purpose and also takes the security a level higher. What if we use an additional mechanism which will help in concealing the existence of the data?

R. Maheswari (✉) · S. Rajarao
VIT Chennai, Vandalur-Kelambakkam Road, Chennai, India
e-mail: maheswari.r@vit.ac.in

S. Sheeba Rani
Sri Krishna College of Engineering and Technology, Coimbatore, India

P. Sharmila
Sri Sairam Engineering College, Chennai, India

© Springer Nature Singapore Pte Ltd. 2019
S. Tiwari et al. (eds.), *Smart Innovations in Communication and Computational Sciences*, Advances in Intelligent Systems and Computing 851,
https://doi.org/10.1007/978-981-13-2414-7_37

1.1 Motivation

Internet users have shifted from using desktops/laptops to mobile devices to carry out different activities. Now mobile devices are a part of our life. We depend on the devices much more than the major advantage of providing a means of communication with people. We use it for simple stuffs like communication via messaging, email to complex activities such as banking and financial transactions because of its exponential growth and various advantages it offers. As it resulted in increased usage and gaining of quick popularity, the malicious and fraudulent activity on mobile platform is growing more quickly than it did on the PC platform. Mobile application developers do not look into security aspects or rather find it difficult to include the best security features since it requires additional research and work to choose the appropriate security mechanism and algorithm, its strength, compatibility, and reliability of it. It will be really appreciated if there is an easy way of including the best security mechanism which will make the app more secure and reliable for the users.

1.2 Difficulties Faced by the Developers

Application developers give maximum attention to the functional requirements and would like to complete it successfully to get the desired output which is a normal practice. It is not that they do not consider about business requirements, but they find it difficult to implement it due to various reasons. Let us take security as a business requirement. An in-depth knowledge in the domain of security is required to make the application more secure and hack-proof which cannot be expected in an application developer. The developer, to contain the situation, has to do a lot of research to find the appropriate security mechanism, and corresponding algorithms and protocols associated with it, the reliability and strength of the selected mechanism and much more.

1.3 Proposed Approach

This work presents a security mechanism using CRLSB Image Steganography for information security. After discussing the problems and difficulties faced by application developers, it is evident that it will be easier for them if there is any ready-to-use implementation of a security mechanism. So the proposed system will be a HTTP API which uses the php web service for posed system will be a HTTP API which uses the php web service for including security as a service in the client application. The principal objective of this work is to design and develop a security web service and an API for mobile applications. The API should be able to provide Security as a Service to its client applications. There as on behind developing an API is that it supports cross-platform calls to the web service no matter what technology is used by it. The objective is to enable different types of applications to access the service and make use of it.

2 System Overview

The proposed system consists of two major components, one is the PHP web service and other component is a mobile API to access the service. The security mechanism that is implemented in PHP is based on enhanced LSB steganography called as CRLSB.

2.1 *Least Significant Bit Steganography*

In this simple method the least significant bits in an image is substituted by bits of secret message. Without any observable changes, a large amount of data can be embedded by LSB. The major advantages of this technique are very effective, easy to implement, takes very less space but it has low imperceptibility. Here is an example of how LSB Method works: Message to be hidden: "E" It is converted into its ASCII equivalent—69. Then it is represented in binary format-01000101. To hide this data, at least 3 pixels are required. Considering the following are the RGB values of three pixels

P1 (10011010(RED) 11000110(GREEN) 10011101(BLUE)),
P2 (10111000(RED) 10000110(GREEN) 10001001(BLUE)),
P3 (10111110(RED) 10000100(GREEN) 11111101(BLUE))

The least significant bits are replaced with data bits:

P1 (10011010 11000111 10011100),
P2 (10111000 10000110 10001000),
P3 (10111111 10000100 11111101).

In this case, only 3 bits are required to change for inserting the character effectively. The message is successfully hidden, since the consequential changes that are created to the least significant bits are too small to be predictable by the ordinary human eye. The advantage of LSB embedding is its simplicity and many techniques use these methods. LSB embedding also allows high perceptual transparency. CRLSB, an enhanced version of LSB, will be further explained in later section. An API is developed to access the web service so that the application which needs to adopt this security mechanism has access to this service. It is not platform-dependent it can be accessed by any mobile application and will be able to serve the requests received. To give an example of how to utilize the service, a simple general-purpose

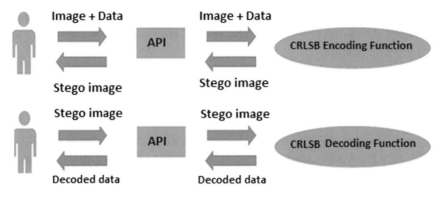

Fig. 1 Preliminary design of the system

steganography android application has been developed which sends requests to the API to consume the security web service which is deployed [3].

Figure 1 depicts the high-level design of the system. A user (user's client application) makes an API call by sending the data to be secured along with the cover image. The request is authenticated and forwarded to the server.

2.2 Literature Review

Since 440 BC, technically steganography is refereed as covered or hidden writing. Although steganography was coined in late15th century, the real usage of it dates back several millennia. In olden days, people use to hide messages on the back of wax writing tables, writing messages on the stomach of rabbits, sometimes messages got tattooed on the scalp of slaves. Spies and terrorist used invisible ink for their confidential work and many times students used those ink for fun. Current steganographic systems uses multimedia objects like image, audio, video, etc., as hidden tool since users often transmit digital pictures over email and other Internet communication.

In a steganography framework the techniques used for embedding the data is kept unknown. In a private key steganography both the sender and receiver knows secret key used for embedding the message, example: the key used to seed the pseudo random number generator. In public key steganography, a private–public key pair is generated and both the sender and receiver will know each other's public key [4]. LSB-based steganographic algorithms are easy to implement, easy to understand and adding data bits to the least significant bit of multimedia objects do not make a remarkable difference to the stego object. Transferring secret message to destination is a challenging task in modern steganography. LSB steganography can be combined with Discreet Cosine Transformation and compression on raw image to enhance security of payload [5]. Filtering based algorithm can be used to hide large data in Bitmap image which uses Most Significant Bit (MSB) for filtering purpose and

Least Significant Bit (LSB) for hiding information in image [6]. LSB steganography can be made more difficult to detect if the data to be embedded into an image is encrypted using cryptography [7]. Jain et al. proposed secret message using DES cryptography with RGB image and Least Significant Bits (LSB). Battisti et al. [8] used digital media with Fibonacci number sequence. Dey et al. [9] proposed LSB data hiding using prime numbers. An improved LSB data hiding using arithmetic and geometric progression was proposed by Satapathy et al. [10] Using the concept of lighter image and darker image creates a new enhanced LSB technique [11]. A pixel is selected to hide the data bit if it satisfies the following condition: If lighter image, MSB contains at least two "1" bits or If darker image, MSB contains at least two "0" bits. If the MSB values and the data bits are same, then no changes are being done to the MSB bits. Modern steganography hides information into digital multimedia files and sometimes hidden messages at the network packet level. Based on the hidden tool used steganography can be categorized into five types: 1. Text Steganography 2. Image Steganography 3. Audio Steganography 4. Video Steganography 5. Protocol Steganography.

3 Proposed System Design

In the proposed method, CRLSB Image Steganography is used to encode the data into the image. After developing and testing the algorithm, we need to develop an API so that the client applications can make use of this algorithm to secure the data. To encode the data, it is converted into binary format. A RGB 24 bit color image is used to form three 22 matrices. These matrices store the Red (R), Green (G), Blue (B) values of all the pixels from the chosen image (A).

$f(p) = P(X, Y) \forall P \subseteq A, X\&Y \in Z+$ where X, Y are the coordinates of pixel P. $R = P(Red), 0 \leq Red \leq 255$ $G = P(Green), 0 \leq Green \leq 255$ $B = P(Blue), 0 \leq Blue \leq 255$. To hide data one can use either of R, G, B matrices by following LSB steganography mechanism. In the proposed approach Red (R) matrix is used for embedding data. The LSB of some elements in the R matrix are replaced by the binary data.

Figure. 2 depicts the architecture of the system. It has two major components, the HTTP API and the web service which is hosted in a web server. End users will use the client applications installed in their mobiles. The application is developed in such a way that it should send request to the HTTP API and receive appropriate response from the same. The API, upon receiving request from client application, will interact with the web service and passes the necessary input data received via POST. When the PHP web service receives the input variables from API, it then executes the requested operations and sends a JSON response to the API. It then forwards the response object to the client application. In client side, the response should be decoded to obtain the expected result. The webserver consists of php class files which performs the encoding and decoding of data using the image constructed from the client application's input. It has two php class files namely crlsb encode into image and crlsb decode from image. As the file name says the first class file

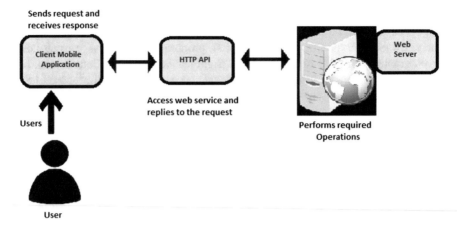

Fig. 2 System architecture design

will take a RGB image and data to be encoded as the input parameters and performs CRLSB encoding and returns the stegoimage as output. The other class file accepts the stegoimage as input parameter and performs CRLSB decoding and returns the original data decoded from the stegoimage.

3.1 Encoding Process

After discussing about how the API works, in this section the encoding process is explained in detail. Without loss of generality a square RGB image of size preferably 1000×1000 (any value greater than 512×512) is used. First, the complement image of the image is generated and the corresponding complement R, G, B values are extracted is from $R' = 255 - R$, $G' = 255 - G$, $B' = 255 - B$.

The complement values obtained from the image will be used to determine the coordinates of the next pixel to encode along with the current green and blue values. The complement matrix undergoes 4 types of rotation in order to get random locations every time even when the same image is used again. The following steps describe the steps involved in encoding process: Input: binary stream of size n.

1. Randomly select a pixel P1 from the cover image.

$$P1 = P(x, y) \forall x \in [0, w] \text{ and } y \in [0, h]$$

2. For every pixel that is selected to hide the data, the pixel coordinates are stored in a 2D array (selected Pixels). This array is maintained so that the selected pixel is checked for redundancy while encoding.
3. The first bit of data is embedded in Red value of pixel P1.

4. Compute the complement image

$G'[i, j] = 255 - G[i, j]$ where $0 \leq i \leq w$
$B'[i, j] = 255 - B[i, j]$ where $0 \leq j \leq h$

5. Based on the server time, perform phase rotation. Phase rotation ensures pseudo randomization in data encoding process. Even if the same image used to hide the same text the sequence of pixel selection will vary.

if hours%4 = 0 then rotation angle = 0
if hours%4 = 1 then rotation angle = 90
if hours%4 = 2 then rotation angle = 180
if hours%4 = 3 then rotation angle = 270

6. Determine the next pixel to embed data using matrices G, B, G', B'.
Let us say X and Y are next pixels, then

$$X = G(P_i) + a \text{ and } Y = B(P_i) + b$$

where a = Green complement value of pixel P and b = Blue complement value of pixel P.

7. Repeat Step 6 until n bits are encoded.

3.2 HTTP API

The HTTP API will act as an interface between the client application and the web server to carry out the requested operations. The encode API accepts file type of image as input and also the data to be encoded and then converts the image into base 64 encoded string data and converts the data into binary format. It then forwards the converted data to the php class files to carry out the encoding process. While decoding, the decode API accepts the file type of image as input (the stego image which has hidden data) and converts it into base 64 encoded string and sends it to the decode class. After performing decoding, it then returns the extracted original data from the stego image to the called application.

3.3 Encoding of Data

To begin with the encoding process an RGB image greater than 512×512 (or bigger dimensions) is taken. First, we need to generate the complement image values for green and blue values as mentioned in step 2. Then, the complement image matrices are rotated corresponding to the phase. These complement values along with green and blue values will be used in determining the position of the next pixel for encoding in a complete random manner (Step-5). Then we use LSB technique to store the data bits in the red values of the randomly selected pixels, one at a time. The various methods involved in encoding process are: (i) Obtain the cover image where data will be encoded. (ii) Select first pixel to encode the data bit. (iii) Embed first data bit into the LSB of red value. (iv) Compute the complement value matrix for the given image. (Complement value = 255 − color value [green & blue]). (v) Rotate the complement value matrix based on the file-sent time.

if hour % 4 = 0, no rotation
if hour % 4 = 1, rotate the matrices by 90°
if hour % 4 = 2, rotate the matrices by 180°
if hour % 4 = 3, rotate the matrices by 270°

where h = hour taken from the file-sent time (Ex: 18:32 = h = 18) (vi) Determine the next pixel's X and Y coordinate to encode the data bit where X will be computed by adding green value of current pixel and its compliment value and Y will be computed by adding blue value of current pixel and its compliment value. (vii) Encode the data bits and repeat until all data bits are encoded.

4 Decoding of Data

The decoding process involves discovering those pixels that hold data bits and arranging the random bits in the proper sequence. In decoding the complement image of G and B matrix for the stegoimage is generated which helps in detecting the proper sequence of the original message. The following steps describe the decoding process.

(i) Obtain the stegoimage and determine the first pixels coordinates to be decoded by using the method followed while encoding. (ii) Decode the data bit from LSB of red value and append it on the data array. (iii) Compute the complement value matrix of the stego image and rotate it as mentioned in encoding process. (iv) Select the next pixel's X and Y coordinates to decode the data bit where X is calculated by adding green value of current pixel and its compliment value and Y is calculated by adding blue value of current pixel and its compliment value. (v) Now decode the data bit from the selected pixel and repeat until all data bits are decoded.

The user needs to input the secret data to be embedded into the cover image. After giving input, the user clicks Encode button which will send a HttpPost Request along with the cover image.

5 Results and Discussions

This part consists of the result so obtained from executing the application and corresponding API. It also discusses about the results obtained and includes algorithm analysis of the implemented CRLSB Image steganography algorithm with existing algorithms. The API has been tested by using Android app and php. Both calls executed successfully and returned the expected result. The results of testing API and php class files manually can be found in Fig. 3.

Fig. 3 Input image for encoding

Measure	CRLSB	BattleSteg	DynamicBattleSteg	BlindHide	HideSeek
(RS)Avg. across all groups/colours	26.47036	26.61133	26.57609	26.65852	26.7294
(RS)Avg. approx. length across all groups/colours	38157.0236	38360.2282	38309.42816	38428.261	38530.4
(SP)Avg. across all groups/colors	27.71236	28.05879	28.04414	27.95601	28.2825
(SP)Avg.approx.length across all groups/colors	39947.36379	40446.7502	40425.63337	40298.586	40769.3
Average Absolute difference	0.004076483	0.00425078	0.004341831	0.0041701	0.00408
Mean Squared Error	0.004076483	0.00516129	0.005070239	0.0083221	0.00409
LpNorm	0.002038241	0.00258065	0.00253512	0.004161	0.00205
Signal to Noise Ratio	76.32224499	75.2975183	75.37481656	73.222793	76.3029
PSNR	81.11587624	80.0911495	80.16844781	78.016424	81.0965
Normalised Cross-Correlation	0.999998981	0.99999911	0.999998894	0.999999	1
Correlation Quality	163.9105129	163.910533	163.9104986	163.91052	163.91

Fig. 4 Comparison of algorithms with respect to image quality metrics

5.1 Design Analysis

The implemented algorithm is compared with similar LSB based steganography algorithms such as Battle Steg, Blind Hide, Dynamic Battle Steg and HideSeek. The algorithm analysis gave an overall positive result when compared. Data to be embedded into the cover image is given in Fig. 4.

Different Image quality metrics used the cover stegoimages obtained from different algorithms are evaluated with PSNR, MSE to evaluate the image quality after modification. PSNR is used in measuring the quality of reconstructed image by computing the ratio between the original data and distortion introduced to it. PSNR and MSE have an inverse relationship and higher the PSNR value, greater the quality of the image is. MSE is an image quality metric used to find difference in images comparing the degree of changes in its color values. Lower MSE value signifies less difference between cover and stegoimages. Correlation refers to the similarity between the cover and stegoimages. Average absolute difference is the average difference between color values of the cover and stegoimages. Histogram analysis is an effective way to find hidden information in an image. Histograms of data samples can be analyzed to detect suspicious parameters in still images. Studies have shown that steganographic embedding is equivalent to low-pass filtering the histogram, characterized by a decrease in the mass center of a characteristic function of the histogram and this decrease can be exploited to identify stego-media.

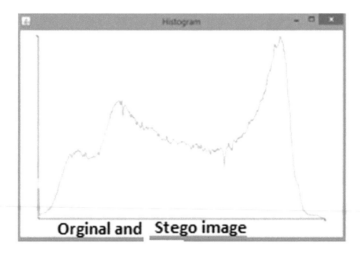

Fig. 5 Histogram analysis (Original and stego image)

The results of the histogram analysis are shown in Fig. 5. Since, there is no difference found in the histogram of original and stegoimage so, it can be concluded that the proposed approach is resistant to histogram analysis attacks.

6 Conclusion and Future Work

An idea of developing security API for mobile has been presented. The proposed approach is successfully implemented and the API is fully functional. The approach was a success since it was planned to develop in one language and provide the service to all kind of applications rather than developing the security mechanism in all the languages. The key feature of this API is that it is platform-independent and can be used by any authenticated client mobile applications and help developers to include a higher level of security mechanism in their application. Even though there is a demo application to show the utilization of this API (a general-purpose steganography android app) presented along with the report, if given more time, instead of presenting it as an idea, it could be completed (a professional messaging application) which will use this API for its security to show how it can be applied in real-time applications and presented it along with the report. As long as the encoding/decoding technique is secured, there will be no threat to this system. Deploying a web service and making it available for use from online secures the code, but makes the process a little complex. This can be improved by creating a downloadable library with class files but after using obfuscator to scramble the code files so that it cannot be reverse engineered easily.

References

1. Mobile Malware-The Rise continues. http://www.mcafee.com/in/security-awareness/articles/mobile-malware.aspx/ (2014)
2. Mobile Malware Threats Fraudsters are still two steps ahead. https://securityintelligence.com/two-steps-ahead/ (2015)
3. An Introduction to APIs. https://zapier.com/learn/apis/chapter-1-introduction-to-apis/
4. Venugopal, K.R., Patnaik, L.M., Raja, K.B., Chowdary, C.R.: A secure image steganography using LSB, DCT and compression techniques on raw images. In: ICISIP, pp. 123–135 (2005)
5. Memon, N., Chandramouli, R., Kharrazi, M.: Image steganography and steganalysis: concepts and practice. In: IWDW, pp. 4–12 (2003)
6. Mandal, Md.P., Islam, R., Siddiqa, A., Delowar, Md.: An efficient filtering based approach LSB image steganography using status bit along AES cryptography. In: ICIEV, pp. 55–60 (2014)
7. Wagner, D.A., Manferdeli, J.L.: Cryptography and Cryptanalysis, pp. 312–319. Springer, Heidelberg (2013)
8. Carli, M., Neri, A., Egiaziarian, K., Battisti, F.: A Generalized Fibonacci LSB Data Hiding Technique. (CODEC2006) TEA, Institute of Radio Physics and Electronics, University of Calcutta, pp. 37–45 (2006)
9. Abraham, A., Sanyal, S., Dey, S.: An LSB Data Hiding Technique Using Prime Numbers, pp. 223–235. IEEE CS Press, UK (2007)
10. Gupta, S., Goel, S., Kaushik, N.: Image steganography least significant bit with multiple progressions. In: FICTA (2015)
11. Cachin, C.: An information-theoretic model for steganography. In: 2nd Information Hiding Workshop, pp. 78–86 (1998)

Classification of Query Graph Using Maximum Connected Component

**Parnika Paranjape, Meera Dhabu, Rushikesh Pathak,
Nitesh Funde and Parag Deshpande**

1 Introduction

The graph classification problem is learning to classify separate, individual graphs in a graph database into two or more categories. Complex data such as chemical compounds, protein molecules, documents, image network, brain network etc. exhibit the graph structure naturally. Therefore, a lot of graph classification methods have been developed in the literature for such complex data [1–7]. Graph classification methods can be broadly categorized into two groups: similarity based methods i.e. graph kernels and subgraph pattern based methods.

Most of the subgraph feature based methods extract discriminative subgraphs to transform the graphs into binary vectors and apply generic classifiers like SVM for classification. The drawback of graph kernels and SVM approach is fixed input dimensionality. Subgraph pattern based methods extract frequent subgraphs at user specified minimum support using several existing frequent subgraph mining (FSG) algorithms like AGM [8], FSG [9], FFSM [10], gSpan [11], Gaston [12] etc. These algorithms require exponential execution time to extract each and every pattern from the graph database for very low minimum support. Storing all these patterns consumes lot of memory. Another problem is matching a query graph with subgraph patterns using graph isomorphism is a NP-hard problem.

To overcome these issues, we propose a novel method to classify query graph using maximum connected component. In this approach, we first construct a DFS tree to store only maximum substructures based on DFS lexicographic order of gSpan. We then parse a query graph and search for the maximum connected component of DFS

P. Paranjape (✉) · M. Dhabu · N. Funde · P. Deshpande
Department of Computer Science and Engineering, Visvesvaraya National Institute
of Technology, Nagpur, India
e-mail: parnika.paranjape@students.vnit.ac.in

R. Pathak
Nutanix Technologies India Pvt. Ltd., Bangalore, Karnataka, India

© Springer Nature Singapore Pte Ltd. 2019
S. Tiwari et al. (eds.), *Smart Innovations in Communication and Computational
Sciences*, Advances in Intelligent Systems and Computing 851,
https://doi.org/10.1007/978-981-13-2414-7_38

tree having highest match with query graph. DFS tree saves a lot of memory by storing only maximum components. Another advantage of DFS tree is quick graph matching which in other cases has to be performed using graph isomorphism which has high time as well as space complexity.

2   Literature Review

In this section we present a brief review of major approaches for graph classification namely similarity based methods, subgraph pattern based methods, and probabilistic substructure approach.

For similarity based methods, a similarity between a pair of graphs can be assessed using graph kernels [13–17] or graph embedding [18]. Weisfeiler Lehman [16] kernel provides faster feature extraction using Weisfeiler Lehman test of graph isomorphism. It transforms the input graph into a sequence of graphs and defines various kernels based on the sequence of graphs. Among these kernels, subtree kernel is most efficient which compares two graphs on the basis of a number of subtrees common to them. Propagation kernels [17] capture the pattern of information spread through a set of graphs. These kernels model the structural information present in node labels, attributes, and edges using early-stage distributions from propagation schemes like random walks. The drawback of these graph kernel approaches is that they work with fixed input dimensionality.

Subgraph feature based methods find out a set of discriminant subgraph features to build a classification model [2, 3, 7, 19]. There are several frequent subgraph mining algorithms like AGM, FSG, FFSM, gSpan, and Gaston. Recently a very fast and memory efficient version of gSpan has been developed called gBolt [20]. Subgraph pattern based methods like [21] use discriminative subgraph features to transform the input graphs into vector space so that generic classifier like SVM can be directly applied. However, COM [2] takes into account the pattern co-occurrence information for graph classification. It considers a group of many weak features as a discriminative feature instead of using a single long pattern for classification. Subdue [22] algorithm is another subgraph pattern based method which performs greedy, heuristic search for subgraphs present in positive examples and absent in the negative examples. Subdue maintains the instances of subgraphs (in order to avoid subgraph isomorphism) and uses graph isomorphism to determine the instances of the candidate substructure in the input graph. The limitation of Subdue is classification decision is based on only best-compressed frequent subgraph patterns.

A principled approach of probabilistic substructure method MaxEnt [23] is to build a global classification model from the local subgraph patterns using the maximum entropy principle. Given underlying subgraph features, this approach learns a conditional probability distribution of the class for a graph by using the support of the features as the constraints imposed on the probability model.

Some limitations of the existing graph classification methods are (1) exponential pattern search space due to implementation based on frequent subgraph mining algorithms (2) high complexity due to graph isomorphism test (3) loss of crucial patterns due to threshold dependency of frequent subgraphs which in many cases might be important.

3 Overview of the Proposed Work

In the proposed work, we construct a DFS tree for each class which stores only maximum substructures belonging to graph dataset of a given class. Based on the approach used by gSpan to extract frequent sub-graphs, we implement DFS lexicographic order for constructing DFS tree. DFS tree offers several advantages over existing frequent sugraph mining algorithms. DFS tree consumes less memory as it stores only maximum substructures as opposed to other methods such as AGM, FSG, FFSM and gSpan which generate all subgraph patterns present in the database satisfying minimum support threshold. When the minimum support is very low even the efficient algorithms like gSpan consumes exponential run time to extract all frequent subgraphs from the database. Another benefit of tree approach is quick graph matching over other approaches using graph isomorphism test which is NP-hard problem.

After constructing DFS trees for all the classes, the query graph and its subgraphs are processed. During parsing of query graph, proposed algorithm generates a score for each graph of a particular class which shares common pattern with query graph. Graph with maximum score generated during parsing of query graph will have highest match with query graph.

In this work, we have focused on binary classification task.

4 Proposed Work

In this section we briefly describe DFS tree construction and classification of query graph using maximum connected component.

4.1 DFS Tree Construction

As stated earlier DFS tree is created for every class and the number of branches of the tree is equal to number of distinct vertex labels of that particular class. All graphs with same class label are grouped in one partition and tree is constructuted for each partition. We will now explain the steps involved in tree construction.

Distinct vertex and edge labels of the given partition are sorted in descending order according to their frequency counts. Based on the frequency counts, new unique labels are assigned to vertices and edges. Let $V_{Old} = vl_1, vl_2, \ldots, vl_p$ be the old ver-

Table 1 Edge-detail table

V_{l1}	V_{l2}	E	(ID, V_{n1}, V_{n2})
0	1	0	(0, 2, 3)

tex labels sorted in decreasing order of counts. Then, the new label set for vertices is, $V_{new} = 1, 2, \ldots, p$. Similarly, for the old edge label set $E_{old} = el_1, el_2, \ldots, el_q$, the new edge label set is $E_{new} = 1, 2, \ldots, q$. After relabeling the vertices and edges of the partition, an Edge-Detail table is created for the given partition to store connectivity information. Each entry contains source and destination vertex labels, edge label, graph ID and corresponding vertex numbers having those vertex labels. The entries are maintained in descending order from most frequent edges to least frequent edges in graph set. We build DFS tree using Edge-Detail table which allows us to check connectivity and consider only those edges which exists in graph database. Structure of an Edge-Detail table is shown in Table 1. Each node of the DFS tree maintains certain information such as vertex label, vertex number from which the pattern originated, edge label of an edge connecting it with its parent vertex, forward edge or backward edge, a list of graph IDs to which particular pattern belongs to, pointer to child nodes. DFS tree algorithm is presented in 1.

Algorithm 1 DFS Tree

Require: Set of graphs along with class labels
Ensure: TS : DFS Tree for each class
 1: $TS = \phi$
 2: $PS \leftarrow$ Create partitions of the graph databse such that the graphs with same label belong to one partition;
 3: **for** each partition $p \in PS$ **do**
 4:    $V_{new} \leftarrow$ New vertex labels as per the frequency count of distinct vertex labels in p;
 5:    $E_{new} \leftarrow$ New edge labels as per the frequency count of distinct edge labels in p;
 6:    $Edge\_Table \leftarrow$ Edge-Detail table to store connectivity information of edges present in p;
 7:    $t \leftarrow$ DFS tree constructed using Edge_Table;
 8:    $TS \leftarrow TS \cup t$;
 9: **end for**
10: **return** TS;

4.2 Finding Maximum Connected Component

Once the DFS trees are constructed for all classes, the matching of query graph to the patterns stored on DFS tree needs to be performed such that the minimum DFS tree of query graph should be matched rather than all patterns belonging to query graph. To achieve this, query graph is relabeled as per the class i for which score is to be generated. After relabeling, an edge detail table of the query graph is built and each edge of the query graph is matched against the edges in the DFS tree. For each match of connected edge, the score of the graph IDs present in the graph list of corresponding nodes is incremented by 1. The proposed algorithm generates the score for each graph in the database which shares common patterns with query graph. Score associated with each graph of a given class is the number of edges matched with query graph. Algorithm for score calculation makes sure that each edge of query graph can occur at the most once in maximum connected component. To ensure this, algorithm 2 produces a disjoint set (DSJ) of graphs along with their scores such that these graphs do not have any query graph edge in common and total score of the set of graphs is less than number of edges in query graph. Let $DSJ = S_1, S_2, \ldots, S_k$ be a disjoint set and $S_i \in DSJ$ represents an entry having graph ID and score then, the final score of the DSJ is computed using (1),

$$DSJ_{score} = \sqrt{(S_1 . score)^2 + \ldots + (S_k . score)^2} \tag{1}$$

Details of the score generation for each graph are summarized in the algorithm 2.

Algorithm 2 Score Generation

Require: Query graph and DFS Tree of a given class i
Ensure: DSJ_i : Disjoint set of graphs and respective scores for class i
1: $Edge\_Table_Q \leftarrow$ Build Edge_Table of query graph after relabeling as per class i;
2: $DFS_{min} \leftarrow$ Minimum DFS tree of query graph using Edge_Table;
3: **for** each edge $e \in DFS_{min}$ **do**
4:     **if** e matches with an edge in DFS_i **then**
5:         Increment the score of graph_IDs present in graph list of corresponding nodes of DFS_i by 1;
6:         $ScoreList \leftarrow$ Append the score with graph IDs to $ScoreList$;
7:     **end if**
8: **end for**
9:
10: $DSJ \leftarrow$ Find a disjoint set of entries from $ScoreList$ such that sum of the scores of entries is less than no. of edges in query graph; **return** DSJ_i;

4.3 Classification Based on Matching Score

Let each class contains N different disjoint sets arranged in the descending order of their scores. We compare only highest DSJ score of each class with each other. If DSJ score ties at any level, we consider next highest DSJ score of respective classes. Let $DSJ_i = DSJ_i^1, DSJ_i^2, \ldots, DSJ_i^N$ denotes disjoint sets for class i. DSJ_i^1 has highest score among all others of class i. We assign a label to query graph based on the class which has highest DSJ_i^1 score.

Table 2 Characteristics of the dataset

Dataset	# Positive examples	# Negative examples	Avg. # vertices	Avg. # edges
Mutagenesis	125	63	17.93	19.79
PTC (MR)	152	192	25.55	25.96
PTC (MM)	129	207	25.047	25.395
PTC (FR)	121	230	26.082	26.527
PTC (FM)	143	206	25.246	25.618

5   Performance Analysis

In this section, we present performance analysis on five real-world datasets Mutagenesis and Predictive Toxicology data (PTC) which are considered as benchmark datasets for graph classification problem. These datasets can be obtained from ChemDB[1] portal. Table 2 shows characteristics of these datasets. All experiments are performed on an Intel(R) Core(TM) 2 Duo machine with 2.20 GHz CPU, 4 GB main memory, and 64 bit Windows platform. The proposed algorithm CMCC is implemented in Java.

We have performed experiments on these datasets with two different cross-validation settings: (1) k-fold cross-validation (2) Leave-one-out cross-validation.

5.1   Analysis Using k-Fold Cross-Validation

We have performed 5-fold cross-validation on PTC datasets and 10 times 10-fold cross-validation on Mutagenesis dataset. Prediction accuracies have been averaged over all runs of the experiment on each dataset.

Accuracy results for PTC datasets in Fig. 1 indicate that the proposed method CMCC significantly outperforms other methods. The results of the methods MaxEnt, Ada, and SVM have been taken from the work [23]. Refer [24] for Subdue results on PTC datasets.

As it can be seen from the Fig. 3, the average accuracy of the proposed method CMCC on Mutagenesis data is competitive with all graph kernel methods except Ramon & Gartner and Shortest path. Ramon & Gartner and Shortest path kernels outperform our proposed method CMCC. Refer [16, 17] for the results of the graph kernel methods.

In Fig. 2, we have also shown the minimum, maximum, and average accuracy of the proposed method CMCC across all runs of the experiment on each dataset. The minimum and maximum accuracies for other established methods are not shown in Fig. 2 as they have not been provided by the authors of these methods in their respective work.

---

[1] http://cdb.ics.uci.edu/cgibin/LearningDatasetsWeb.py.

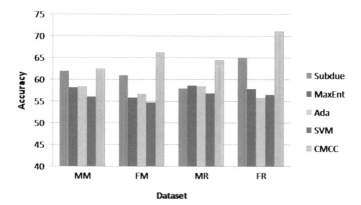

Fig. 1 Prediction accuracy comparisons on PTC datasets

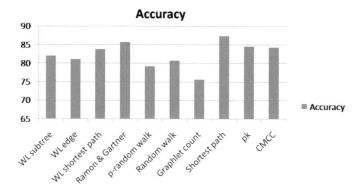

Fig. 2 Prediction accuracy comparisons on Mutag dataset

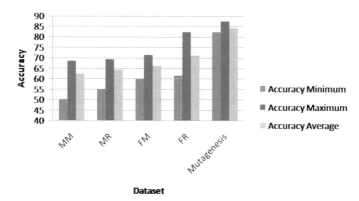

Fig. 3 Minimum, maximum, and average accuracy of CMCC on all datasets

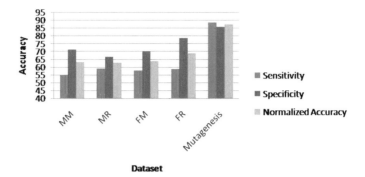

Fig. 4 Performance of CMCC on all datasets using leave-one-out cross-validation

5.2 Analysis Using Leave-One-Out Cross-Validation

In this case, we have performed experiments on all datasets using Leave-one-out method of cross-validation. For this setting, comparison with other established methods is not shown since their experimental setup is similar to k-fold cross-validation setting. The results in the Fig. 4 show specificity, sensitivity and normalized accuracy of the proposed method CMCC on all five real-world datasets.

6 Conclusion and Future Scope

In this work, we proposed a Maximum Connected Component approach for graph classification which uses a DFS tree for storing maximum subgraphs based on DFS lexicographic order used by gSpan. Storing of patterns in DFS tree not only saves memory as compared to frequent subgraph algorithms but the execution time is also reduced because of quick graph matching. Experimental results on PTC datasets demonstrate that the proposed method significantly enhances the accuracy over all established methods used in the experiments. Though the proposed method could not perform better than Ramon & Gartner and Shortest path graph kernels on Mutagenesis dataset, its overall performance is competitive with other remaining graph kernels.

The size of DFS tree grows exponentially for the graphs with high connectivity and large number of vertices with distinct vertex labels. This results in huge memory requirements and query graph matching becomes time consuming task. To alleviate these issues, we plan to enhance the proposed method for storing only discriminative CMCC patterns which will significantly reduce the memory requirements and improve the classification accuracy.

References

1. Deshpande, M., Kuromochi, M., Wale, N., Karypis, G.: Frequent substructure-based approaches for classifying chemical compounds. IEEE Trans. Knowl. Data Eng. **17**(8), 1036–1050 (2005)
2. Jin, N., Young, C.: Graph Classification Based on Pattern Co-occurrence. CIKM, Hong Kong, China (2009)
3. Jin, N., Young, C.: Graph Classification Using Evolutionary Computation. SIGMOD, USA (2010)
4. Ranu, S., Hoang, M., Singh, A.K.: Mining discriminative subgraphs from global-state networks. In Proceedings of the 19th ACM International Conference on Knowledge Discovery and Data Mining, pp. 509–517 (2013)
5. Dang, X.H., Singh, A.K., Bogdanov, P., You, H., Hsu, B.: Discriminative subnetworks with regularized spectral learning for global-state network data. In: Joint European Conference on Machine Learning and Knowledge Discovery in Databases, pp. 290–306. Springer, Berlin, Heidelberg (2014)
6. Dang, X.H., You, H., Bogdanov, P., Singh, A.K.: Learning predictive substructures with regularization for network data. In: IEEE International Conference on Data Mining (ICDM), pp. 81–90 (2015)
7. Pan, S., Zhu, X., Zhang, C., Philip, S.Y.: Graph stream classification using labeled and unlabeled graphs. In: IEEE 29th International Conference on Data Engineering (ICDE), pp. 398–409 (2013)
8. Inokuchi, A., Washio, T., Motoda, H.: An apriori-based algorithm for mining frequent substructures from graph data. In: Principles of Data Mining and Knowledge Discovery, pp. 13–23 (2000)
9. Kuramochi, M., Karypis, G.: Frequent subgraph discovery. In: Proceedings of the IEEE International Conference on Data Mining ICDM, pp. 313–320 (2001)
10. Huan, J., Wang, W., Prins, J.: Efficient mining of frequent subgraphs in the presence of isomorphism. In: IEEE 3rd International Conference on Data Mining, ICDM, pp. 549–552 (2003)
11. Yan, X., Han, J.: gspan: graph-based substructure pattern mining. In: Proceedings of the IEEE International Conference on Data Mining, ICDM, pp. 721–724 (2002)
12. Nijssen, S., Kok, J.N.: A quickstart in frequent structure mining can make a difference. In: Proceedings of the 10th ACM SIGKDD International Conference on Knowledge discovery and data mining, pp. 647–652 (2004)
13. Schölkopf, B., Tsuda, K., Vert, J.P. (eds.): Kernel Methods in Computational Biology. MIT press (2004)
14. Mahé, P., Ueda, N., Akutsu, T., Perret, J.L.: Extensions of marginalized graph kernels. In: Proceedings of the 21st ACM International Conference Machine Learning, p. 70 (2004)
15. Seeland, M., Karwath, A., Kramer, S.: A structural cluster kernel for learning on graphs. In: Proceedings of the 18th ACM SIGKDD International Conference on Knowledge discovery and data mining, pp. 516–524 (2012)
16. Shervashidze, N., Schweitzer, P., Leeuwen, E.J.V., Mehlhorn, K., Borgwardt, K.M.: Weisfeiler-lehman graph kernels. J. Mach. Learn. Res. 12, 2539–2561 (2011)
17. Neumann, M., Garnett, R., Bauckhage, C., Kersting, K.: Propagation kernels: efficient graph kernels from propagated information. Mach. Learn. **102**(2), 209–245 (2016)
18. Riesen, K., Bunke, H.: Graph classification by means of Lipschitz embedding. IEEE Trans. Syst. Man Cybern. Part B (Cybernetics) **39**(6), 1472–1483 (2009)
19. Kong, X., Philip, S.Y.: gMLC: a multi-label feature selection framework for graph classification. Knowl. Inf. Syst. **31**(2), 281–305 (2012)
20. Zhou, K.: gspan algorithm in data mining. https://github.com/Jokeren/DataMining-gSpan. Accessed 20 Nov 2017
21. Zhu, Y., Yu, J.X., Cheng, H., Qin, L.: Graph classification: a diversified discriminative feature selection approach. In Proceedings of the 21st ACM International Conference on Information and Knowledge Management, pp. 205–214 (2012)

22. Ketkar, N.S., Holder, L.B., Cook, D.J.: Subdue: compression-based frequent pattern discovery in graph data. In: Proceedings of the 1st international workshop on Open Source Data Mining: Frequent Pattern Mining Implementations, pp. 71–76 (2005)
23. Moonesinghe, H.D.K., Valizadegan, H., Fodeh, S., Tan, P.N.: A probabilistic substructure-based approach for graph classification. In: 19th IEEE Conference on ICTAI 2007, vol. 1, pp. 346–349 (2007)
24. Ketkar, N.S., Holder, L.B., Cook, D.J.: Empirical comparison of graph classification algorithms. In: IEEE Symposium on Computational Intelligence and Data Mining (CIDM), pp. 259–266. IEEE (2009)

All Domain Hidden Web Exposer Ontologies: A Unified Approach for Excavating the Web to Unhide Deep Web

Manpreet Singh Sehgal and Jay Shankar Prasad

1 Introduction

It has been turned out that web pages make up only 1% (approximately) [1–3] of the entire repository, whereas rest of the 99% which is more valuable is stored in the databases. The former is termed as PIW (Publically Indexed Web) whereas the latter is called Hidden Web or deep web. The irony of the matter is that one can view hidden web when one knows where it is situated (exact url of the hidden web) Considering the huge volume and decentralized behavior of this online hidden repository there is a need of one platform where one can mention the area of interest and the data/information/knowledge from this huge repository is automatically flown to the user. Our objective in this paper is to discuss the possibility of such automation. This paper critically analyses the requirements of information (hidden web data) extraction and make appropriate assumptions and set certain constraints for setting up the environment which is conducive to extract hidden web data for all the domains. Section 2 presents related work in this field. Section 3 discusses the proposed algorithm. Experiments and Results are empirically analyzed in Sect. 4. Section 5 concludes this paper with the future directions to the implementers of this algorithm. At the end a set of references are listed which are referred in this journey of knowledge research.

M. S. Sehgal (✉) · J. S. Prasad
MVN University, Palwal, Haryana, India
e-mail: manpreet.sehgal@asu.apeejay.edu

J. S. Prasad
e-mail: jayshankar.prasad@mvn.edu.in

© Springer Nature Singapore Pte Ltd. 2019
S. Tiwari et al. (eds.), *Smart Innovations in Communication and Computational Sciences*, Advances in Intelligent Systems and Computing 851,
https://doi.org/10.1007/978-981-13-2414-7_39

2   Related Work

Fetching hidden web data [1, 2] for almost every possible domain is always a much needed and sought after research area. This paper is the algorithmic attempt to discuss the means and ways to implement an ontology-based tool that can automatically sense the fields in the hidden web page and generate a wrapper form that is filled and submitted to extract data hidden behind the search interfaces. This section briefly describes attempts to unhide the hidden web data of specific domains.

There has been a considerable rise in the number of methods that attempted to uncover deep web [15], as the approach named HWPDE to extract data from structured hidden web page [3, 4]. This technique introduced new ways called data sponger for the extraction, data squeezer for data pouring and ADTD for Automatic Data type detection mechanisms. This tool automatically creates a database and table in it at the client side after harvesting and processing hidden data from a hidden web page.

Liu and Grossman [5] proposed the technique for mining the data records from the Web page. They called this technique MDR. Contiguous and non-contiguous data records were the target of MDR and MDR proved successful in mining both. Experimental results showed a remarkable improvement against existing techniques.

In [6] the proposal named VIPS (VIsion-based Page Segmentation) in the form of algorithm was made so as to extract the semantic structure in the web page. This kind of semantic structure is a hierarchical where each node corresponds to a block which is assigned a value called degree of coherence.

The method to identify templates of web page along with the tag structures of a document so that structured data can be extracted from deep web sources and can be presented as the results using templates on demand of the user query [7].

A machine learning approach was introduced by Wang and Hu in [8] that is able to identify data rich tables in the web page among many of them present on the page as web designers may use <table> tag for the easy layout purpose.

The use of ontology driven and dynamic ontologies for information extraction and representation is a prominent field [9–11].

In [12] Proposed the method for eliminating noises from web pages for the purpose of improving the accuracy and efficiency of web content mining. Methods to crawl hidden objects using kNN queries are the main thrust area of [13]. Researchers also conducted comparative studies to effectively evaluate the search engines [14]. Several methods have been proposed so far to crawl the hidden web page [7–15] yet the domain independent searching is fascinating to handle, hence this paper proposes an approach to fetch them.

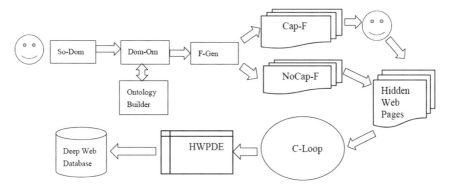

Fig. 1 The conceptual diagram of ADHWENT

3 Proposed Work

In this paper domain independent algorithm to extract hidden web data is proposed so that information seeker does not need to look for domain specific hidden web crawlers thereby reducing the time spent in knowledge discovery from the repository which is rooted deep into the www away from the accessibility of surface web search engines. Following subsections represents the conceptual architecture and the details of its components.

3.1 The Conceptual Architecture

Figure 1 shows the conceptual diagram of the All Domain Hidden Web Exposer Ontologies in short ADHWENT. The process of excavating World Wide Web is mentioned as eight step approach, starting from selecting or typing domain in the search box to inputting the extracted deep webpages to HWPDE [4] if necessary to store the data from the webpages into the relational databases for further mining and analysis.

Following sections describe the details of this process.

So_Dom (Statement of domain). This step is the input of the user into the free form text field, which is equipped with autocomplete feature to aid in selecting the words from already filled up file.

Dom_Om (Domain Ontology Mapper). This paper assumes the presence of the master ontolgy file that lists exhaustive set of ontologies, we will refer this file as M-Ont in further discussion. The keyword from the SoDom is picked up and matched against the keywords (along with their corresponding synonymous words) from M-Ont. In case there is no match found than an automated email is sent to the ontology designer notifying about the need of an ontology for the domain concept mentioned in So_Dom, and user is directed to "Coming Soon" page. On the other hand in case of a match, the ontology is browsed and traversed and as a result the hash map is prepared for the later assignments of labels and their values in the search forms.

Ontology Builder. This component is only used in case there is no matching ontology mention available in the M-Ont and is beyond the scope of this paper.

F_Gen stands for Form Generator. This phase prepares the empirical form with concrete labels gathered from the UI forms (from the urls of websites mentioned in "individuals by class" instances in the master ontology) and their values filled in from the values portion of hash map after finding appropriate match for labels from the hash map designed in the Dom-Om step.

Cap_F and NoCap_F. There are two categories of User Interface Forms, One that requires CAPTCHA to prove non-robotic (Cap_F) behavior, and others without this restriction (NoCap_F). In case of Cap_F the forms are presented to the user to fill in the CAPTCHA asked for whereas in other case form values are directly submitted to generate hidden web pages.

C-Loop. This phase is called consumption loop as its objective is to prepare generated dynamic hidden webpages into the form acceptable by HWPDE [4], which is another tool designed to sponge HTML data from HTML Pages and squeeze the same after appropriate processing to the relational database.

HWPDE (Hidden Web page Data Extractor). This application is used to sponge the hidden web data from the processed pages by C-Loop and then to squeeze the data into the relational database for the purpose of knowledge analysis and decision making.

Deep Web Database. The output of the ADHWENT is in the form of relational database which captures the tabular information presented in the hidden web page.

3.2 Algorithms of the Components and Their Descriptions for the Black Boxes

Algorithm: So_Dom.

1. *Get the statement of Domain from user in the text box D*
2. *Return the value typed in D to Dom_Om(D)*

So_Dom looks for the typed search item in the textbox and provides the same to Dom_Om in D

Algorithm: Dom_Om(D).

1. *Get the domain returned from Sodom*
2. *Search Master Ontology for the domain retrieved from Sodom.*
3. *If relevent ontology is found then return in "dbag" else*

   a. *Refer dictionary or thesaurus for D.*
   b. *Search Master Ontology for all the aliases found in dictionary.*

c. *if relevent ontology is found than return in "dbag" else make a note to ontology builder to build missing ontology.* (beyond the scope of this paper)

The above algorithm finds a match between the returned domain from SoDom (with the help of dictionary/thesaurus if required) and the ontology from the master ontology file. In case of unsuccessful search for the relevant ontology the ontology builder is marked for the creation of ontology for future searches. The implementation of Ontology builder is beyond the scope of this paper.

Algorithm F_Gen (dbag)

1. *Extract all the urls and processing file links from the dbag in address_table.*
2. *Generate a listbox in HTML and display the urls of hidden websites poured in address_table (Step 1 above).*
3. *Create a blank form, name it as FRM.*
4. *for each url in the listbox generated in step 2*

   a. *update dbag to contain form elements of the url file.*
   b. *for each item in the updated dbag*
      (1) *if the key of item is "captcha" Generate image control in html and place CAPTCHA image and a text box for the CAPTCHA entry.*
      (2) *else follow steps c, d and e*
      (3) *Create a Lable and assign Key of the item to it*
      (4) *Generate a combo box (if the domain is finite) and input box (for the free form text value) and populate the values in the combo box with the values associated with the key of the current item in the dbag.*
      (5) *Create a new line.*
   c. *Create a Submit Button.*
   d. *Update the action attribute of the form to contain the absolute address (url/processing_file) of processing file.*
   e. *Save the generated form in FRM*

5. *return FRM*

F_Gen routine takes the dbag generated by Dom_Om (Domain) for fetching the domain bag (dbag). This task is accomplished by pouring the list of addresses from the dbag extracted from master ontology on to relational table called address_table. A list box is created to display all the hidden websites referred for the creation of key value pairs. For every other non CAPTCHA item in the domain bag (dbag) an assignment is made to the label from key of the item and to the input from the value of the key. All of these pairs are separated by new lines. At the end of the control placing loop a link is created through submit button so that the values can be transferred to the actual user interface forms. It is notable that the action attribute of the form needs to be an absolute address so that the processing file is correctly targeted in the event of form submission.

Algorithm NoCap_F (FRM).

1. *Create empty hash-map H*
2. *Fetch matched ontology key value pairs into the H*
3. *Initiate the label detection process on the form*

   a. *For each detected label on the form,*
      (1) *Select matching key from H.*
      (2) *Assign value of the selected key (in a above) into the input element*
      (3) *Submit the form and return*

NoCap_F is an attempt to assign values to the form that does not ask for CAPTCHA or to the form where user has already filled in CAPTCHA to proceed. This code will retrieve the values in a hash map H from the matched ontology and will fill the values in the form (CAPTCHA-less) after label detection process. It first fetches the key value pairs from the matched ontology into the Hash Map, and then assigns values from this hash map against the input fields of the detected labels that closely match with the keys. Once the query is formed in the form of filled in fields it is fired and the result page is fed to HWPDE [4] which consequently will create a database of hidden web information. The details of C-loop, HWPDE and the format of relational database construction has been already explained in [4].

4   Performance Metrics and Results

As the success of any idea depends on how effectively it is designed as algorithm, the performance metrics for ADHWENT has been chosen as speed, i.e., how long ADHWENT takes to produce its result. At present the speed of ADHWENT depends upon the availability of candidate ontology, if one exists than the time that ontology builder takes to build the candidate ontology gets saved thereby remarkably improving the performance and speed. Other essential component is the choice of data structures used in algorithm. A special care has been taken in this regard and data structures like Hash maps (Used in ADHWENT for fetching knowledge from Ontologies), graphs (ontology is modeled as a graph, where nodes are the entities and edges are the relationships in the entities.) are used extensively in algorithmic details of ADHWENT. Following subsections discusses the time complexities of the various processes used in the proposed work.

4.1   So_Dom Performance

The Process of So_Dom is to fetch user input which is possible in a constant time as shown in Eq. 1.

$$T(So\_Dom) = O(1) \tag{1}$$

4.2 Dom_Om Performance

This component searches the ontology match in the master ontology M-Ont (this is a kind of Meta ontology that describes other ontologies) and as the ontology is organized as a graph where nodes are objects and edges are the relationships causing properties to be associated to objects. The search process can be either Breadth first Search (BFS) or Depth First Search (DFS). Assuming the ontology graph is implemented using Adjacency lists and the search criteria is BFS or DFS. The time complexity of Dom_Om is shown in Eq. 2.

$$T(Dom\_Om) = O(V + E), \tag{2}$$

where V is the set of Objects in Master Ontology and E is the set of Edges connecting various objects. However the above complexity captures the essence of one successful search of a master ontology. For the unsuccessful search, the thesaurus (assumed as implemented as a hash map) is searched using a binary search with the Time Complexity $\theta(\lg n)$ and then Master Ontology is searched for with the same $O(V + E)$ so in general the time complexity of Dom_ Om can be written as Eq. 3.

$$T(Dom\_Om) = p \cdot \theta(\lg n) + k \cdot O(V + E) \tag{3}$$

where $P \in \{0|1\}$ and $k \geq 1$.

4.3 F_Gen Performance

The major activity of F-gen is the multiple traversals (say k) of the graphs associated with the urls of the domain, where each node of the dbag is processed and displayed on the form along with the possible values. The traversal is similar to the search algorithm where the last item in the graph is always the item to be searched as the motive is to linearly traverse the entire graph. Hence the time complexity of this major step of F_Gen is the worst case complexity as shown in Eq. 4 of any of the search method (BFS or DFS) in addition to the cost of updating graph and creating labels and other form elements is

$$T(F\_Gen) = k * O(V + E), \tag{4}$$

where k is the number of urls for the selected domain in the M-Ont (Master Ontology).

Table 1 Comparison of proposed work against domain specific extraction mechanism

Criteria	Domain specific hidden web extractor algorithm	Proposed ADHWENT algorithm
Domain	Specific	Independent
Complexity	Domain restricted	General
Coverage	Limited	Broad

4.4 NoCap_F Performance

The performance measure of this algorithm is dependent on the label detection process that undergoes comparisons with the keys stored in the hash map. Hash map has an advantage that the search time is $O(1)$. Suppose there are k values on an average associated with any key (for any detected label) and there are n form elements on the form, than combining all of these factors as given in Eq. 5.

$$T(NoCap\_F) = O(1) + O(n * k) \tag{5}$$

4.5 Results

ADHWENT presents improved results when compared with the domain specific hidden web extractor algorithms [7–9] in many ways. Table 1 depicts the improving trends.

5 Conclusion and Future Scope

The proposed work of fetching hidden web data with the help of ontologies which is independent of domain is much needed as there are different ontologies available in abundance for specific domains. This paper attempted to shed light on the algorithmic details of the use of such master ontology and in the process it is realized that by using proper data structures, objects, and making use of libraries of standard functions, remarkable results can be obtained.

ADHWENT is the innovative idea in the field of extraction of deep web data for all domains provided the ontology structure is available. The future scope lies in the process of designing algorithm for Ontology Builder for missing ontologies in the master ontology. As a note to the future researchers, a novel implementation of the components discussed in the proposed work is anticipated. As a limitation ADHWENT cannot handle CAPTCHA forms and user intervention is sought for to counter the CAPTCHA.

References

1. Bergman, M.K.: The Deep Web: Surfacing Hidden Value (2000)
2. Lawrence, S., Giles, C.L.: Searching the World Wide Web. Science **280**(5360), 98–100 (1998)
3. Lawrence, S., Giles, C.L.: Accessibility of information on the web. Nature **400**, 107 (1999). https://doi.org/10.1038/21987
4. Sehgal, M., Anuradha.: HWPDE: novel approach for data extraction from structured web pages. Int. J. Comput. Appl. (0975–8887), **50**(8), 22–27 (2012)
5. Liu, B., Grossman, R., Zhai, Y.: Mining data records in web pages. In KDD 03: Proceedings of the ninth ACM SIGKDD International Conference on Knowledge Discovery and Data Mining, pp. 601–606 (2003)
6. Cai, D., Yu, S., Wen, J.R., Ma, W.Y.: VIPS: a Vision-based page segmentation algorithm. Microsoft Tech. Rep. MSR-TR-2003-79 (2003)
7. Anuradha, Sharma, A.K.: A novel technique for data extraction from hidden web databases. Int. J. Comput. Appl. **15**(4), 45–48 (2011)
8. Wang, Y., Hu, J.: A machine learning based approach for table detection on the web. In: Proceedings of the 11th International Conference on World Wide Web, pp. 242–250 (2002)
9. Yildiz, B., Miksch, S.: OntoX—a method for ontology-driven information extraction. In: Proceedings of the International Conference on Computational Science and its Applications, pp. 660–673 (2007)
10. McDowell, L., Cafarella, M.J.: Ontology-driven information extraction with OntoSyphon. In Proceedings of the 5th International Semantic Web Conference, pp. 428–444 (2006)
11. Hwang, C.: Incompletely and imprecisely speaking: using dynamic ontologies for representing and retrieving information. In: Proceedings of the 6th International Workshop on Knowledge Representation Meets Databases, pp. 29–30 (1999)
12. Sivakumar, P.: Effectual web content mining using noise removal from web pages. Wirel. Pers Commun. 84–99 (2015)
13. Yan, H., Gong, Z., Zhang, N., Huang, T., Zhong, H., Wei, J.: Crawling Hidden Objects with kNN Queries. IEEE Trans. Knowl. Data Eng. **28**(4), 912–924 (2016)
14. Song, D., Luo, Y., Heflin, J.: Linking heterogeneous data in the semantic web using scalable and domain-independent candidate selection. IEEE Trans. Knowl. Data Eng. **29**(1), 143–156 (2017)
15. Schafer, R.: Accurate and efficient general-purpose boilerplate detection for crawled web corpora. Lang Resour. Eval. **51**(3), 873–889 (2017)

Data Stream Classification Using Dynamic Model for Labeling Strategy

Nitesh Funde, Meera Dhabu and Parnika Paranjape

1 Introduction

Classifying data stream is one of the challenging problems in real-time applications because of its operational constraints such as limited memory and time constraints [1]. The importance of data stream mining technique has greatly increased in many real-time commercial applications such as credit card fraud detection [2], intrusion detection, internet traffic management [3], and sensor-based monitoring systems. Data streams have three properties: concept drift, infinite length, and concept evolution [4]. Since data stream is of infinite length, traditional multiscan data mining algorithms are not capable of processing it because they would need infinite training and storage time. Concept drift happens when underlying data distribution changes over the time. Hence, classification model must have to be updated progressively which reflects the most recent concept. Concept evolution is the arrival of a novel class in the real-time data stream classification applications such as fault detection, intrusion detection, and social networks. These challenges have acquired attention of many researchers since the last two decades that has led to the publications of classifying data streams. These publications are as follows: G-eRules [5] , Hoeffding trees [6], and very fast decision rules (VFDR) [7] need only single scan of data where they have adapted a concept drift in real-time environment.

However, most of the existing approaches are based on supervised learning which are trained with labeled data for classification of data streams. But in real-time streaming environment, where large volume of data arrives at a high speed, only limited labeled data may be available for classification. Thus, we have proposed a classification model which build from both small amount of labeled and unlabeled data. The classification model is built by creating submodels on the data stream chunks. The micro-clustering technique is used for training submodel which contains partially labeled data. The process of classifying or labeling the test instances also needs to be addressed for improving the classification accuracy within these constraints.

N. Funde (✉) · M. Dhabu · P. Paranjape
Department of Computer Science and Engineering, Visvesvaraya National Institute
of Technology, Nagpur 440010, India
e-mail: nitesh.funde@gmail.com

© Springer Nature Singapore Pte Ltd. 2019
S. Tiwari et al. (eds.), *Smart Innovations in Communication and Computational
Sciences*, Advances in Intelligent Systems and Computing 851,
https://doi.org/10.1007/978-981-13-2414-7_40

The two different approaches for classification are presented and experimented. First approach is max weighted-sum inter-submodel, and second is high-frequency max-weight intra-submodel. Assigning a label to a test instance is entirely based on how the submodels have been built during training phase. The different challenges such as concept evolution and concept drift are also addressed for data stream classification.

This paper is organized as follows. Section 2 discusses the related work of data stream classification algorithm. Section 3 presented the proposed dynamic classification algorithm. Section 4 provides experimental details, results, and discussion, and at last, Sect. 5 gives conclusion.

2   Related Work

Researchers have effectively addressed the concept drift, infinite length, concept, and feature evolution problems in data stream classification techniques [8, 9]. Different incremental approaches are studied for infinite length and concept drift problem [10–12].

In single incremental model, the data stream classification is done using only one model which is updated over the time [10]. Ensemble-based classification consists of different models which can be used for data stream classification [11, 12]. The models are updated to reflect the recent instances and to accommodate in memory. The model with lowest classification accuracy is discarded for concept drift. The ensemble-based models are having good classification accuracy than the single incremental-based approach.

Existing literature have considered the problem of concept drift and infinite length. Authors assume that the number of classes in stream data is fixed. Recently, Masud et al. [11] addressed the concept evolution problem in classifying data streams. An arrival of a novel class in data streams indicates concept evolution. In real-world applications such as fault detection and intrusion detection, new class may occur at any time in the data stream. Existing classifiers find outliers from test instances of data streams due to noise and concept drift [13]. But, if these outliers are found with large volume, then it will be likely due to emergence of a novel class, i.e., concept evolution. It avoids misclassification of test instances as outlier. Tennant et al. [14] proposed a parallel version of the data stream classifier, i.e., MC-NN: Micro-cluster nearest neighbor which handles concept drift and scalability issue in the real-time application.

3   Proposed Data Stream Classification Algorithm

This section discusses a data stream classification algorithm. The definitions required for classification are given in Table 1.

Table 1 Definitions

Term	Description
Centroid [4]	The mean of data instances in the particular cluster of a submodel. It is the attribute of the cluster which is the measure of central location in the space
Radius [4]	The radius of a cluster is the maximum distance between all the instances and the centroid. It gave the idea of how vast the cluster is spread out
Weight [4]	Weight attribute defines the weight of each label (given in actual labeled dataset) in the cluster. It is the total number of data instances of a particular label in that cluster
Proportion	It is in reference to the labels in the cluster. It is calculated as the weight of the label divided by the size of the cluster (total number of data instances in that cluster.) Weight and proportions help us to find out the accuracy of the submodels built

In this paper, ensemble of p submodels is used for classification. Each submodel consists of C_i clusters. The data stream instance (test instance) can go into one of these clusters of the submodels. The data stream instance is either classified as an existing class or a new class. Let $\{M_1, M_2..., M_m\}$ represent the m submodels. Algorithm 1 and Algorithm 2 illustrate the proposed data stream classification.

Algorithm 1 Preprocessing Data Chunks

Input: Equal size n data chunks
Output: Current ensemble of M classifiers with their statistics of cluster summary information
1: **for** $j = 1$ to p do **do**
2:     Apply k-mean clustering algorithm to each of the chunks;
3:     Obtain K clusters;
4:     Compute the cluster summaries, i.e., Centroid C_i , Radius R_i, Weight W_i, and Proportion P_i;
5:     M: current ensemble of p models $\{M_1, M_2..., M_p\}$;
6: **end for**

Algorithm 2 Data Stream Classfication

Input: Current ensemble of M classifiers, x_k: test instance
Output: Current ensemble of M classifiers with their statistics of cluster summary information
1: **for** $x_k = 1$ to n **do**
2:    **if** $D(x_j, C_i(s)) \leq R_i(s)$ **then**
3:       x_j is an existing class instance in particular cluster;
4:       Labeling strategy 1: Max weighted-sum inter-submodel;
5:       Labeling strategy 2: High-frequency max-weight intra-submodel;
6:    **else**
7:       x_k is an oulier;
8:       Insert x_k in buf //enqueue;
9:       **if** (buf.length $> n$) **then**
10:          create new model;
11:          Add model;
12:          Again apply Algorithm1 for preprocessing;
13:       **end if**
14:    **end if**
15: **end for**

3.1 Training Phase

At first, data stream is divided into equal size frames; we called it as chunks. The size of each data chunk is set to n data points. Algorithm 1 is for preprocessing the data chunks. K clusters are obtained by applying k-mean clustering algorithm to each chunk. The statistical summaries of each cluster are calculated such as centroid C_i, radius R_i, weight W_i, and proportion P_i. The micro-cluster is a popular term [11] for representing the statistical summary of each cluster. Remaining raw data stream instances are discarded. We represent K-micro-clusters as a classification model as these micro-clusters are used for classifying unlabeled data. The number of classification model forms ensemble. Therefore, preprocessing is done for data stream classification, i.e., the ensemble of p models is now ready for classifying test instances.

3.2 Overlap Check

After the clusters are built, each cluster in the particular submodel checks for overlapping. For checking if the two clusters overlap or not, we can use the following formulae:

Suppose, R_1 and R_2 are the radius of the two clusters which needs to be checked. Distance between both the centroids is calculated as $dCentroid$

1. Touching clusters: $(R_1 + R_2) = dCentroid$
2. Overlapping clusters: $(R_1 + R_2) > dCentroid$
3. Far apart: $(R1 + R2) < dCentroid$.

3.3 Testing Phase

Algorithm 2 illustrates the data stream classification process. If x_k is a data stream instance to be tested, then it is individually checked with all the clusters of all the submodels. Distance of the point x_k from the centroid of that cluster $PCdistance$ is calculated and is checked with the radius of that cluster. If $PCdistance \leq Radius$, then the point lies inside the cluster. This is checked for all the clusters in the submodel and similarly for all the submodels. If it is inside cluster, then we can classify that instance with existing class using the following two labeling strategies.

3.3.1 Labeling Strategy

- Labeling strategy is of two types:

1. Max weighted-sum inter-submodel
2. High-frequency max-weight intra-submodel.

Considering X_k as a data instance to be tested, if the distance of the point from the centroid C_i of that cluster is less than or equal to the radius R_i of the cluster, then the point lies inside the cluster. For each cluster, we have maintained the proportion list of each label in that cluster. When the point lies inside the cluster, this proportion list is kept track of. The test data instance x_k is checked with all the clusters in a particular submodel. We assume seven clusters in each of six submodels. The proportion list of all such clusters in a single submodel in which the point lies is added up attribute-wise. Finally, we get a single proportion list with seven attributes for a single submodel. Let it be represented as $PL_1 = \{l_1, l_2 \ldots l_7\}$ where li represents each label proportion.

3.3.2 Max Weighted-Sum Inter-submodel

In this method, the point is tested with all the six submodels in the similar manner. Hence, we get six PL_i. Now we add all the six PL_i attribute-wise. Hence, we get a single proportion list for a test point x_k. Now, we find which label has the maximum proportion in the summed proportion list. The index with the max proportion will be given as a label for that test instance X. For example, if the final proportion list after summing is $\{a, b, c, d, e, f, g\}$ and b is the max of all, then the label given is 2.

3.3.3 High-Frequency Max-Weight Intra-submodel

In this method, after we get six proportion lists one for each submodel, we do not sum them again. We find out the label according to all the six proportion lists by finding out indexes where the max is located in each proportion list. After getting six labels for a particular test point, we give the label whose frequency is highest among all. This is how the label is decided for each test instance X.

4   Experimental Results and Discussion

The experiments are performed on an Intel Core i3 machine with 4GB memory and 2.20 GHz processor CPU in a Java environment with 64-bit windows operating system.

4.1   Dataset

We used Forest Cover Dataset [12]. It is available at UCI Machine Learning Repository. It contains geospatial characteristics of various types of forests. There are 7 classes, 54 features, and approximately 581,000 instances. The dataset characteristics are multivariate, and attribute characteristics are categorical and integer. The problem is to predict forest cover type.

The algorithm is tested on an experimental set of real data for different values of chunk size n. The experiments have considered the following parameters:

(1) Number of submodels in the ensemble = 6.
(2) Number of clusters in each submodel = 7.
(3) Values for n = 500, 1000, 1500, 2000.

4.2   Results and Discussion

The error rate is evaluated for different sizes of chunks using two labeling strategies, i.e., MWSISM and HFMWISM. The error rate indicates the percentage of data wrongly classified. Figure 1 shows that the error rate of HFMWISM is less compared to MWSISM. We can observe that increasing the size of chunk reduces the error rate.

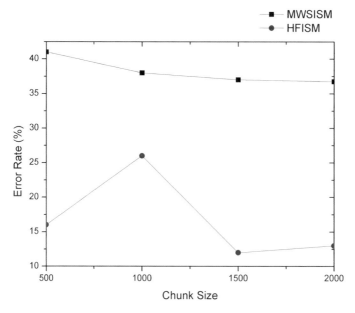

Fig. 1 Comparison of error rate using labeling strategies

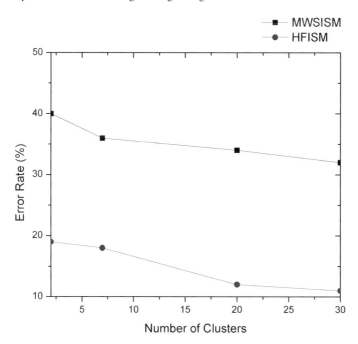

Fig. 2 Comparison of error rate using different number of clusters

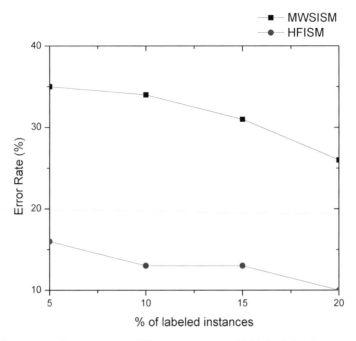

Fig. 3 Comparison of error rate using different percentages of labled training data

The sensitivity analysis of different parameters has been done. The error rate is evaluated with variable number of clusters and percentage of labeled instances in the submodel. We can observe that, as there is increase in number of clusters in the submodel, error rate is reduced using both the labeling strategies. It is shown in Fig. 2.

The impact of percentage of labeling instances in the submodel is shown in Fig. 3. We can observe that as there is an increase in the percentage of labeled data, the error rate decreases.

5 Conclusion

The growing importance of real-time applications leads to generation of data streams to a great extent. This paper presented data stream classification technique by building different models of chunked data streams. These models are built as micro-clusters, and k-nearest neighbor algorithm is used for data stream classification. An ensemble of these models is viable for classification of the test stream instances. The two labeling strategies, i.e., MWSISM and HFMWISM, are used for labeling the test instances. Comparison is performed for two approaches of classifying data streams. Experimental results justify that the HFMWISM is better than MWSISM. We will consider a data stream classification technique for network traffic as our future work. The algorithm can be further optimized for real-time applications.

References

1. Han, J., Pei, J., Kamber, M.: Data Mining: Concepts and Techniques. Elsevier (2011)
2. Salazar, A., Safont, G., Soriano, A., Vergara, L.: Automatic credit card fraud detection based on non-linear signal processing. In: 2012 IEEE International Carnahan Conference on Security Technology (ICCST), pp. 207–212, Oct 2012
3. Gama, J.: Knowledge Discovery from Data Streams. CRC Press (2010)
4. Masud, M., Gao, J., Khan, L., Han, J., Thuraisingham, B.M.: Classification and novel class detection in concept-drifting data streams under time constraints. IEEE Trans. Knowl. Data Eng. **23**(6), 859–874 (2011)
5. Le, T., Stahl, F., Gomes, J.B., Gaber, M.M., Di Fatta, G.: Computationally efficient rule-based classification for continuous streaming data. In: Research and Development in Intelligent Systems XXXI, pp. 21–34. Springer (2014)
6. Domingos, P., Hulten, G.: Mining high-speed data streams. In: Proceedings of the Sixth ACM SIGKDD International Conference on Knowledge Discovery and Data Mining, KDD '00, (New York, NY, USA), pp. 71–80, ACM (2000)
7. Gama, J.A., Kosina, P.: Learning decision rules from data streams. In: Proceedings of the Twenty-Second International Joint Conference on Artificial Intelligence—Volume Volume Two, IJCAI'11, pp. 1255–1260. AAAI Press (2011)
8. Masud, M.M., Chen, Q., Gao, J., Khan, L., Han, J., Thuraisingham, B.: Classification and novel class detection of data streams in a dynamic feature space. In: Joint European Conference on Machine Learning and Knowledge Discovery in Databases, pp. 337–352, Springer (2010)
9. Spinosa, E.J., de Leon, A.P., de Carvalho, F., Gama, J.: Olindda: a cluster-based approach for detecting novelty and concept drift in data streams. In: Proceedings of the 2007 ACM symposium on Applied computing, pp. 448–452, ACM (2007)
10. Aggarwal, C.C., Han, J., Wang, J., Yu, P.S.: A framework for on-demand classification of evolving data streams. IEEE Trans. Knowl. Data Eng. **18**(5), 577–589 (2006)
11. Masud, M.M., Gao, J., Khan, L., Han, J., Thuraisingham, B.: Classification and novel class detection in data streams with active mining. In: Pacific-Asia Conference on Knowledge Discovery and Data Mining, pp. 311–324. Springer (2010)
12. Yang, Y., Wu, X., Zhu, X.: Combining proactive and reactive predictions for data streams. In: Proceedings of the eleventh ACM SIGKDD International Conference on Knowledge Discovery in Data Mining, pp. 710–715. ACM (2005)
13. Gehrke, J., Garofalakis, M., Rastogi, R.: Data Stream Management: Processing High-speed Data Streams (2016)
14. Tennant, M., Stahl, F., Rana, O., Gomes, J.B.: Scalable real-time classification of data streams with concept drift. Futur. Gener. Comput. Syst. **75**, 187–199 (2017)

Regulatory Framework for Standardization of Online Transactions Using Cryptocurrencies

Astitva Narayan Pandey and Himanshu Gupta

1 Introduction

This paper proposes a regulatory framework for the use of cryptocurrencies in online transactions without compromising the core characteristics that make up the said cryptocurrency. Cryptocurrencies are a relatively new concept that helps decentralize the online currency exchange by shifting the trust factor from a third-party financial institution to cryptographic trust, implementing cryptographic concepts such as digital signatures, block-chain and peer-to-peer networking successfully into a robust, yet currently flawed, self sufficient financial system. The core concepts of cryptocurrencies such as total anonymity, while necessary to a point, also make it easier for the said currencies to be used for illegal purposes. The proposed regulatory framework suggests ways to diminish the illicit potential of the said currencies, as well as ways to bring them under tax boundaries of the law, while maintaining partial anonymity of its users, without compromising the core characteristics that are at the foundation of cryptocurrency.

2 Inception and Growth of the "Bitcoin"

On October 31, 2008, an unidentified programmer/a group of programmers released a paper named "Bitcoin: A Peer-to-Peer Electronic Cash System" [1] in the cryptography mailing list at Metzdowd.com. This paper would go on to become the

A. N. Pandey (✉) · H. Gupta
Amity Institute of Information Technology, Amity University
Uttar Pradesh, Noida, Uttar Pradesh, India
e-mail: pandey.astitva@gmail.com

H. Gupta
e-mail: himanshu_gupta4@yahoo.co.in

© Springer Nature Singapore Pte Ltd. 2019
S. Tiwari et al. (eds.), *Smart Innovations in Communication and Computational Sciences*, Advances in Intelligent Systems and Computing 851,
https://doi.org/10.1007/978-981-13-2414-7_41

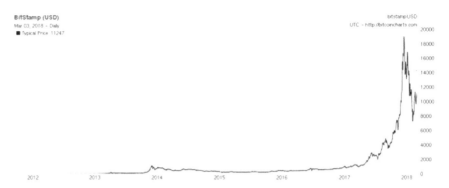

Fig. 1 Exponential growth in price of Bitcoin versus US$. Data collected from 2012 till March 2018. *Source* Bitcoincharts [3]

foundation on which cryptocurrencies would build upon. In January 2009, Bitcoin v0.1 was released, a software infrastructure for the creation and spending of bitcoins.

It was not until the next year though that these virtual coins would hold any real-world value. In May 2010, the first purchase was made by the new currency, a pizza. February of the next year would mark the Bitcoin reaching the value of 1 US$. In September 2012, the Bitcoin Foundation [2] was formed to standardize, protect and promote the use of Bitcoin. In April 2013, Bitcoin peaked to the value of 250 US$, and to 1200 US$ by the November of the same year. Ever since this virtual currency has seen an exponential growth (Fig. 1).

3 Working

A cryptocurrency is a pure peer-to-peer online currency which is independent of any financial institution. The goal of a cryptocurrency is to shift the trust factor in a transaction from a third party to cryptographic hash-based proof-of-work. In this paper, we shall discuss the working of Bitcoin as detailed in the 2008 paper by Satoshi Nakamoto, as it is the most popular cryptocurrency in the world at present.

3.1 Transactions

A Bitcoin is defined as a "*chain of digital signatures*" [1]. During each transaction, a hash of previous transaction is digitally signed by the owner and the public key of the next owner is added to the coin. To avoid any double-spending, all transactions are publically announced over the network with a timestamp (Fig. 2).

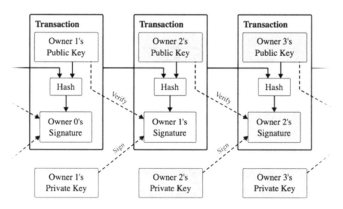

Fig. 2 Block-chain of transactions of a Bitcoin with each transaction adding a block with a hash of the previous transaction and public key of the new owner. *Source* "Bitcoin: a peer-to-peer electronic cash system" by Nakamoto [1]

3.2 Proof-of-Work

When hashed, the chain must start with a number of zero bits. This proves that some computational power has been put into calculating that hash, and the chain with most computational power put into it can be trusted as the computational power put in by the honest nodes must always exceed the short-lived computation by an attacker or a group of attackers. The average work required in a number of zero bits is exponential and can easily be verified by a single hash. Each CPU gets one vote in determining the majority decision, which is represented by the longest chain, with most amount of work put into it.

3.3 Incentive

Cryptocurrencies rely on *mining* and *miners* to introduce new currency in the system. Analogous to gold mining, miners of cryptocurrencies work and spend CPU time and electricity to mine out new coins, which marks the first transaction of the block of that coin. Another incentive can be found in transaction fees. Once a predetermined number of coins have entered circulation, no new coins would be minted and the system would entirely rely on transaction fees, making it inflation free. It is projected that the last Bitcoin would be minted in the year 2020.

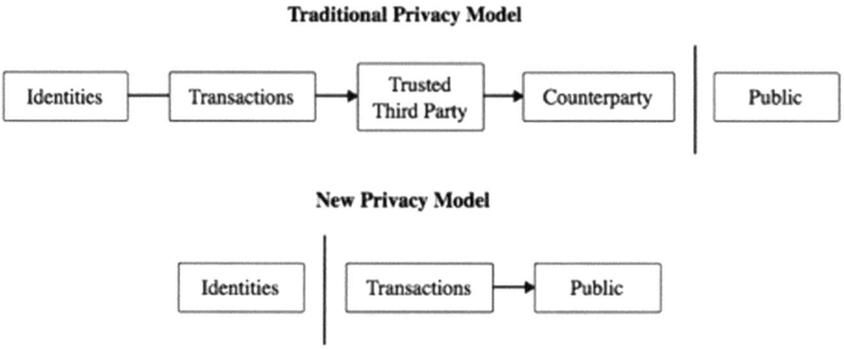

Traditional Privacy Model

New Privacy Model

Fig. 3 Comparison between traditional privacy model where a trusted third party does not make the transaction information public, versus the privacy model adapted by Bitcoin where in transactions are public, but the identities of the parties are not. *Source* "Bitcoin: a peer-to-peer electronic cash system" by Nakamoto [1]

3.4 Privacy

Bitcoin achieves a level of privacy by keeping the public keys in the block anonymous. It is visible to all that someone is spending an amount, because all transactions are publicly announced, but that spending information cannot be linked to anyone (Fig. 3). Additionally, a new pair of keys is used for each transaction, further keeping the anonymity of the spender intact. This is one of the key concerns relating to the use of cryptocurrencies, as it cannot be determined if the spending is done for lawful reasons or not.

3.5 Block-Chain

Bitcoin Protocol has an ordered, back-linked block-chain data structure. Blocks link back to the previous block, hence forming a chain. Each block in the chain is identifiable by a hash generated using the SHA256 algorithm on the header of the block (Table 1). Since the previous block's hash is included in the header of the current block, it influences the hash of the current block, making it so that work for the entire chain up till the current block needs to be redone in order to change even a single block.

The 80 Bytes block header further consists of three sets of metadata. One to connect to the parent block, second to store difficulty and timestamp, and third the merkle tree root. Table 2 details the structure of the block header.

Table 1 Structure of a block (Bitcoin) [4]

Size	Field	Description
4 Bytes	Block size	The size of the block, in bytes, following this field
80 Bytes	Block header	Several fields form the block header
1–9 Bytes	Transaction counter	How many transactions follow
Variable	Transactions	The transactions recorded in this block

Table 2 Structure of a block header (Bitcoin) [4]

Size	Field	Description
4 Bytes	Version	A version number to track software/protocol upgrades
32 Bytes	Previous block hash	A reference to the hash of the previous (parent) block in the chain
32 Bytes	Merkle root	A hash of the root of the Merkle tree of this block's transactions
4 Bytes	Timestamp	The approximate creation time of this block (seconds from Unix Epoch)
4 Bytes	Difficulty target	The proof-of-work algorithm difficulty target for this block
4 Bytes	Nonce	A counter used for the proof-of-work algorithm

4 Benefits of Cryptocurrencies

Being decentralized and having an open ledger, cryptocurrencies radically reduces the possibility of fraud due to a need of reconciliation. The near real-time nature of the cryptocurrencies also enables them to cut the possible losses due to the foreign exchange volatility. It also enables the participating banks and financial institutions to reduce pressure of the treasury management.

The open distributed nature of the system also accommodates the need for redundancy. Since any and all nodes in the system can participate in validating a transaction, even if some nodes are unavailable due to some calamity or downtime, transactions still can be approved by the remaining nodes. Since a copy of the ledger is maintained at all the nodes in the network, it ensures availability.

The parallel nature of the transmission of a block through the network, in contrast to the linear nature of a financial institution handling a transaction, analogous to an assembly line, enables all the nodes to update the information simultaneously, reducing the effective delay in processing of a transaction due to decision-making, ultimately reducing the overall cost of processing, and enhances transparency of decisions in the system.

Some other benefits of cryptocurrencies include

1. Facilitate access to financial transactions to population with no access to financial institutions.
2. Avoids pitfalls of managed or commodity-based monetary systems.
3. Self-enforcing contracts that are independent of any trusted financial institution.

5 Concerns Surrounding Cryptocurrencies

Decentralization while having some obvious benefits discussed above, still has come concerns regarding it. There is a keen need for regulation regarding the block-chain technology and cryptocurrencies, as there still exist some fundamental flaws in the very core of these currencies.

Providing total anonymity enable anyone to have multiple wallets, making it further hard for countries or authorities to track bulk movement of currencies, making it even harder for them to track illegal activities.

To further add to this, since the currencies are completely decentralized, the jurisdiction relating to any transaction is ambiguous, rendering regulations and taxonomy un-implementable. There also are no Know Your Customer implementation for these currencies, which are mandatory for financial systems and institutions in many countries.

Finally, there is a big concern regarding the ambiguity of the role of cryptocurrencies as many a people treat them as currency, commodity, security, property, loan, deposit, derivative or foreign contract; making it virtually impossible for them to be considered under any law or regulation.

6 Countries' Stand

See Table 3.

7 Proposed Framework

We can address the concerns of the nations around the globe surrounding the current state of cryptocurrencies by a United Nations level regulation. It is necessary that all nations work together in encouraging and regulating this new, untamed technology with great potential.

It must be mandated for all cryptocurrencies/software clients to only issue wallets and allow the use of their currency after the user has gone through proper KYC verification. After the verification, the customer must receive a unique ID or code, through which the wallet holder shall be identifiable, along with the country to which he belongs to. As an example, the culprits in the infamous WannaCry ransomware attack that demanded ransoms as Bitcoins could have easily been tracked by the use of this unique ID. Any currency that refuses to follow this requirement shall not be allowed in the country for any of their operations, be it use of the currency or mining.

Each block generated by a transaction must also have the ID of the recipient embedded into it. This simple sacrifice in partial privacy of one of the two parties involved in the system would enable countries to easily track bulk transactions within

Table 3 Different countries' approach towards cryptocurrencies

Stance	Country	Description
Complete prohibition	Russia	In February of 2015, Russian Prosecutor General's Office said that "(Bitcoin) cannot be used by individuals or legal entities." [6]
	Iceland	The Icelandic Central Bank said "it is prohibited to engage in foreign exchange trading with the electronic currency bitcoin, according to the Icelandic Foreign Exchange Act." [7]
	China	On December 5, 2013, Chinese Central Bank prohibited financial institutions from dealing with Bitcoin transactions, although private parties and individuals can legally trade Bitcoin [8]
Protection from illicit activities	Singapore	Intermediaries must verify the identities of their customers and report suspension
	United States of America	Bitcoin exchanges and miners must monitor transactions and report
Taxation	United States of America	Bitcoin is taxable if used in the real-world economy
	Japan	Gains from trading Bitcoins, purchases made using Bitcoins and revenue of transactions are taxable
	Finland	Capital gains when profit is in Bitcoin after is obtained as payment is taxable
	Germany	Mining profits and trading are taxable unless hoarded for at least one year
Ambiguous	Israel	Most national financial and fraud prevention institutions issued a warning against the risks of cryptocurrencies
	India	The Reserve Bank of India and the Finance Minister have issued warnings against the potential risks of cryptocurrencies

Source: An Analysis of the Cryptocurrency Industry [5]

their boundaries, making it easier to close in on money laundering suspects as well as terrorism funding. This system would still not compromise much of the privacy of the receiving party as only the government of residence has access to the data to be able to link the unique ID to a particular person, which only shall be used by proper authorities in certain circumstances. This would also enable them to levy a tax automatically during that transaction. The tax value can be set by the local law of the country the recipient resides in and can be used as a tool to either encourage or discourage the use of a certain currency, as per the country's policy. A system for only levying tax on certain bulk transactions can also be implemented as per the country's policy.

Mining plants would also require a license to operate in a country, along with some strict use of power regulations to help reduce the negative impact of cryptocurrencies

version	02000000
previous block hash (reversed)	17975b97c18ed1f7e255adf297599b553 30edab87803c8170100000000000000
Merkle root (reversed)	8a97295a2747b4fla0b3948df3990344c 0e19fa6b2b3a19c8e6badc141787
timestamp	258b0553
bits	535f0119
nonce	48750833
transaction count	63
UID	**091-818d3fdecadb9**
coinbase transaction	
transaction	
...	

Fig. 4 64-bit UID (User ID) embedded in a block of a Bitcoin transaction, enabling us to identify the spender in the transaction, along with the country they belong to

on the environment. Reports indicate that the global consumption of electricity for mining Bitcoin is equivalent to 3.4 Million US homes, and it has been observed that many mining companies set up around power plants and places with low power prices (such as Sichuan and Yunnan of China). Since the mining plants need to be registered with the local government, enforcing which currencies can be mined in the country also becomes easier.

8 Detailed Architecture

At the core of this proposal is a unique ID for all the users that is attached to the block when someone receives some currency. This unique 64-bit key can be embedded in the block header, extending the size of the block header from 80 to 88 (Fig. 4).

Use of a 64-bit number implies that 9,223,372,036,854,775,807 unique IDs can be issued, making it sufficient for use even in the near future. Additionally, a mere 8 Byte difference in the size of the block is unlikely to cause any latency.

Out of these 64-bits, 12 bits can be used to uniquely identify the country the user belongs to, we assume it to be the same as the country's telephone extension number when written in hexadecimal in this paper, giving us 4096 possible country codes and 4503599627370496 possible users for each country (Fig. 5).

When a transaction is made in a cryptocurrency client software, it shall automatically determine the destination country (of the recipient) by analyzing the UID, and tally the taxation laws of the said country from a central database available online to split the transaction into two, namely one from the original owner to the destination, and second to the treasury wallet maintained by the country.

091	818d3fdecadb9
Country identifier	User identifier

Fig. 5 Structure of the UID field detailed with Country ID (12-bits/3 hexadecimals) and User ID (52-bit/13 hexadecimals)

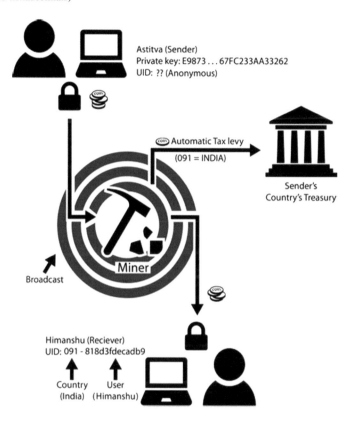

Fig. 6 Detailed flowchart of the transaction process inclusive of a unique ID for the recipient (UID) with a country stamp which would enable the said country to levy appropriate tax automatically, and determine the jurisdiction of the said transaction when needed for legal purposes

9 Working Principle

1. All wallet services must mandate KYC as per countries' regulations to further use the services.
2. The KYC would generate a unique 64-bit key (UID) which would denote the country of KYC with the first 12 bits, and the user with the remaining 52 bits.
3. A transaction is made through a cryptocurrency software client called X.
4. X checks the first 12 bits of the UID of the recipient and determines the country he belongs to.

5. X goes online to a database maintained by X, updated as per the current cryptocurrency specific tax laws of different countries and retrieves the tax formula and the details of the treasury account for the country the recipient belongs to.
6. X then divides the transaction into two, one that transfers to the original recipient, and second which transfers tax to the country (which includes the UID of the recipient as well), and sends these transactions to be mined (Fig. 6).
7. The country can then track how much tax it received and what was the initial amount being transferred using this information and can track bulk transfers within its boundaries. These transactions then can be linked to the original recipient with the help of the UID, only by the government of the country he belongs to, with access to the KYC records.

   This maintains the anonymity of the recipient, as only the government can know his true identity; still maintaining the decentralized nature of the cryptocurrencies by keeping it independent of any financial institution, yet bringing it under the boundaries of law.

10   Challenges

1. The framework proposed in this paper could only come to fruition if it is implemented at a global level with a UN level resolution. A point can be raised that we must stop the use of cryptocurrencies in terror funding to gain traction with different countries.
2. Since the tax levied by the countries would be in form of the cryptocurrency the transaction was held in, the countries would have to maintain a treasury specific to these different currencies. How can this currency be used by the country, and how can this currency be brought back into circulation?
3. This model calls for a massive shift in the system, what incentives do the countries have to bring about this huge change?
4. Can this mode of spending, in the near future with IOT devices taking over, when everything is connected to every other thing and the cloud at all times, replace the physical currency we are used to today? What would be the economic implications if it does end up happening?

11   Conclusion

As new technologies emerge, they bring in new potential for a better world. We must embrace the positives of these new technologies, while putting together our collective intelligence towards mitigating and resolving the potential shortcoming or issues the come with, via the help of regulatory as well as technical changes. No new thing is perfect and we cannot be closed to the potential and possibilities that they bring to the table.

Bitcoin, and all other cryptocurrencies such as Ethereum [8], Ripple [9], Litecoin [10], etc. shook the world in a big way, which led to the spread of the scare relating to these currencies, but regulators and researchers around the world are hard at work to figure out how the seemingly infinite potential of this technology can be tamed, giving way to a revelation that is bound to change our economic society similar to how the coin did.

The framework detailed in this paper is not perfect in any definition of the word but is a stepping stone towards the vast research that can be done on this topic, further improving upon our solutions, for the greater good of the world.

References

1. Nakamoto, S.: Bitcoin: a peer-to-peer electronic cash system (2008). https://bitcoin.org/en/bitcoin-paper
2. Bitcoin Foundation: Non-profit corporation founded to "standardize, protect, and promote the use of bitcoin cryotographic money for the benefit of users worldwide." (2012–present). https://bitcoinfoundation.org/
3. Bitcoincharts, provider for financial and technical data related to the Bitcoin network (2018). https://bitcoincharts.com/charts/bitstampUSD
4. Mastering Bitcoin: Programming the Open Blockchain, Chapter 9, The Blockchain (2017). http://apprize.info/payment/bitcoin_3/9.html
5. Farell, R.: An analysis of the cryptocurrency industry, University of Pennsylvania (2015). https://repository.upenn.edu/wharton_research_scholars/130
6. Hamburger, E.: The Verge (2014). http://www.theverge.com/2014/2/9/5395050/russia-bans-bitcoin
7. Schwartz, D., Youngs, N., Britto, A.: The Ripple protocol consensus algorithm. Ripple Labs Inc. (2014). https://ripple.com/files/ripple_consensus_whitepaper.pdf
8. Ethereum: One of the largest cryptocurrency in the world (2015–present). https://www.ethereum.org/
9. Ripple: One of the largest cryptocurrency in the world (2014–present). https://ripple.com/
10. Litecoin: One of the largest cryptocurrency in the world (2011–present). https://litecoin.com/

Autonomics of Self-management for Service Composition in Cyber Physical Systems

Swati Nikam and Rajesh Ingle

1 Introduction

Cyber Physical Systems have gained lot of popularity since its inception. Lot of research work is seen on abstraction, architecture, verification, validation, certification, robustness, safety, security, reliability, co-design tools, model-based development, service composition, autonomics and resource management [1]. Major work done by various researchers along with different approaches and challenges in service composition, resource provisioning and autonomics is summarized in [2]. Service composition means to combine more than one service when an individual service is not sufficient to satisfy the requirement but after combining it satisfies the requirement. Service Composition is very well studied in the context of Web Service, Cloud computing, Wireless sensor Network, Pervasive Computing, Opportunistic Networking and Internet of Things. Different phases of service composition are Service Discovery, Service Selection, Service Composition, Service Deployment and Service Execution. Out of which we have focused on service selection phase which is a sub-problem of service composition.

The paper is organized as follows. Section 2 gives an overview of related work done in service composition in CPS domain. Section 3 discusses two MADM methods as potential solution approach for solving Optimal service selection problem. Section 4 describes OSS problem. Section 5 gives insights to implementation details along with results and Sect. 6 gives concluding remarks.

S. Nikam (✉)
Department of Computer Engineering, Dr. D. Y. Patil Institute of Technology, Pune, India
e-mail: swatinikam3@gmail.com

R. Ingle
Department of Computer Engineering, PICT, Pune, India
e-mail: ingle@ieee.org

© Springer Nature Singapore Pte Ltd. 2019
S. Tiwari et al. (eds.), *Smart Innovations in Communication and Computational Sciences*, Advances in Intelligent Systems and Computing 851,
https://doi.org/10.1007/978-981-13-2414-7_42

2   Related Work

Tremendous literature is found on service composition in various domains like Web service [3], Pervasive Computing [4], Opportunistic network [5], Cloud Computing [6], Internet of things [7]. Similarly some efforts towards service composition in CPS domain is also found which can be summarized as follows.

Wang [8] have presented efficient context-aware service composition framework along with algorithm of atomic service filtering algorithm. Service composition problem is discussed as two phase context-sensitive service composition optimization problem [9] which is solved using Particle Swarm Optimization Method.

Some efforts are seen towards service naming [10] and proposing ontology for CPS [11]. Huang [12] have discussed context-sensitive resource-explicit service model and for service composition AI planning techniques is used. Wang [13] have discussed service composition problem in cyber physical social systems using mixed integer programming approach but the service characteristics are like a web service.

To summarize the literature study, the work related to service composition is either focused on proposing a framework [8] or theoretical discussion which is limited to ontology [11] description with physical entity consideration but very few authors talk about results. Also the existing methods which are present in literature for service selection either focus on any one or two selected attributes to select optimal service or they assign the weights [3] to different attributes to express the importance of one attribute over the other. It also uses skyline operator [13] to prune the redundant services. No concrete methods are found which focus on all the attributes to select the optimal service. And one more challenge is that, the attributes are combination of positive and negative attributes which needs to be considered simultaneously. So we reviewed the literature related to Multi-Attribute Decision-Making (MADM) methods as the potential solution approach and then we compared performance of selected MADM methods to check their feasibility for solving the service selection problem.

3   Potential Solution Options for OSS Problem

Multi-Criteria Decision-making (MCDM) methods are divided into Multi-Objective Decision-Making (MODM) and Multi-Attribute Decision-Making (MADM) methods. Various MADM methods [14] present are SAW (Simple Additive Weighting Method), WPM (Weighted Product Method), TOPSIS (Technique for Order Preference by Similarity to Ideal Solution), PSI (Preference Selection Index), AHP (Analytic Hierarchy Process), PROMETHEE (Preference ranking organization method for Enrichment Evaluation Method), VIKOR, Entropy, Fuzzy Based methods, WEBDA (Weighted Euclidean distance based approach), Grey Relational Analysis, ANP (Analytical Network Process).

We analyzed few significant methods of MADM as a potential solution approach. Vyas and Chetan [15], Amin and Johansson [16] have compared few methods of MADM. These methods are applied in different domains like Web service ranking [17], Ranking cloud services [18], Automated network selection in wireless environment [19], to rank Bayesian network options [20], for machine selection in manufacturing cell [21]. But to the best of our knowledge, no literature is found of application of MADM methods in the CPS domain, so we have focused on performing feasibility study of selected MADM methods for solving OSS problem in service composition in cyber physical systems.

Looking at the nature of CPS service selection problem and considering inherent characteristics of CPS and after studying the literature present on MADM methods, it was found that PSI and TOPSIS seems to be more relevant. Because TOPSIS makes use of full information of attributes and does not require attribute preference to be independent. Also PSI was chosen because there is no requirement of computing the weights of the attributes.

The steps of PSI and TOPSIS algorithms are as follows.

(i) **PSI (Preference Selection Index)**: The algorithm steps [22] are as follows.

Step (1) Define the Problem: Identify the alternatives along with criteria.

Step (2) Formulate the decision matrix: Construct a decision matrix as M alternatives as column and N criteria as rows and the element are X_{ij}.

Step (3) Normalize the data: To make the attributes dimensionless. Transform it into 1 and 0.
For positive attribute

$$N_{ij} = X_{ij}/X_{jmax}$$

And for negative attributes

$$N_{ij} = X_{jmin}/X_{ij}$$

Step (4) Compute the means value of normalized data P.

Step (5) Computer the preference variation for every attribute.

$$\Phi_j = \sum_{i-1}^{n} ([N_{ij} - P])^{\wedge}2$$

Step (6) Determine the variation in preference value.

$$\Omega_j = [1 - \Phi_j]$$

Step (7) Compute the overall preference value.

$$\omega_j = \frac{\Omega_j}{\sum_{j=1}^{m} \Omega_j}$$

overall preference value that is denominator should be 1.

Step (8)   Compute the preference selection index for each attribute.

$$\Theta_j = \sum_{i=1}^{M} x_{ij} * \omega_j$$

Step (9)   Select the appropriate result from ranked results.

(ii) **TOPSIS Algorithm**: The steps [23] are shown in Fig. 1.

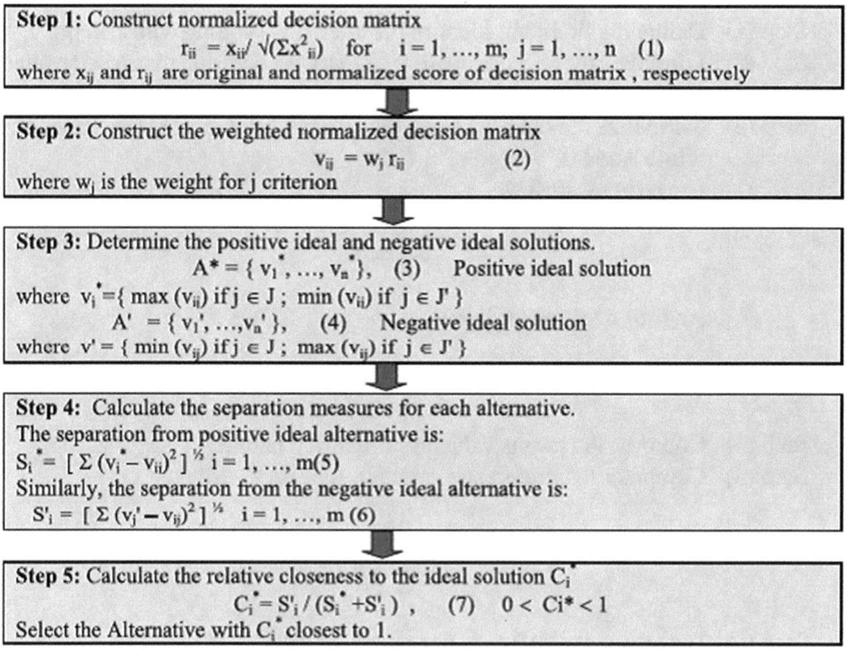

Step 1: Construct normalized decision matrix
$$r_{ij} = x_{ij}/\sqrt{(\Sigma x^2_{ij})} \quad \text{for} \quad i = 1, ..., m; \ j = 1, ..., n \quad (1)$$
where x_{ij} and r_{ij} are original and normalized score of decision matrix, respectively

Step 2: Construct the weighted normalized decision matrix
$$v_{ij} = w_j \, r_{ij} \qquad (2)$$
where w_j is the weight for j criterion

Step 3: Determine the positive ideal and negative ideal solutions.
$$A^* = \{ v_1^*, ..., v_n^* \}, \quad (3) \qquad \text{Positive ideal solution}$$
where $v_i^* = \{ \max (v_{ij}) \text{ if } j \in J ; \ \min (v_{ij}) \text{ if } j \in J' \}$
$$A' = \{ v_1', ..., v_n' \}, \quad (4) \qquad \text{Negative ideal solution}$$
where $v' = \{ \min (v_{ij}) \text{ if } j \in J ; \ \max (v_{ij}) \text{ if } j \in J' \}$

Step 4: Calculate the separation measures for each alternative.
The separation from positive ideal alternative is:
$$S_i^* = [\Sigma (v_i^* - v_{ij})^2]^{\frac{1}{2}} \ i = 1, ..., m (5)$$
Similarly, the separation from the negative ideal alternative is:
$$S_i' = [\Sigma (v_j' - v_{ij})^2]^{\frac{1}{2}} \ i = 1, ..., m (6)$$

Step 5: Calculate the relative closeness to the ideal solution C_i^*
$$C_i^* = S_i' / (S_i^* + S_i') , \quad (7) \quad 0 < C_i^* < 1$$
Select the Alternative with C_i^* closest to 1.

Fig. 1 TOPSIS algorithm steps [23]

4 Problem Description

Let us consider an application with a set of n tasks and each task t can be fulfilled with any individual service among available services S which consists of multiple concrete services which are same in functionality but different in quality of service.

Let $Q = (q_1, q_2 \ldots q_k)$ is set of QoS attributes of each concrete service.

$$Q = \{EC, R, RT, F, SER, CC, A\}$$

where

EC Execution Cost,
R Reputation,
RT Response Time,
F Frequency,
SER Successful Execution Rate,
CC Communication Cost,
A Availability.

The attributes can be categorized as positive attributes (if high then good) and negative attributes (if low then good). Here we have considered all the attributes except Availability. So if the value of Availability $= 1$, i.e., True then only this service is considered for MADM selection else it is considered that service is not available.

The overall objective of OSS problem is to select optimal CPS service for a given task from the set of services using MADM methods.

5 Implementation Details and Results

As discussed earlier optimal service selection is a subpart of our main service composition problem. Computer used for experimentation is Intel i5, 2.00 GHz CPU and 8 GB RAM, 500 GB hard disk. We have used Java for implementation. Service alternatives along with their QoS criteria are captured and stored in .XML files. We have done performance analysis of PSI and TOPSIS on accuracy and execution time.

(1) Accuracy: To check the accuracy of PSI and TOPSIS algorithms, we have first tested 14 test cases which were specifically designed by varying one or more QoS criteria from all positive criterias, all negative criterias, few negative and few positive criterias. As this design of test cases was required to see the effect of positive and negative criteria when the service is selected. In this case the actual optimal results were known to us. We have shown a sample test case 1 in Table 1. Similarly the criteria instances were collected from our prototype for remaining 13 test cases too. Because of space limitation we have not shown the other test cases.

Optimal result $=$ Alternative 12
PSI Result $=$ Alternative 12
TOPSIS Result $=$ Alternative 12

Table 1 Sample test case 1 with 12 service alternatives and 6 QoS attributes (Criteria)

Alternatives no	Criteria					
	EC	R	RT	F	SER	CC
1	955.256	2.268	0.827	2.146	0.033	518.552
2	996.065	2.758	0.505	2.344	0.011	759.870
3	910.938	1.672	0.616	3.695	0.041	752.818
4	823.916	1.569	0.707	4.934	0.456	658.022
5	832.830	2.660	0.581	4.760	0.376	696.325
6	716.072	1.919	0.551	4.075	0.408	607.399
7	536.255	1.411	0.578	3.418	0.258	721.573
8	614.476	1.325	0.770	1.267	0.154	526.308
9	769.310	2.836	0.548	3.788	0.449	542.232
10	980.028	2.657	0.722	2.159	0.479	961.472
11	300.000	4.000	0.400	6.000	0.600	400.000
12	200.000	5.000	0.300	7.000	0.700	300.000

Table 2 Comparison of successful achievement of optimal service election using PSI and TOPSIS

Test cases	Successful achievement of optimal service selection achieved using PSI	Successful achievement of optimal service selection achieved using TOPSIS
Case 1	Yes	Yes
Case 2	Yes	Yes
Case 3	Yes	No
Case 4	Yes	Yes
Case 5	Yes	No
Case 6	No	No
Case 7	Yes	Yes
Case 8	No	No
Case 9	Yes	Yes
Case 10	Yes	Yes
Case 11	Yes	Yes
Case 12	Yes	No
Case 13	Yes	Yes
Case 14	Yes	No

i.e. If both method PSI and TOPSIS gives optimal service selection result correctly, then in Table 2, we write as YES in both PSI and TOPSIS result. Similarly successful achievement of optimal service selection for remaining test cases using PSI and TOPSIS is recorded in Table 2. We calculated accuracy of PSI and TOPSIS algorithm with all test cases which is as follows.

Table 3 Result of PSI and TOPSIS

Number of alternatives (Services)	Execution time of PSI (ms)	Execution time of TOPSIS (ms)
100	1.534	1.515
200	2.194	3.29
300	4.52	5.187
400	5.107	5.074
500	6.682	10.626
600	7.275	10.592
700	1.312	1.989
800	1.239	2.335
900	1.726	2.779
1000	2.038	1.236

Accuracy of PSI = 12/14 * 100 = 85%
Accuracy of TOPSIS = 8/14 * 100 = 57%.

(2) Execution Time: We increased number of service alternatives from 12 to consecutively 100, 200, 300, 400, 500, 600, 700, 800, 900, 1000 Services. Then we recorded their execution time which is given in Table 3.

Figure 2 results indicate that the execution time of PSI is minimum as compared to TOPSIS in most of the cases. So we observed that as the number of alternatives increases then the execution time of TOPSIS algorithm increases, so TOPSIS may not be suitable from scalability point of view. PSI seems to be potential candidate for solving OSS problem from accuracy, execution time and scalability point of view.

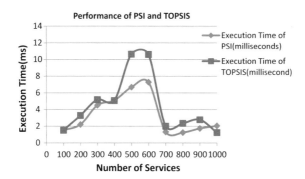

Fig. 2 Performance of PSI and TOPSIS

6 Conclusion

We have compared performance of PSI and TOPSIS for the optimal service selection problem, which is a subset of our main service composition problem. Thus MADM methods considers all the positive and negative attributes while choosing the optimal service. Comparative analysis is done based on accuracy, execution time and scalability. PSI shows 85% accuracy and execution time is also minimum in almost all the cases. So PSI seems to be most suitable solution approach for solving Optimal service selection problem. These MADM methods ranks all the alternatives and hence topmost service is selected which can be used in autonomic service composition. If topmost service is not available then the next service from the ranked service list will be used as a substitute for service composition.

Acknowledgements I am thankful to Dr. D. Y. Patil Institute of Technology for being facilitator to carry out this research.

References

1. Wan, J., Yan, H., Suo, H., Li, F.: Advances in cyber physical systems research. KSII Trans. Internet Inf. Syst. **5**(11) (2011)
2. Nikam, S., Ingle, R.: Survey of research challenges in cyber physical systems. Int. J. Comput. Sci. Inf. Secur. **15**(11), 192–199 (2017)
3. Jatoth, C., Gangadharan, G.R., Fiore, U., Buyya, R.: QoS-aware big service composition using MapReduce based evolutionary algorithm with guided mutation. Futur. Gener. Comput. Syst. (2017)
4. Kalasapur, S., Kumar, M., Shirazi, B.A.: Dynamic service composition in pervasive computing. IEEE Trans. Parallel Distrib. Syst. **18**(7), 907–918 (2007)
5. Tamhane, S.A., Kumar, M., Passarella, A., Conti, M.: Service composition in opportunistic networks. In: 2012 IEEE International Conference on Green Computing and Communications (GreenCom), pp. 285–292. IEEE (2012)
6. Wu, T., Dou, W., Chunhua, H., Chen, J.: Service mining for trusted service composition in cross-cloud environment. IEEE Syst. J. **11**(1), 283–294 (2017)
7. Nikam, S., Ingle, R.: Comparative study of service composition in CPS and IoT. In: International Conference on Advances in Cloud Computing, pp. 1–7 (2014)
8. Wang, T., et al.: Automatic and effective service provision with context-aware service composition mechanism in cyber-physical systems. Int. J. Adv. Inf. Sci. Serv. Sci. **4**(11), 151–160 (2012)
9. Wang, T.: A two-phase context-sensitive service composition method with a workflow model in cyber-physical systems. In: Proceedings 17th IEEE International Conference on Computational Science and Engineering, pp. 1475–1482 (2014)
10. Hellbrück, H., et al.: Name-centric service architecture for cyber-physical systems. In: Proceedings IEEE 6th International Conference on Service-Oriented Computing and Applications, pp. 77–82 (2013)
11. Huang, J., et al.: Extending service model to build an effective service composition framework for cyber-physical systems. In: IEEE International Conference on Service-Oriented Computing and Applications, pp. 130–137 (2009)
12. Huan, J., et al.: Towards a smart cyber physical space—context sensitive resource explicit service model. In: 33rd International IEEE Conference on Computer Software and Application, pp. 122–127 (2009)

13. Wang, S., Zhou, A., Yang, M., Sun, L., Hsu, C.-H.: Service composition in cyber-physical-social systems. IEEE Trans. Emerg. Topics Comput. (2017)
14. Tzeng, G.H., Huang, J.-J.: MADM Methods and Applications. CRC press
15. Vyas, G., Chetan, M.: Comparative study of different multicriteria decision-making methods. Int. J. Adv. Comput. Theory Eng. **2**, 9–12 (2013)
16. Amin, K., Johansson, R.: Utilization of multi attribute decision making techniques to integrate automatic and manual ranking options. J. Inf. Sci. Eng. **30**, 519–534 (2014)
17. Umm-e-Habiba, Asghar, S.: A survey on multi-criteria decision making approaches. In: IEEE International Conference on Emerging Technologies (2009)
18. Garg, S.K., Versteeg, S., Buyua, R.: A framework for ranking of cloud services. Futur. Gener. Comput. Syst. **29**, 1012–1023 (2013)
19. Leung, V.: Automated network selection in a heterogeneous wireless network environment. IEEE Netw. (2007)
20. Karami, A.: Utilization and comparison of multi attribute decision making techniques to rank Bayesian network options (2011)
21. Jian, S., et al.: PSI for machine selection in flexible manufacturing cell. In: Proceedings of MATEC Web of Conference, 139 (2017)
22. Maniya, K., Bhatt, M.G.: A selection of material using a novel type decision-making method: Preference selection index method. Mater. Des. **31**(4), 1785–1789 (2010)
23. Behzadian, M., et al.: A state-of the-art survey of TOPSIS applications. Expert Syst. Appl. **39**(17), 13051–13069 (2012)

Ranking-Based Sentence Retrieval for Text Summarization

Abhishek Mahajani, Vinay Pandya, Isaac Maria and Deepak Sharma

1 Introduction

Recently, information is considered as a vital asset which can have different applications running from measurable purposes to learning portrayal purposes. With the amount of data collected on a daily basis, there is a plethora of information available at our disposal. In such circumstances, it is highly imperative to extract all the important information while disregarding the repetitive and less significant information. Subsequently, the task of summarization becomes a very challenging task since it has to capture all the significant information and compile it into a document, which is semantically and etymologically correct while taking into consideration the rationality of the information coalesced. It is exceptionally troublesome for people to fathom and decipher the substance of the content especially when it is very large. A solution to this problem is having an efficient summarization technique capable of condensing the information in a concise manner while preserving the semantics. Information can be extracted from all kinds of documents such as newspapers, magazines, novels, and exploration papers, which are used by different people with various thought processes. So, what does an effective summarization mean? The most accurate summary produced can be defined as the one which takes into account all the aspects of the content, prioritizes the most eminent

A. Mahajani (✉) · V. Pandya · I. Maria · D. Sharma
Department of Computer Engineering, KJSCE, Mumbai, India
e-mail: a.mahajani@somaiya.edu

V. Pandya
e-mail: vinay.hp@somaiya.edu

I. Maria
e-mail: isaac.m@somaiya.edu

D. Sharma
e-mail: deepaksharma@somaiya.edu

© Springer Nature Singapore Pte Ltd. 2019
S. Tiwari et al. (eds.), *Smart Innovations in Communication and Computational Sciences*, Advances in Intelligent Systems and Computing 851,
https://doi.org/10.1007/978-981-13-2414-7_43

information in the document, and creates a document which is significantly smaller than the original content. This paper proposes a novel methodology for generating summary using a scoring algorithm, which extracts the most significant words from the document and ranks the sentences by calculating the amount of homogeneity amongst the sentences and words extracted. The scoring algorithm utilized is a combination of cosine similarity and maximal marginal relevance and to identify the most significant words we are using location-based and TF–IDF statistical information retrieval method. Here, the framework endeavored to enhance the nature of summary phonetically, by changing the strategy for discovering the similarity and pertinence parameters to recognize catchphrases.

The paper proposes and discusses a text summarization framework in the following sections. In Sect. 2, related work is discussed. In Sect. 3, the proposed methodology and the scope of the system are highlighted. Sections 4 and 5 discuss the implementation and results of the proposed system. Finally, conclusion and future scope are discussed in Sect. 6.

2 Literature Survey

Advancements in text summarization emerged in the early 50s. Text summarization can be broadly characterized into two categories, viz., extractive text summarization and abstractive text summarization. Extractive text summarization is the method of extracting content from the document and combining it to form a text smaller in size. Abstractive text summarization takes summarization to a stride further. It is capable of depicting information by creating new sentences. Abstractive summarization can be divided into structured and semantic approaches. Each of these classifications can be subdivided into subcategories based on various methods [1].

Structured approach fundamentally encodes the most indispensable information from the document(s) through mental blueprints like layouts, extraction principles, and elective structures like tree, ontology, rule, and graph structure.

- In tree-based approach [2], sentences from multiple documents are clustered according to the themes they represent. Secondly, these themes are re-ranked, selected, and ordered according to their significance. This is followed by identification of common information using syntactic trees. The syntactic trees formed are subsequently merged using fusion lattice computation to assimilate information from different themes. Linearization is carried out for the formation of sentences from the merged tree using tree traversal.
- Ontology-based method [3] is based on the predefined knowledge base to create summaries belonging to a particular domain, i.e., it is domain specific. In this approach, the domain ontology for generating Chinese news articles is laid out by the area specialists. The archive preprocessing stage creates the important terms from the news corpus and furthermore the Chinese news lexicon. The important

terms generated are classified using term classifier and the classified meaningful terms are passed to fuzzy inference mechanism to generate fuzzy ontology.

- Rule-based method [4] is based on random forest classification and feature scoring. The scoring is based on the constraints laid down by the user. The rules can be set in many ways such as: using verbs and nouns which are related to each other; keywords and syntactic constraints; domain constraints.

- Graph-based Approach [5, 6] uses the graph data structure for language representation. Here, every word unit is represented by a node and the structure of the sentences is determined by directed edges. These edges represent the relationship between any two words. The underlying feature of this method is that it uses the shortest path algorithm to find the smallest sentences with a considerable amount of information. The sentence formation is subjected to constraints such as it is mandatory to have a subject, verb, and predicate in it. Along with this, a compendium is used for linguistic and summary generation purposes.

  Alternatively, [7] a graph is created in which a vertex represents a sentence. Sentences are connected to each other by an edge. Weights are assigned to the edges on the basis of similarity between the sentences. Finally, PageRank algorithm is executed on the graph and the sentences with the highest PageRank scores are extracted.

In the semantic method, etymology outline of document(s) is utilized to feed into the NLG framework. This strategy has experience in recognizing expressions and verb states by processing and handling etymological information.

- Information item set [8] is said to be the smallest element of coherent information in a sentence. Text entries, its attributes, and predicates are identified in this method. Similar to extractive text summarization methods, frequency-based models are used for item set selection. The sentences are generated by combining them and ranking of sentences is done. Out of these sentences, the highly ranked content is selected to be a part of the summary.

- Semantic text representation model [9] is based on aims to analyze input text using the semantics of words rather than the structure of the text. Here, the abstractive summarization is accomplished as a semantic portrayal of supply records. Content determination is finished by selection of the most relevant predicate contention structures. At last, the summary is created by utilizing a dialect apparatus. However, the framework does not deal with comprehensive semantics in the summarization technique.

General notion in extractive text summarization is to weight the sentences of a document as a function of high-frequency words, disregarding the very-high-frequency common words. Extractive summarization framework later, in addition to the above-mentioned method (i.e., recurrent dependent weights), also used the following approaches for deciding the sentence weight [10]:

- Cue method [11] is based on the theory that the importance of a sentence is calculated by the presence or absence of certain cue (hint) words in the cue dictionary.

This is similar to attention mechanism, wherein the focus is on words which draw our attention and are very impactful which helps in understanding the context.

- Title-based method [12] takes into consideration the heading and subheading in a document. Generally, the heading tags represent the whole idea in a document or a paragraph in a few words. Following this methodology will help in understanding a broad view of the context. However, the information assimilated using this method is very limited and can be misleading.
- Location method [13] exploits the idea of finding relevant information in certain part of the document. It is a perceived notion that the text at the starting or any document can be treated as the introduction and that it will give a general idea about what the document is based on. Likewise, the ending is considered as the conclusion of what is discussed in the document and can help in understanding the overall outcome of the document. This method skips the detailed information mentioned in the body part of the document. This is because the body of any context consists of detailed information elaborating the given ideas.

Besides this, sentence extraction can be done using neural network architectures. One of these methods is the [14] classify architecture which involves traversing the document sequentially, and deciding whether to add the sentence into the summary. Select architecture involves picking sentences in an arbitrary manner. Classifier architecture has been proven to be better than select architecture after various experimental analyses. The other method [15] focuses on the side information of the document. The essence of the content lies in the side information such as title, image caption, twitter handle. Thus, the given framework summarizes a document by making use of document encoder and attention mechanism over side information.

3   Proposed Model

The insights collected from the literature survey gave an overview of the advantages and disadvantages of different summarization methods. The proposed framework went for centering and consolidating the upsides of different frameworks and minimalizing the disadvantages. The system uses term frequency and inverse document frequency methods to identify the most relevant words in an archive with the help of a threshold for important content retrieval. However, before the extraction of noteworthy words, preprocessing of input data was done to bring it into a standardized format. Besides this, the preprocessing helped in disposing of the irregularities.

The extracted top 'n' most significant words were combined together as a sentence, which can essentially be called as a query. This query is used for finding the best sentences which can be used for the generation of the summary. The system calculates the similarity between the sentences and the query. Cosine similarity is used for calculating the resemblance between the query and the sentences. Conversion of sentences into a vector format was done in advance. The glove was used to generate these embedding. Cosine similarity measures the similarity between two vectors. If

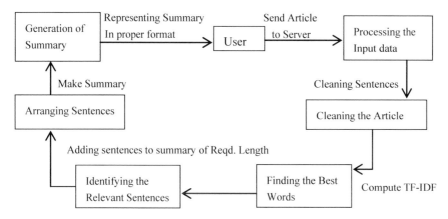

Fig. 1 Data flow diagram of the proposed model illustrating the flow of document through different phases for generation of summary

the cosine similarity is one then the 2 vectors lie on the same line which means that they are very similar to each other. Using this, the system was able to find the most significant sentences in the archive.

During the process of assigning the rank, the position of sentences in the document was also taken into consideration. The position additionally affects the significance of sentences in the substance. These sentences are given higher ranks when contrasted with the sentences which are present in the body part consisting of numbers, technicalities, and detailed information. This aggregate scoring helped the system in getting a consolidated list of sentence ranks.

Once the sentence ranks were generated, the method for selection of sentences is developed. The system used a Maximal Marginal Relevance (MMR) algorithm to select the sentences. It helps in quantifying the extent of dissimilarity between the sentences taken into consideration and those which are already read. The dissimilarity helps the system in choosing sentences, which are not related. This results in a diversified summary generation. Higher MMR score means the sentence is both important to our summary and has large dissimilarity. Various phases for the generation of summary in the system can be seen in the Fig. 1.

4   Implementation

The algorithm used by the system is explained in the following steps:

1. Data preprocessing is the first step which is executed. Preprocessing basically involves cleaning the data by lemmatizing it with the stemmer and then extracting the sentences from it using regex. The process involved removal of the HTML tags and mathematical operators for achieving consistent results.

2. Calculation of term frequency for each word. This function takes the sentence and finds the term frequency of words in that sentence. Term frequency measures how frequently a word occurs in the text. This returns a dictionary of words with their respective term frequency scores [16].

$$TF(t) = \frac{(Number\ of\ times\ term\ (t)\ appears\ in\ a\ document)}{(Total\ number\ of\ terms\ in\ the\ document)} \qquad (1)$$

3. Calculation of inverse document frequency for each word. This method finds the inverse document frequency value of the word in the sentence. This is an important factor which is used to assign importance to a word. While calculating Term Frequency (TF) all words are assigned equal importance but there are certain stop words such as 'as', 'that', 'is', etc., which occur frequently but do not give much valuable information. Hence, it is necessary to weigh down certain terms but scale up some rare meaningful words [16].

$$IDF(t) = \frac{[\log_e(Total\ number\ of\ documents)]}{(Number\ of\ documents\ with\ term\ t\ in\ it)} \qquad (2)$$

4. Calculate the TF–IDF scores for every word in the document by multiplying the outputs obtained from *Step II* and *Step III* for every word [16].

$$Value(t) = TF(t) \times IDF(t) \qquad (3)$$

5. Building a sentence query consisting of the most significant words. The TF–IDF scores were used to find the most significant words. The query built can be of the desired number of words. The method is used for building a query sentence consisting of 'n' best words.
6. Here, the system uses cosine similarity to calculate the similarity between the query and the sentences in the system. These similarities are taken into consideration, since it helps in understanding the sentences which consist of the most relevant words [17].

$$sim(D_1, D_2) = \frac{\sum_i t_{1i} t_{2i}}{\sqrt{\sum_i t_{1i}^2} \times \sqrt{\sum_i t_{2i}^2}} \qquad (4)$$

where t_i is the term weight for a word w_i.
7. After the best sentences are obtained by calculating the cosine similarity, the location of the sentences is taken into consideration. The system gives higher ranks to the sentences at the beginning and at the end of the paragraph. These sentences generally have a higher density of information per number of words. These ranks are averaged out and final ranks for the sentences are achieved.
8. The final step consists of calculating the Maximal Marginal Relevance between the sentences in the document. This helps the system in identifying sentences with relevant and manifold ideas which have 0 correlations between them and

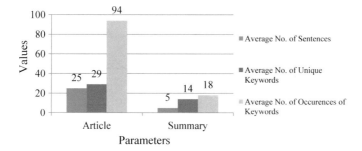

Fig. 2 Graph depicts the various parameters, mentioned in legend, of articles used for testing and the summaries generated during the testing

are important at the same time. The number of words, in summary, is chosen dynamically based on the domain, length, and content in the article [18].

$$MMR(D_i) = \lambda \times Sim_1(S_i, Q) - (1 - \lambda) \times \max_{D_i \in S} Sim_2(S_i, S_j) \qquad (5)$$

where $\lambda \in [0, 1]$ is an adjusting factor for emphasizing the relevance of the sentence and avoiding redundancy; S_i is the candidate sentence; Q is the query; Sim_1 and Sim_2 are the cosine similarity measures.

The system was based on Android and webapp platform, which used Django for python backend. The accepted the article and on submitting the article is sent as a POST request for processing and generating the summary.

5 Results

The implementation was followed by a testing phase for validation and verification purposes. The system was tested using a two-step strategy which is as follows.

- Objective Testing

Here, the system was tested against a set of 100 articles of different sizes and pertaining to different domains. Different parameters were evaluated to check the effectiveness of the summarizing system. The following graphs (Fig. 2) represent different characteristics of input data and output data.

While generating summaries, ratio of count of words in summary and article; ratio of count of keywords in summary and article; ratio of count of occurrences in summary and article were calculated. The preliminary results generated from the testing are shown in Table 1.

These ratios indicated that the algorithm focused on extraction of relevant data, which can be inferred from the ratio of keywords. Consequently, the higher ratio of occurrences of keywords in comparison with ratio of words indicated that the

Table 1 Parameters obtained on conducting the objective testing for 100 articles and their corresponding summaries

Parameters	Article	Summary	Ratio (Summary: Article)
Count of words	501	111	0.22
Count of unique keywords	29	18	0.62
Count of occurrences of keywords	94	20	0.21

Note The testing was done for a series of (λ) values but λ = 0.5 produced better keywords-oriented summaries

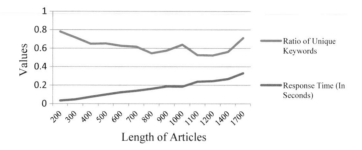

Fig. 3 Graph represents the ratio of unique keywords (Summary: Article) and response time generated of system during objective testing

density of keywords representing important information is higher in the summary as compared to the article. The following two graphs indicate the ratio of unique keywords obtained in summary to article and the response time of the system for different word lengths. This can be observed in Fig. 3 stated subsequently.

The system was tested with 100 samples of articles and the corresponding results indicate that the system gave an accuracy of 65% for matching the number of unique keywords obtained from articles to summary. Furthermore, language semantics were also checked at random to test whether the flow of sentences does not violate the rules of grammar. It was found that the flow and sentence formation was semantically correct.

- Subjective Testing

Subjective testing was carried out to understand the amount of context covered by the summary. To test this, a group of 10 people was chosen and given 20 different articles and their summaries such that no person has the same summary article pair. Each person was asked to read the article and summary. The person who has read the article had to ask a question to the person having the corresponding summary about what the article revolved around. The central idea behind this testing was to check how the algorithm fared in capturing and representing all the ideas represented in

Table 2 Parameters obtained on conducting the subjective testing for 20 articles

Parameters	Values
Percentage of questions answered	80%
Percentage of grammatically correct summaries	100%
Percentage of meaningful correct summaries	70%
Average percentage of context covered by summary	55%
Average ratings received by the summary	4.3

the article in a concise format. The statistics of the results obtained from the testing can be observed in Table 2 propounded below.

6 Conclusions

In this paper, an algorithm in view of extractive techniques was actualized to accomplish summarization. The outcomes acquired demonstrate that the algorithm can provide the user with succinct information having the least repetition. The framework fared well against the extrinsic and intrinsic testing techniques. In this paper, accentuation was laid on numerous strategies for leading synopsis as talked about in the literature review. MMR scoring strategy alongside TF–IDF information retrieval has given a valuable and a recipient technique to direct summarization. This conclusion is bolstered by the outcomes got subsequent to testing.

Nonetheless, there still remains an overwhelming test in the field of summarization to condense records from unstructured or semi-structured information in the correct path as indicated by user specifications. There are numerous issues not tended to in the field of summarization in full degree. Future research involves chipping away at broadening this usefulness utilizing abstractive methodologies including natural language processing and deep learning. It will help in delivering a balanced summary, which is exceedingly coherent, cognizant, less repetitive outline, and data rich.

Acknowledgements The system was developed by students as a part of final year project. The project was conceivable due to guidance from the project mentor, Dr. Deepak Sharma. The infrastructure provided by the college, K. J. Somaiya College of Engineering, India, helped in the successful development of the project.

References

1. Moratanch, N., Chitrakala, S.: A survey on abstractive text summarization. In: 2016 International Conference on Circuit, Power and Computing Technologies (ICCPCT), pp. 1–7, Mar 2016. IEEE
2. Barzilay, R., McKeown, K.R.: Sentence fusion for multidocument news summarization. Comput. Linguist. **31**(3), 297–328 (2005)
3. Lee, C.S., Jian, Z.W., Huang, L.K.: A fuzzy ontology and its application to news summarization. IEEE Trans. Syst. Man Cybern. Part B (Cybernetics), **35**(5), 859–880 (2005)
4. John, A., Wilscy, M.: Random forest classifier based multi-document summarization system. In: 2013 IEEE Recent Advances in Intelligent Computational Systems (RAICS), pp. 31–36, Dec 2013. IEEE
5. Ganesan, K., Zhai, C., Han, J.: Opinosis: a graph-based approach to abstractive summarization of highly redundant opinions. In: Proceedings of the 23rd International Conference on Computational Linguistics, pp. 340–348, Aug 2010. Association for Computational Linguistics
6. Lloret, E., Palomar, M.: Analyzing the use of word graphs for abstractive text summarization. In: Proceedings of the First International Conference on Advances in Information Mining and Management, Barcelona, Spain, pp. 61–66 (2011)
7. Barrios, F., López, F., Argerich, L., Wachenchauzer, R.: Variations of the similarity function of textrank for automated summarization (2016). arXiv:1602.03606
8. Genest, P.E., Lapalme, G.: Framework for abstractive summarization using text-to-text generation. In: Proceedings of the Workshop on Monolingual Text-To-Text Generation, pp. 64–73, June 2011. Association for Computational Linguistics
9. Khan, A., Salim, N., Kumar, Y.J.: A framework for multi-document abstractive summarization based on semantic role labelling. Appl. Soft Comput. **30**, 737–747 (2015)
10. Gupta, V., Lehal, G.S.: A survey of text summarization extractive techniques. J. Emerg. Technol. Web Intell. **2**(3), 258–268 (2010)
11. Zhang, J., Sun, L., Zhou, Q.: A cue-based hub-authority approach for multi-document text summarization. In: Proceedings of 2005 IEEE International Conference on Natural Language Processing and Knowledge Engineering, IEEE NLP-KE'05, pp. 642–645. IEEE (2005)
12. Ferreira, R., Freitas, F., de Souza Cabral, L., Lins, R.D., Lima, R., França, G., Favaro, L.: A context based text summarization system. In: 2014 11th IAPR International Workshop on Document Analysis Systems (DAS), pp. 66–70, Apr 2014. IEEE
13. Zhang, P.Y., Li, C.H.: Automatic text summarization based on sentences clustering and extraction. In: 2nd IEEE International Conference on Computer Science and Information Technology, ICCSIT 2009, pp. 167–170. IEEE, Aug 2009
14. Nallapati, R., Zhou, B., Ma, M.: Classify or select: neural architectures for extractive document summarization (2016). arXiv:1611.04244
15. Narayan, S., Papasarantopoulos, N., Lapata, M., Cohen, S.B.: Neural extractive summarization with side information (2017). arXiv:1704.04530
16. Ramos, J.: Using TF-IDF to determine word relevance in document queries. In: Proceedings of the First Instructional Conference on Machine Learning, vol. 242, pp. 133–142, Dec 2003
17. Xie, S., Liu, Y.: Using corpus and knowledge-based similarity measure in maximum marginal relevance for meeting summarization. In: IEEE International Conference on Acoustics, Speech and Signal Processing, ICASSP 2008, pp. 4985–4988. IEEE, Mar 2008
18. Carbonell, J., Goldstein, J.: The use of MMR, diversity-based reranking for reordering documents and producing summaries. In: Proceedings of the 21st Annual International ACM SIGIR Conference on Research and Development in Information Retrieval, pp. 335–336, Aug 1998. ACM
19. Allahyari, M., Pouriyeh, S., Assefi, M., Safaei, S., Trippe, E.D., Gutierrez, J.B., Kochut, K.: Text summarization techniques: a brief survey (2017). arXiv:1707.02268

Design of Dmey Wavelet Gaussian Filter (DWGF) for De-noising of Skin Lesion Images

Ginni Arora, Ashwani Kumar Dubey and Zainul Abdin Jaffery

1 Introduction

Image distortion is a prevalent occurrence, several interferences from the various types of noises [1]. The fundamental forms of this disturbance are through various forms of noises like Gaussian, Speckle, Poisson, Salt and Pepper, etc. Gaussian noise is one of the common and most prevailing types of noise [2]. The distortion may be caused due to a noise source in the surroundings, where an image is captured or may be the consequence of an intrinsic flaw in the camera device used. For instance, lenses may be wrongly aligned, improper focal length may exist, other disadvantageous environmental conditions may also be there, etc. These factors render rigorous noise analysis and close noise estimation as a crucial aspect of image de-noising, which leads to selecting proper noise model for image processing systems an important task. Thus, de-noising becomes an in dispensable preprocessing method for processes like image segmentation or edge detection. The prominent goal of de-noising algorithm lies in reducing noise without losing image characteristics [3]. Currently, there is a multitude of such algorithms available, but they are inefficient as they do not attain better measures like signal to noise and retain the original image traits.

The objective of this paper is to develop a de-noising method based on the performance measures, peak signal noise ratio and mean square error measures. This de-noising method is basically for the removal of Gaussian type of noise which is

G. Arora (✉) · A. K. Dubey
Amity University Uttar Pradesh, Noida, UP, India
e-mail: garora@amity.edu

A. K. Dubey
e-mail: dubey1ak@gmail.com

Z. A. Jaffery
Department of Electrical Engineering, Jamia Millia Islamia, New Delhi, India
e-mail: zjaffery@jmi.ac.in

© Springer Nature Singapore Pte Ltd. 2019
S. Tiwari et al. (eds.), *Smart Innovations in Communication and Computational Sciences*, Advances in Intelligent Systems and Computing 851,
https://doi.org/10.1007/978-981-13-2414-7_44

most common among all types of noise. This proposed method is said as Dmey Wavelet Gaussian Filter.

The paper has been divided into following sections: Sect. 2 shows the literature survey in this field. Section 3 delineates study of various image restoration methods, which further describes wavelet transformation, low pass filters, and related noise. Section 4, highlights the proposed method for de-noising. Section 5 shows experimental results and discussion. The execution is assessed by figuring the measures—Mean Square Error and Peak to Signal Noise Ratio (PSNR). Section 6 is the conclusion of paper.

2 Literature Survey

In the recent years, many researchers have been attracted towards automatic detection or diagnosis of skin diseases. This automatic detection needs various phases to focus on which starts from preprocessing to final classification of skin disease. The initial phase which removes noise from an image plays a vital role during the whole process. Various filters with different combinations have been examined in relevant literature on medical image analysis.

Rani [1] and Elfouly et al. [4] displays an investigation of different picture de-noising methods (BIOR, SURELET, HAAR, CURVELET) utilizing wavelet changes and different combinations have been applied in which Curvelet change giving better outcomes under various fluctuation conditions for shading pictures. Sridhar et al. [2] also worked with different wavelets and compression metrics but it concluded about compression depends upon the size of an image. Filters plays a major role while de-noising which has been focused in Hosyar et al. [3, 5] and Durai [6] which uses various types of filters on different noises like Gaussian, Poisson, Salt and Pepper and Speckle.

Rao [7] lightened about different types of wavelet transforms and their way of implication and also Janani et al. [9] and Aktar et al. [10] different fusion rules applied on wavelets. Arora et al. [12, 13] discussed about role of preprocessing till classification phase of skin disease in medical image processing. Kumar et al. [14] and Ruikar [15] thrown light in which case there is a need to go for advance wavelet filtering with specific type of noise.

3 Image Restoration Methods

Image Restoration methods are used to improve the quality of an image either by removing noise or by enhancing the quality by different techniques and algorithms. Wavelet transforms are one of the sources to be used for reducing noise. These are of different types namely Haar, Daubechies (Db), Symlet (Sym), Coiflet (Coif), Biorthogonal (Bio), Meyer and Discrete Meyer (Dmey) wavelet [3]. Haar is the most fundamental wavelet sort. When contrasted with the Haar wavelet, the Daubechies wavelet is more convoluted. Symlets are practically symmetrical wavelets, exhibited by Daubechies, as alterations to the Db family. Coiflets are practically symmetric. Biorthogonal wavelets are a piece of the group of orthogonal wavelets. They have applications in flag and picture handling. Dmey wavelet is Meyer wavelet work in discrete shape [4].

Filters play a major role in process of de-noising in image processing. There is variety of filters like low pass filters categorized as Median, Mean, Gaussian, Wiener, Laplacian and Prewitt filters. Median is among the most eminent and productive channels, which has simple execution while mean has a more simple structure when contrasted with a Median channel. Gaussian is that specific channel, which is known for obscuring and stifling the commotion.

The Wiener channel gives a straight estimation of the authentic picture [5]. To recognize edges in a picture, we can utilize the Laplacian channel [6]. The Prewitt channel manager is used as a piece of picture planning, particularly inside edge area estimations.

Noise is the disturbing element for which filters are required. There can be various forms of noises like Gaussian, Salt and Pepper, Poisson and Speckle noise [7]. Gaussian noise is that noise, which affects each pixel value. It is the most common noise in occurrence. The monochromatic spot develops in the image due to Salt and Pepper noise. The reason for this cause is image sensor and errors in data transmission. Poisson noise is also termed as shot noise and is an electronic noise. Speckle noise increases in number of degree [5].

4 Proposed Method: Dmey Wavelet Gaussian Filter (DWGF)

In this method, skin image has been taken as an input which is noised with Gaussian type of noise. This image has been preprocessed by various wavelet transforms and low pass filters. MSE and PSNR are calculated for each type of wavelet and low pass filters. Then on the basis of low MSE and high PSNR, best wavelet and low pass filter is selected [5]. Dmey Wavelet performed best among wavelets and Gaussian filter performed best among filters. This method gives better performance known as Dmey Wavelet Gaussian Filter (DWGF). Block diagram has been shown in Fig. 1. This method has been tested on 100 images dataset taken from International Skin Imaging Collaboration (ISIC Archive) [8]. Some sample images are shown in Fig. 1.

In this proposed method, DWGF is Dmey Wavelet with Gaussian Filter is used for de-noising of an image. In this, Dmey wavelet which is a discrete form of Meyer wavelet is working with Gaussian low pass filter for better removal of noise on the basis of quality measures.

The block diagram work flow of proposed method is shown in Fig. 2. Skin diseases image is passed from preprocessing phase in which two categories of filters have been

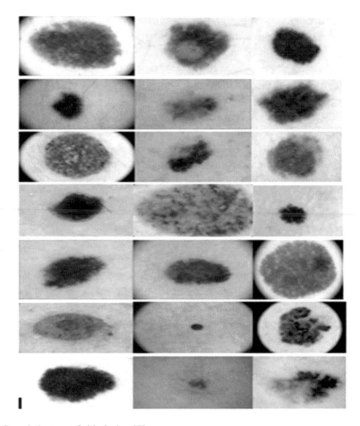

Fig. 1 Sample images of skin lesion [8]

taken. The first category consist of Wavelet transform filters like Haar, Daubechies, Symlet, Coiflet, Biorthogonal, Meyer, and Dmey. Each type of wavelet filter has been applied on images and for each PSNR and MSE has been calculated. The second category consists of low pass filters like Median, Mean, Gaussian, Weiner, Laplacian and Prewitt. Each type of low pass filter has been applied on images and for each PSNR and MSE has been calculated. Finally from each category of filters, best filter have been identified with high PSNR and low MSE which comes to be as Dmey Wavelet and Gaussian Filter acting as Dmey Wavelet Gaussian Filter which proves to be better for de-noising skin lesion image.

5 Implementation and Results

Implementation has been done for DWGF method and other filters by using MAT-LAB on Gaussian type of noise. Experiments were conducted on skin images for

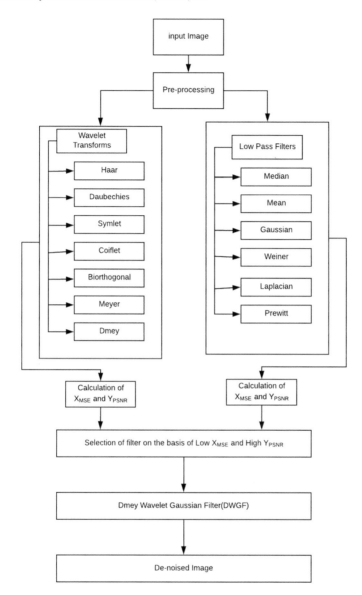

Fig. 2 Block diagram of DWGF working

demonstrating the results. In this paper, a very common skin lesion sample image, as shown in Fig. 3, has been taken as an example to illustrate and deploy the proposed de-noising method.

To test the exactness of filtering algorithms, the following steps are put into effect in the below order:

(a)   **(b)**

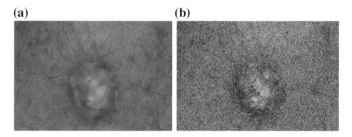

Fig. 3 **a** Original RGB and **b** Gray scale image

   i. A non-corrupt skin image is taken as input.

  ii. The image is transformed into a gray scale image.

 iii. Gaussian noise is infused with 10% noise density into the image.

 iv. The wavelet transformation filtering and low pass filters are applied for reproduction of image.

  v. To check the performance of filters, for noise removal and reconstructed image quality, the MSE and PSNR image quality measures are computed.

 vi. On the basis of quality measures, i.e., Low MSE and High PSNR, our proposed method proves to be better for de-noising.

5.1 MSE Quality Measure

Mean Square Error [1] is given by (1):

$$X_{MSE} = \frac{1}{mn} \sum_{i=0}^{m-1} \sum_{j=0}^{n-1} [L(i, j) - K(i, j)]^2, \tag{1}$$

where M and N are the aggregate number of pixels in the flat and the vertical axes of the picture, g indicates the Noise, and f means the filtered picture. The most minimal mean square blunder speaks to the best quality picture.

5.2 PSNR Quality Measure

The Peak Signal to Noise Ratio given by [1] is as shown in (2):

$$Y_{PSNR} = 20 \log_{10}(MAX) - 10. \log_{10}(X_{MSE}) \tag{2}$$

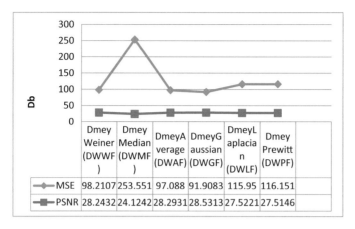

Fig. 4 Analysis of wavelet transforms filters

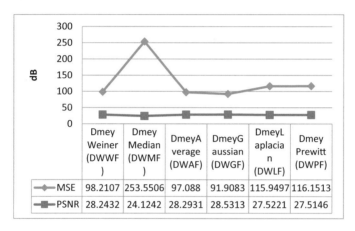

Fig. 5 Analysis of low pass filters

In a nutshell, PSNR value should be high and MSE value should be less [5, 9] for better de-noising effect. Figure 3 shows the original RGB lesion image and converted gray scale image on which the preprocessing have been performed.

Table 1 illustrates the value of PSNR and MSE performance measure calculated for types of wavelet transform filters, low pass filters and wavelet transform with low pass filters. The high PSNR and low MSE for each type of filter is highlighted in bold.

Figure 4 shows the analysis of different wavelet transforms filters with their PSNR and MSE values. Figure 5 shows the analysis of different low pass filters and Fig. 6 shows the analysis of wavelet transform with low pass filters with their PSNR and MSE values.

Table 1 Methods with performance measure and resulted image

S. no.	Methods	MSE	PSNR	Result
1.	Haar wavelet	203.3042	25.0833	
2.	Db wavelet	101.3589	28.1062	
3.	Symlet wavelet	99.4106	28.1905	
4.	Coif wavelet	101.4786	28.1011	
5.	Bior wavelet	99.4904	28.1870	
6.	Rbio wavelet	99.1732	28.2009	
7.	Dmey wavelet	99.1016	28.2040	
8.	Weiner filter	99.6649	28.1794	
9.	Median filter	90.2898	28.6084	
10.	Average filter	100.1405	28.1587	
11.	Gaussian filter	96.0412	28.3402	
12.	Laplacian filter	117.4150	27.4676	
13.	Prewitt filter	115.6477	27.5334	

(continued)

Table 1 (continued)

S. no.	Methods	MSE	PSNR	Result
14.	Dmey Weiner (DWWF)	98.2107	28.2432	
15.	Dmey Median (DWMF)	253.5506	24.1242	
16.	Dmey Average (DWAF)	97.0880	28.2931	
17.	Dmey Gaussian (DWGF)	91.9083	28.5313	
18.	Dmey Laplacian (DWLF)	115.9497	27.5221	
19.	Dmey Prewitt (DWPF)	116.1513	27.5146	

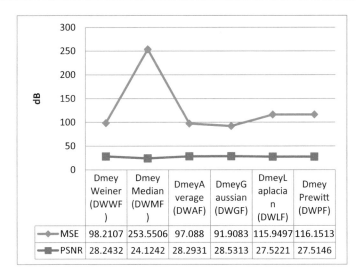

Fig. 6 Analysis of DWGF method

The experimental results show that the Dmey Wavelet Filter fused with Gaussian Low Pass Filter is best suited for de-noising of skin disease input image. Gaussian filter is also beneficial for further analysis of skin lesion [11].

6 Conclusion

In this paper, a Dmey Wavelet Gaussian Noise Filter (DWGF) is proposed for denoising of skin lesion images. This method proves to be better in terms of removal of unwanted noise by measuring its performance through quality measures. MSE and PSNR are tending to be low and high for effective image quality. Also, comparative analysis of an array of image de-noising techniques using wavelet transforms and low pass filters has been performed. From the MSE and PSNR values mentioned in the aforementioned tables and figures, it is observed that our proposed method (DWGF) gives better results. Further, this method can be integrated with segmentation for efficient and accurate analysis without loss of essential features.

References

1. Rani, S.H., Premi, G.: Comparative analysis of various wavelets for denoising color images. ARPN J. Eng. Appl. Sci. **10**(9), 3862–3867 (2015)
2. Sridhar, S., Rajesh Kumar, P., Ramanaiah, K.V.: Wavelet transform techniques for image compression—an evaluation. Int. J. Image Gr. Signal Process **2**, 54–67 (2014)
3. Hoshyar, N., Al-Jumailya, A., Hoshyar, A.N.: The beneficial techniques in preprocessing step of skin cancer detection system comparing. Procedia Comput. Sci. **42**, 25–31 (2014)
4. Elfouly, F.H., Mahmoud, M.I., Dessouky, M.M., Deyab, S.: Comparison between Haar and Daubechies wavelet transformations on FPGA technology. Int. J. Electr. Comput. Energ. Electron. Commun. Eng. **2**, 96–101 (2008)
5. Hoshyar, A.N., Al-Jumailya, A., Hoshyar, A.N.: Comparing the performance of various filters on skin cancer images. Procedia Comput. Sci. **42**, 32–37 (2014)
6. Durai, R., Thiagarasu, V.: A study and analysis on image processing techniques for historical document preservation. Int. J. Innov. Res. Comput. Commun. Eng. **2**(7), 5195–5202 (2014)
7. Rao, R.M., Bopardikar, A.J.: Wavelet Transforms—Introduction to Theory and Applications, 1st edn. Pearson Education, New Delhi, India (2008)
8. International Skin Imaging Collaboration, https://isic-archive.com/
9. Janani, P., Premaladha, J., Ravichandran, K.S.: Image enhancement techniques: a study. J. Sci. Technol. **8**, 1–12 (2015)
10. Aktar, M.N., Lambert, A.J., Pickering, M.: An automatic fusion algorithm for multi-modal medical images. J. Comput. Methods Biomech. Biomed. Eng. Imaging Vis. 1–15 (2017)
11. Udupi, V.R., Raghvendra, A.S., Inamdar, H.P.: Computer vision method for biomedical image analysis. J. IETE Tech. Rev. **18**(5), 365–373 (2015)
12. Arora, G., Dubey, A.K., Jaffery, Z.A.: Performance measure based segmentation techniques for skin cancer detection. Int. Conf. Recent Dev. Sci. Eng. Technol., 226–233 (2018)
13. Arora, G., Dubey, A.K., Jaffery, Z.A.: Classifiers for the detection of skin cancer. Int. Conf. Smart Comput. Inf., 351–360 (2017)
14. Kumar, A., Tiwari, R.N., Kumar, M., Kumar, Y.: A filter bank architecture based on wavelet transform for ECG signal denoising. Int. Conf. Signal Process. Comput. Control 21–23 (2017)
15. Ruikar, S.D., Doye, D.D.: Wavelet based image denoising technique. Int. J. Adv. Comput. Sci. Appl. **2**(3) (2011)

Medical Image Watermarking in Transform Domain

Harsh Vikram Singh and Ankur Rai

1 Introduction

The fast expansion and innovation in high-speed computer systems and high-speed Internet have made imitation and circulation of digital data easier than ever before. This is due to the advancement in information and communication technologies which have provided a new means to access but such advancement has some pros and cons which has been discussed in the paper [1, 2].

The medical information is also compromise with their safety counter to illegal access and manipulation. Several diseases can be diagnosis by extracting information from the medical images. Therefore, hiding information into the medical image requires more attention and care, so that it must not mislead the doctors while extracting data from the medical image for the diagnose. Patient's essential data ought to be kept unaffected and undistorted [3]. The patient's sensitive personal information requires more confidentiality and integrity against unauthorized access and must be robust against several image processing attacks. Data integrity and source authentication of information are the basic pillars of security are major requirement for health data management and distribution.

In the field of telemedicine, there are several issues such as management of patient's information system, hiding diagnosis part from unauthorized access, information retrieval and data integrity can be solved by means of steganography or digital image watermarking [3, 4]. Therefore, embedding the secure and essential patient's data necessitates more attention while embedding it without affecting the diagnosis region.

H. V. Singh (✉) · A. Rai
Kamla Nehru Institute of Technology, Sultanpur 228118, India
e-mail: harshvikram@gmail.com

A. Rai
e-mail: ankur.rai143@gmail.com

© Springer Nature Singapore Pte Ltd. 2019
S. Tiwari et al. (eds.), *Smart Innovations in Communication and Computational Sciences*, Advances in Intelligent Systems and Computing 851,
https://doi.org/10.1007/978-981-13-2414-7_45

In past, few years digital data hiding is shown a significant area for research in integrity, security and confidentiality. The essential end exclusive information is escaped in digital audio, digital music or in digital photographs by means of watermarking [5]. The term Digital Watermarking was first presented in 1993, when Tirkel proposed two watermarking methods to conceal the essential data in the images [6]. Owing to the robust features of watermarking against several image processing attacks it drawn a much attention of researcher towards copyright protection, illicit access in telemedicine and health data management system etc.

In this proposed work the cover image is marked into two regions known as ROI and RONI. Watermark data is hidden into the RONI region without affecting the ROI region. The purpose of this work is to reduce degradation to our medical image. The result achieved in this simulation reveal that the PSNR of host image is achieved greater than 35 dB.

2 Frequency Domain-Based Hiding Techniques

Frequency domain modulation methods are commonly applied for watermarking. In frequency domain technique, the spectral coefficients can easily observe the distinctiveness of Human Visual System (HVS). To make an equilibrium among robustness and imperceptibility, utmost watermark algorithms choose to embed information in the mid-range frequencies. The most often used transform domain techniques are discussed here.

2.1 Discrete Cosine Transforms (DCT)

The frequency domain-based methods are much robust against attacks based on signal processing than the spatial domain techniques [7]. In this type of transform domain data are represented in terms of frequency domain rather than spatial domain. This technique classifies the whole image data into three class of frequencies, known as low, middle, and high frequency band. This watermarking algorithm splits the host image into 8×8 of non-overlapping blocks for inserting the watermark. It also revels imperceptibility property of watermarking. DCT follows the lossy compression technique, in which the hidden data in higher frequency band is often discarded. Therefore, lower and middle frequency are best for embedding watermark in host image. Generally, the middle frequency band is chosen for embedding watermark, since this band has high capacity than lower and higher band [8].

2.2 *Discrete Wavelet Transform*

Wavelet transform based watermarking technique can be used in several applications such as data hiding, compression as well as in image processing. Owing to the properties like spatial localization and multiresolution, DWT is more favorable for watermarking. The information of an image can be easily extracted using such property [8]. The partitioned data are embedded into distinct frequency components. DWT has several advantages over other traditional transform techniques, which can be studied using its discontinuous [9].

One dimensional DWT used for fragmenting the signal into two parts, i.e., low-level frequency and high-level frequency band using low pass filter as well as high pass filter respectively [10, 11]. The high frequency band mostly included the edges of a signal whereas low-level frequency band is furthermore separated into low frequency level and high-level frequency sub-bands. This course can be recurrent till the chosen number of partition level.

3   Proposed Watermarking Model

In the proposed method, medical image is divided into two regions; ROI and RONI, where ROI contains patient's essential and useful information whereas RONI region can be used for embedding purpose [12–14].

In watermarking process, we applied DWT-SVD hybrid method to provide imperceptible and potentially robust method. The RONI region is used for embedding, keeping the ROI part unaffected from any modification. The hiding and removal algorithm is given below;

Watermarking embedding

1. Image I is decomposed into four sub-bounds LL, HL, LH, and HH Using DWT.
2. IDWT Applied to high frequency sub-band and an image of high frequency.
3. SVD is apply to I^k:

$$I^h = U^h \, S^h \, V^h$$

4. SVD is applied to W:

$$W = U^w \, S^w \, V^w$$

5. Alter $S^I : S^{*^I} = S^I + S^w$
   S is known as scaling factor
6. Acquire the altered high-frequency image I^*

$$I^{*^h} = U^h S^{*^h} V^h$$

7. DWT is applied to I^{*^h} and attain altered high frequency component HH*

$$HH^* = DWT\left(I^{*^h}\right)$$

8. The watermarked image I* is recovered using modified HH* and unmodified HL, LH, and HH and IDWT at the end.

Watermarking extraction

1. Hidden watermarked image I* is decomposed into four sub-bounds LL, HL, LH, and HH* Using DWT
2. IDWT is applied to modified HH* and obtain altered high frequency image I.
3. SVD is performed to I^{*^h}:

$$I^{*^h} = U^{*^h} S^{*^h} V^{*^h}$$

4. Again, SVD is performed to watermarking W:

$$W = U^w S^w V^w$$

5. Singular values are extracted using

$$S^{w'} = (S^{*^h} - S^h)$$

6. Recover the watermarking

$$w' = U^w S^{w'} V^w$$

4 Result and Discussion

The proposed scheme is tested on a gray scale medical image of size 256×256 from MathWorks DICOM image and different size of watermarks is used. The value of scaling factor is 0.10. The original image, watermark image, hidden watermarked, and recovered watermark are shown in Fig. 1.

The performance parameter such as PSNR and SSIM are measured for evaluating the quality proposed algorithm. The table shows the measured value of PSNR and SSIM for different images such as MRI, CT Scan and USG image (Table 1).

From above table we can say that PSNR values decreases with increasing the size of watermark. It happens because we are embedding more values of pixel in original image and that's why MSE is increasing and PSNR is decreasing.

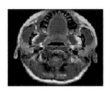 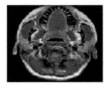

(a) Original Image   (b) Watermark Image   (c) Watermarked Image   (d) Extracted Watermark

Fig. 1 Images of proposed model [26]

Table 1 SSIM and PSNR values with size of watermark

Watermark size	Original image	PSNR (original image and watermarked image) (dB)	PSNR (watermark image and extracted image) (dB)	SSIM (original image and watermarked image)	SSIM (watermark image and extracted image)
16×16	MRI	56.12	42.17	0.952	0.912
	CT scan	48.44	42.13	0.934	0.945
	Ultra sound image	48.41	41.47	0.934	0.952
32×32	MRI	47.76	45.30	0.966	0.973
	CT scan	45.43	41.14	0.924	0.931
	Ultra sound image	43.23	36.31	0.959	0.926
64×64	MRI	46.24	44.71	0.975	0.945
	CT Scan	42.14	40.54	0.957	0.954
	Ultra sound image	40.32	37.15	0.956	0.912

Table 2 PSNR of different image by Gaussian noise attack

Gaussian watermarked image	Watermark image (64 × 64)	PSNR between watermark image and extracted image (dB)
MRI image	Hospital logo	37.73
CT scan image	Hospital logo	41.74
Ultrasound image	Hospital logo	47.73

PSNR is also calculated after performing several signal processing operations during hiding, such as Gaussian noise. Table 2 shows the calculated values of PSNR and SSIM for different images.

In the experiment, we used several 256 × 256 images like MRI, CT scan and Ultrasound, while changing the size of watermarks. We define ROI portion used for diagnosis purpose as a rectangular in middle of the image. The remaining portion is called RONI, where we have to embed the secret image. For watermarking we used DWT-SVD method. The scaling factor is taken 0.10.

For measuring the quality of watermarking different parameters are used, such as SSIM, PSNR. We watermarked and measure the PSNR as well as SSIM for different images. We also add attack such as Gaussian noise and measure the PSNR.

5 Requirement of Medical Image Security

Digital image watermarking can be applied for several applications. Some of them are summarized below.

5.1 Robustness and Protection

The demands on vigorous and safe data embedding are increasing with increase in number of patient information records and digital imaging modalities and therefore, the supervision of digital images with safeguarding of patient record becoming an important issue [15].

5.2 Broadcast Monitoring

Broadcast Monitoring illegal transmission of copyright content can be detected by using this application. It can authenticate whether the copyright content is really advertised or not [16].

5.3 Tamper Detection

Tamper recognition can be done using weak watermark. If the embedded data is ruined or altered, it specifies the existence of meddling and henceforth digital data cannot be reliable [17].

5.4 Authentication and Integrity Verification

Content authentication can perceive any change in digital content. Fragile or semi-fragile watermark can be used for such application since it has little robustness to variation in an image [18–20].

5.5 Security

Watermarking upsurges safety and robustness against geometric and numerous signals processing spells [21–23].

5.6 Covert Communication

For such application, it is very fruitful, where transfer of secret information took place. The transmitted watermark data must be unnoticed and only be recovered by the authentic recipient [24, 25].

6 Conclusion

The complete and essential information for watermarking has been discussed in this paper, which will assist the novel researchers to get the maximum information in this domain. Various aspects for medical image watermarking methods have been carried out. Apart from it a hybrid watermarking model is presented in the paper, where advantages of DWT & SVD are used to form a robust watermarking model. A relative investigation of watermarking methods is offered with their advantages and limitations.

After reviewing several watermarking algorithms, it can be concluded that watermarks embedded using wavelet transforms with SVD are robust against several different attacks as well as they are more suitable for diagnosis purpose. Therefore, they can potentially useful for medical image watermarking.

Acknowledgements The proposed scheme has been tested on a gray scale medical image of size 256×256 accessed and taken from MathWorks DICOM image [26] which are freely available for researchers to access and use.

References

1. Chao, H.M., Hsu, C.M., MiaouS, G.: An authentication data based method with n, incorporation, and concealment for electronic patient records. IEEE Int. Evid. Knowl. Biomed. **7**(2), 46–53 (2002). https://doi.org/10.1109/4233.992161
2. Rai, A., Singh, H.V.: SVM based robust watermarking for enhanced medical image security. Multimed. Tools Appl., 1–14 (2017). https://doi.org/10.1007/s11042-016-4215-3

3. Rong, Z., Keng, P.: Watermarks medicinal indications for telemedicine. IEEE Commun. Evid. Expert. Med. **6**(4), 185–212 (2011). https://doi.org/10.1101/4234.945292
4. Rai, A., Singh, H.V.: Machine learning-based robust watermarking technique for medical image transmitted over LTE network. J. Intell. Syst. **27**(1), 105–114 (2017). ISSN (Online): 2191-026X; ISSN (Print): 0334-1860, https://doi.org/10.1515/jisys-2017-0068
5. Kumar, B., Singh, H.V., Singh, S.M., Mohan, K.: Protected extended-spectrum embedding for telemedical submissions. Sci. Res. J. Inf. Secur. **2**, 91–98 (2011). https://doi.org/10.4236/jis.2011.22009
6. Schyndel, R.G., Tirkel, A., Osborne, C.F.: A Digital Watermark, Reports of the 4th Discussion on Image Processing, ICIP-1994, pp. 86–90 (1994)
7. Cox, I., Miller, M., Bloom, J., Fridrich, J., Kalker, T.: Digital Watermarking and Steganography, 2nd edn. ISBN: 978-0123725851
8. Brancock, E., Meeks, M., Harrmison, R.: Processer discipline subdivision Indigo State University steganography with wavelets: easiness leads to Robustness. Southeast on, IEEE, pp. 588–593, 4–7 (2009)
9. Bouridane, G., Ibrahim, M.K.: Digital video watermarking based on unbalanced multilevel wavelets. IEEE Oper. Signal Dispens. **54**(4), 1519–1536 (2006)
10. Cox, I.J., Miller, M.L., Bloom, J.A.: Digital Watermarking, Morgan Kaufmann (2001)
11. Poldar, V.M., Han, S., Ching, E.: A review of digital video watermarking procedures. In: 4th IEEE Global consultation on Industrial Informatics (INDIN) (2005)
12. Kakarala, R., Ogunbona, P.O.: Signal investigation using a multidimensional form of the singular values decomposition. IEEE Trans. Image Process. **10**, 724–735 (2001)
13. Chang, C.C., Tsai, P., Lin, C.C.: SVD-based digital video watermarking model. Pattern Recognit. Lett. **26**, 1577–1586 (2005)
14. Singh, A.K., Dave, M., Mohan, A.: Multilevel encrypted text watermarking on medical images using spread-spectrum in DWT domain. Wirel. Pers. Commun. (2015)
15. Singh, H.V., Gangwar, S.P., Yadav, R.: Study and analysis of wavelet based image compression techniques. Int. J. Eng. Sci. Technol. **4**, 1–7 (2012)
16. Giakoumaki, A., Pavlopoulos, S., Koutsouris, D.: A medical image watermarking scheme based on wavelet transform. In: Proceedings of the 25th Annual International Conference of the IEEE-EMBS, Cancun, Mexico, pp. 856–859 (2003)
17. Singh, H.V., Gangwar, S.P., Yadav, R.: Emerging trends in transformed based image compression-a review. Int. J. Inf. Sci. Appl. **2**, 591–595 (2010)
18. Zhou, Z., Tang, B., Liu, X.: A block svd based image watermarking method. In: Proceedings of the 6th World Congress on Intelligent Control and Automation, Dalian, China (2006)
19. Giakoumaki, A., Pavlopoulos, S., Koutsouris, D.: Multiple image watermarking applied to health information management. IEEE Trans. Inf. Technol. Biomed. **10**(4) (2006)
20. Hyung, K.L., Hee, J.K., Ki, R.K., Jong, K.L.: ROI medical image watermarking using dwt and bit-plane. In: Asia-Pacific Conference, Communications, pp. 512–515 (2005)
21. Singh, A.K., Kumar, B., Dave, M., Mohan, A.: Multiple watermarking on medical images using selective discrete wavelet transform coefficients. J. Med. Imaging Health Inf. **5**, 607–614 (2015)
22. Chiang, K.H., Chien, K.C.C., Chang, R.F., Yen, H.Y.: Tamper detection and restoring system for medical images using wavelet-based reversible data embedding. J. Digit. Imaging **21**, 77–90 (2008)
23. Eswaraiah, R., Reddy, E.S.: A fragile ROI-based medical image watermarking technique with tamper detection and recovery. In: Fourth International Conference on Communication Systems and Network Technologies (2014)
24. Lai, C.C., Tsai, C.C.: Digital image watermarking using discrete wavelet transform and singular value decomposition. IEEE Trans. Instrum. Meas. **59**(11) (2010)

25. Sharma, A., Singh, A.K., Ghrera, S.P.: Robust and secure multiple watermarking for medical images. Wirel. Person. Commun.; Podilchuk, C.I., Delp, E.J.: Digital watermarking: algorithms and applications. IEEE Ind. Dispens. Mag. 33–46 (2016)
26. https://in.mathworks.com/company/newsletters/articles/accessing-data-in-dicom-files.html

Denoising of Brain MRI Images Using a Hybrid Filter Method of Sylvester-Lyapunov Equation and Non Local Means

Krishna Kumar Sharma, Dheeraj Gurjar, Monika Jyotyana and Vinod Kumari

1 Introduction

Medical imaging has contributed significantly in the diagnosis of disease noninvasively. Medical image processing is encouraging far beyond the boundary of a particular subject and combining the ideas of Medical science, chemistry, radiology, computer science, mathematics and physics. Magnetic resonance imaging measures the spatial distribution of nuclear spins of the proton in the body. The relaxation time of proton spins are the most important parameters in MRI, they are called as $T2^*$, $T2$, $T2$ [1–3]. MRI is very useful for diagnosing soft tissues, and for diagnosing brain it provides images from any projection in multiple sections without moving patient that provides advantages for diagnosis and surgical treatment. A brain MRI is useful for detecting number of brain conditions, including: aneurysms, multiple sclerosis, spinal cord injuries, hydrocephalus, stroke, infections, tumours, cysts, swelling, hormonal disorders, etc. [4].

Medical image denoising algorithms are designed to improve the quality of image by attenuating the noises. There are many algorithms to denoise the images according to types of noise in MRI. Modeling of noise in images depends upon the statistical properties of noise in the image [5]. The nature of noise will be gaussian if noise is additive and there are various techniques to denoise. Magnitude images are common

K. Kumar Sharma (✉)
Department of CSI, University of Kota, Kota, India
e-mail: krisshna.sharma@gmail.com

D. Gurjar
Department of CSE, Madhav Institute of Technology & Science, Gwalior, India

M. Jyotyana
Department of CS, CURAJ, Kishangarh, Ajmer, India

V. Kumari
University of Kota, Kota, India

© Springer Nature Singapore Pte Ltd. 2019
S. Tiwari et al. (eds.), *Smart Innovations in Communication and Computational Sciences*, Advances in Intelligent Systems and Computing 851,
https://doi.org/10.1007/978-981-13-2414-7_46

in MRI because they overcome the problem of phase artifacts by discarding the phase information. In magnitude images, noise behaves in multiplicative form and display rician noise model. Rician noise [6] is similar to rayleigh noise if pixel intensity in the absence of noise is equal to zero.

Impulse noise occurs where quick transients, and bi-polar impluse noise is called salt-and-pepper noise. The uniform density is useful for numerous random number generators that are used in simulation. Exponential and gamma densities are useful in laser imaging [7].

In MRI the acquired imaginary and real data after fourier reconstruction is having noise in resultant image. Noise in this resultant image has gaussian distribution. But the operation on MRI is non-linear so it convert noise from Gaussian PDF to Rician PDF. Most denoising techniques are following Gaussian PDF, wherever other PDFs can also be used. Gaussian PDF is most general because, when singal-to-noise ratio (SNR) is more then MRI data shows behaviour similar to normal PDF. If SNR is low then data PDF deviate from Gaussian to Rician PDF [8].

In the medical image processing, clinical features of the image should not be disturbed from operations, if anatomical features are changing then diagnosis will become difficult. There are two ways to denoise images one is at the time of acquisition of data at the hardware level and another is after acquisition or by post processing method. First method is cost effective and increases the acquisition time. There are available various post processing techniques and generally is classified in the three categories; spatial domain, transform domain and statistical methodology [9, 10]. The rest paper has been discussed as follows: The state-of-the-art works have been elaborated in the second part of the paper. In the third part, proposed denoising method has been presented. In Sect. 4, comparative analysis and result have been discussed using brainweb dataset and original dataset. In the Sect. 5, paper has been concluded.

2 Methods

The local smoothing filter aims to reduce noises and it reconstruct the geometrical features. In filtering, local smoothing filter does not preserve the fine features, structures and textures. Spatial techniques process on all functional features of image as noise and that remove details and fine features of image.

2.1 Non-local Means (NLM) Filter

In [11], the NLM algorithm have been discussed using similar windows of a same image. Periodicity assumption is generalized for study of MRI. Given input is a noisy image in the discrete form $v = \{v(i)|i \in I\}$. Weighted average of image is computed using the all pixels in image as following equation,

$$NL(v)(i) = \sum_{j \in I} w(i, j)v(j) \tag{1}$$

where, $w(i, j)$ is estimated using the distance between i and j. $w(i, j)$ follows certain conditions as; $0 \le w(i, j) \le 1$ and $\sum_j w(i, j) = 1$. Distance between i and j rely on the vectors $v(N_i)$ and $v(N_j)$ of gray level intensity. L2 distance is a reliable method to compare image windows.

$$\mathbf{E}\|v(N_i) - v(N_j)\|_{2,a}^2 = \|u(N_i) - u(N_j)\|_{2,a}^2 + 2\sigma^2 \tag{2}$$

In Eq. (2), u is the input image and v is a noisy image. σ^2 denotes the noise variance. Equality in Eq. (2) preserves similarity between pixels [12]. Weights related to quadratic distances are calculated by

$$w(i, j) = \frac{1}{Z(i)} \exp^{-\frac{\mathbf{E}\|v(N_i) - v(N_j)\|_{2,a}^2}{h^2}} \tag{3}$$

where, $Z(i)$ denotes the mormalizing factor as:

$Z(i) = \sum_j \exp^{-\frac{\mathbf{E}\|v(N_i) - v(N_j)\|_{2,a}^2}{h^2}}$. The exponential function is managed by the value of h. Similarity between image windows can be computed using various methdos as discussed in further methods.

2.2 Optimized Block Wise Nonlocal Means (ONLM) Filter

Redundancy in image is used as information to denoise the 2-D image in the NLM filters. But if same algorithm has been extended to 3-D image then it will be cost effective [13]. In block-wise execution of the NL-means approach is performed in steps. Partitioning the volume in to blocks with overlapping supports: 3-D MRI image having volume Ω^3 in overlapping blocks B_{i_k} such as $\Omega^3 = U_k B_{i_k}$ is partitioned. Here B_{i_k} is not empty block and each block has center at x_{i_k}, which belongs to the volume Ω^3. To display the continuity of block, overlapping of block has to be non-empty. In this ONLM filter, we used the previous NLM filter method to calculate the value of pixel as follows:

$$NL(u)(B_{i_k}) = \sum_{B_j \in V_{i_k}} \omega(B_{i_k}, B_j)u(B_j), \tag{4}$$

where V_{i_k} are the neighbourhood voxels of B_j and weight between block and voxel can be calculated by the following expression:

$$\omega(B_{i_k}, B_j) = \frac{1}{Z_{k(i)}} \exp -\frac{\| u(B_{i_k} - u(B_j)) \|_{2,a}^2}{2\beta\sigma^2|N_i|} \tag{5}$$

where, $Z_{k(i)}$ is a normalizing constant and $\sum_{B_j \in V_{i_k}} \omega\left(B_{i_k}, B_j\right) = 1$. $NL(u)(x_i)$ denotes a intensity which is restored block-wise. Block-wise restoration of intensity is estimated as follows:

$$NL\,(u)\,(x_i) = \frac{1}{|A_i|} \sum_{p \in A_i} A_i\,(p) \tag{6}$$

This algorithm reduces the complexity of the Non-local means technique.

2.3 Adaptive Nonlocal Means Filter (ANLM)

When we solve the nonstationary noise and uses the global noise variance then this results suboptimal. To overcome that problem local noise estimation is introduced [14]. Estimation of local noise can be calculated using euclidean distance between two patches as follows:

$$d(N_i, N_j) = \mathbf{E}\|u(N_i) - u(N_j)\|_2^2 = \mathbf{E}\|u_0(N_i) - u_0(N_j)\|_2^2 + 2\sigma^2 \tag{7}$$

where, u_0 is the noise free image.
Local variance in the image can be calculated as:

$$\sigma^2 = min(d(R_i, R_j)), \ \forall j \neq i \tag{8}$$

R can be calculated by substracting the u and $\psi(u)$. $\psi(u)$ can be calculated as the average of $3 \times 3 \times 3$ 3D region around the voxel.
Behavior of Rician noise is asymmetric due to biasness of resultant weighted average. To avoid this bias Manjon et al. used enhanced method, NLM filter for handling rician noise. Unbiased intensity can be restored as follows:

$$NL_R(u)(x_i) = \sqrt{max\left(\sum_{x_j \in V_i} \omega(x_i, x_j)u(x_j)^2 - 2\sigma^2, 0\right)} \tag{9}$$

The same approach can be applied to the block-wise technique also as follows:

$$NL_R(u)(B_i) = \sqrt{max\left(\sum_{B_j \in V_i} \omega(B_i, B_j)u(B_j)^2 - 2\sigma^2, 0\right)} \tag{10}$$

2.4 Adaptive Multi Resolution Nonlocal Means Filter (AMRNLM)

A multi resolution framework adapts the different filtering features by incorporating in the transform space over the various space-frequency resolutions. Hard wavelet subbands mixing: In NL-based restoration, selection of filtering parameters plays an important role. Procedure of hard wavelet subband mixing will be as follows [15]: Noise MR image can be denoised using two set of filtering features: S_u modified to preserve image features and S_0 adapted to remove noisy components in the image. Resultant image will be respectively I_u and I_0. I_u and I_0 will be decomposed in to subbands using 3-D discrete wavelet transform (DWT) at first level decomposition. Where, LLL represents a low subband. Where as high wavelet subbands are represented by $LHL, LLH, HLL, HHL, HLH, LHH, HHH$. LLL, LHL, LLH and HLL of I_u and the HHL, HLH, LHH and HHH of I_0 have been selected for further.

Inverse $3D - DWT$ have been used to reconstruct a resultant image I. In this filter, adaptive mixing procedure is used to handle noise level and spatial information. Using threshold technique, comparison of coefficients is done, below threshold coefficient are considered noise and above data information. First, I_n, I_u and I_0 are calculated using $3D - DWT$. For each subband of I_n, the signed distance $d_{k,n}$ between noisy wavelet coefficients is computed and a value for thresholding is computed. Similarly, signed distance $d_{k,u}$ and $d_{k,0}$ are computed between denoised wavelet coefficients. Another popular threshold estimation technique BayeShrink can be used for calculation of threshold.

2.5 Sparseness and Self-similarity Based Nonlocal Means Filter (SSNLM)

A collaborative denoising technique that works on two principles sparseness and self-similarity is discussed. Sparseness is the ability to represent image using small number of base functions [16]. Self-similarity uses information of similar pattern available in a image.

Initially, image is denoised using oracle based discrete cosine transform (DCT) filter ($ODCT3D$).

Second, rotationally invariant nonlocal means filter ($RI - NLM3D$) is applied on denoised image to compute similarity between voxels and patches.

Image is denoised by taking average of similar realizations in a signal. Voxels with similar neighbourhood have similar values and can be obtained using following expression:

$$\hat{x(i)} = \frac{\sum_{j \in \Omega} \omega(i, j) y(i)}{\sum_{j \in \Omega} \omega(i, j)} \tag{11}$$

where $\omega (i, j) = \exp -1/2 \frac{(y(i)-y(j))^2+3(\mu N_i - \mu N_j)}{2h^2}$. Average values of patches around voxels i and j is denoted by N_i and N_j.

3 Proposed Method

Proposed method is divided in to two steps [17] firstly, denoising by preserving edges and second denoising using NLM based on Sparseness and self-similarity (EPSSNLM).

- Denoising with Edge Preserving Priors: In this step, image is smoothed by slowly varying regions by preserving the transitions. This step provides a improved information related to medical diagnosis. This type of priors allows a better representation of anatomical details. Image is processed by MAP solution of Sylvester-Lyapunov equation [18]. Let X is a image restored from a image Y, and image X has NxM size. In this optimization problem, solution of X is estimated using principle of maximum a Posteriori (MAP) as follows:

$$X' = \frac{argmin}{X} E(X, Y) = \underbrace{E_Y(X, Y)}_{Data\,Fidelity\,Term} + \underbrace{E_X(X)}_{Prior\,Term} \tag{12}$$

where, $\underbrace{E_Y(X, Y)}_{} = -\log p(YX)$ & $\underbrace{E_X(X)}_{} = -\log p(X)$ denotes data fidelity term. Energy function of the images X and Y are defined according to MM algorithm to solve quadratic optimization problem. MM algorithm iteratively minimizes $E(X, Y)$, a energy function as descibed by following equation:

$$E(X, Y) = \sum_{i,j=0}^{N-1,M-1} d(x(i, j), y(i, j)) + \frac{1}{2} \sum_{i,j=0}^{N-1,M-1} \left[\omega_v(i, j)\,\delta_h^2(i, j) + \omega_h(i, j)\,\delta_v^2(i, j) \right] \tag{13}$$

Energy equation $E(X, Y)$ can be obtained as in Eq. (13). In Eq. 13, $d(x, y) = -x \log p(y|x)$. Weights are represnted using the $\omega_v(i, j)$ and $\omega_h(i, j)$ in each iterations and estimated as follows:

$$\omega(\delta) = -\frac{1}{\delta p(\delta)} \frac{dp(\delta)}{d\delta}, \tag{14}$$

where ω stands for ω_v & ω_h and $p(\delta) = K \exp -\alpha\rho(\delta)$. In this Eq. (14), K is the condition and $\rho(\delta)$ represents potential function. The partial derivative of $E(X, Y)$ with respect to $x(k, l)$ will be as Eq. (15).

$$\frac{\delta E(X, Y)}{\delta x(k, l)} = \frac{\delta(x(k, l), y(k, l))}{\delta x(k, l)} + X_1 - X_2 \tag{15}$$

where X_1 is $[\omega_v (k,l) \delta_v (k,l) + \omega_h (k,l) \delta_h (k,l)]$ and X_2 is $\omega_v (k+1,l) \delta_v$ $(k+1,l) + \omega_h (k,l+1) \delta_h (k,l+1)$. Therefore the stationary point of $E(X,Y)$ can be computed by solving Eq. (16).

$$\frac{1}{\omega (k,l)} \frac{d}{dx (k,l)} d (x (k,l), y (k,l)) + [VD + HD] = 0, \ 0 \le k,l \le M, N$$
$$(16)$$

where VD is $2x (k,l) - x (k-1,l) - x (k+1,l)$ and HD is $2x (k,l) - x (k,l-1) - x (k,l+1)$. After solving Eq. (16) recursively we obtained using $\odot$ hadammard operator (component wise multiplication) as:

$$\sum_{t-1} \odot (X_{t-1} - X^{ML}) + (\theta_N X + X \theta_M) = 0 \tag{17}$$

If we solve above equation in iteration then it is a Sylvester equation.

$$\theta_N X + X \theta_M + C_{t-1} = 0, \ where \ C_{t-1} = \sum_{t-1} \odot (X_{t-1} - X^{ML}) \tag{18}$$

In each iteration, we find out the Sylvester equation and compute this equation and calculate eigenvalues and they are unique and stable.
- Apply Sparseness and self-similarity based NLM to denoise multiplicative noise from the medical images and priori technique preserves edges.

4 Result Analysis

Efficiency of denoising algorithms can be measured based on quantitative performance by peak signal-to-noise ratio (PSNR) and structured similarity index (SSIM). Performance can be measured based on visual assessment, but sharp curves and small useful information cannot be evaluated. Brainweb dataset of simulated MR images [19] was used to conduct experiment. This simulated dataset contains $T1$ weighted, $T2$ weighted and PD weighted images. Above discussed denoising algorithms are applied on a synthetic dataset and on a real MRI images. In Brainweb dataset, noise is varying by 10, 20 and 30% in T1-Weighted and T2-Weighted images.

In Fig. 1, T1-Weighted MR images are used for comparison of above discussed denoising methods. In Fig. 1, above discussed methods are compared using PSNR metric. From Fig. 1, we see that level of noise increases as the value of PSNR decreases. In comparison, proposed method performed better than other denoising methods for PSNR metric.

In Fig. 2, T1-Weighted MR images are used for comparison of above discussed denoising methods. In this comparison, SSIM metric is used. From Fig. 2, we find that when level of noise increases then value of SSIM decreases and proposed method performed better than other denoising methods for SSIM metric.

Fig. 1 Comparison of
various denoising methods
for T1-weighted MRI images
based on PSNR metric

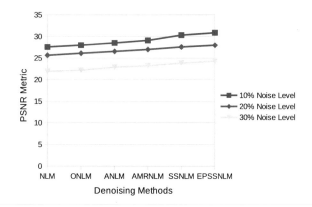

Fig. 2 Comparison of
various denoising methods
for T1-weighted MRI images
based on SSIM metric

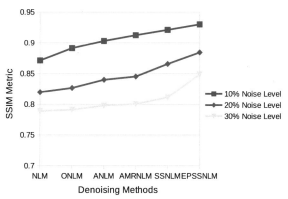

Fig. 3 Comparison of
various denoising methods
for T2-weighted MRI images
based on PSNR metric

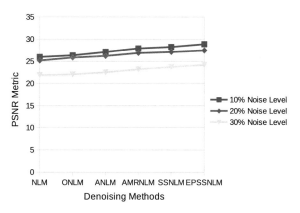

In Fig. 3, T2-Weighted MR images are used for comparison of above discussed denoising methods. In this comparison, PSNR metric is used. From Fig. 3, we find that when level of noise increases then value of PSNR decreases and proposed method performed better than other denoising methods for PSNR metric.

Fig. 4 Comparison of various denoising methods for T2-weighted MRI images based on SSIM metric

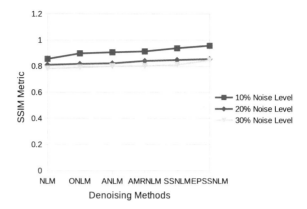

In Fig. 4, T2-Weighted MR images are used for comparison of above discussed denoising methods. In Fig. 4, above discussed methods are compared using SSIM metric. From Fig. 4, we see that level of noise increases as the value of SSIM decreases. In comparison, proposed method performed better than other denoising methods for SSIM metric.

Figures 1, 2, 3 and 4 concludes that Sylvester-Lyapunov Equation and Non Local Means based denoising method gives better results and also preserves the clinical information in the data. NLM uses similar information available in the image to denoise and also preserves the features. As the algorithm is designed with adaptive and improved with sparseness and self-similarity features, then obtained results are improving. High value of PSNR and SSIM are describing the efficiency of the algorithm. Comparison of these denoising filters is also conducted on the real MRI images of the brain and proposed technique performed well as compare to others.

5 Conclusion and Future Work

This work described several state-of-the-art denoising methods based on nonlocal means techniques. In this paper, we proposed a hybrid method using based on sylvester-lyapunov equation and non local means to denoise the brain mri images. We discuss the comparative analysis of propsed method and state-of-the-art methods using PSNR and SSIM coefficients. This comparison has been conducted at different noise levels on the synthetic data to verify the efficiency of proposed method. This experiment is conducted on brainweb dataset and real brain mri images. Our extensive experiment shows effectiveness of proposed method.

In future work, we will implement our proposed method on other medical images than MRI such as CT, endoscopy, etc. An efficient method will be designed to analyse these preprocessed images and that will help in diagnosis.

Acknowledgements We want to thank UGC for funding this work under the category of Minor Research Project. We would like to thank University of Kota, Kota for supporting and providing resources to complete this work. We also thank to dr. dheeraj sharma, medical officer, Govt. of Rajasthan, India for helping in this work.

References

1. Jost, G., Lenhard, D.C., Sieber, M.A., Lohrke, J., Frenzel, T., Pietsch, H.: Signal increase on unenhanced T1-weighted images in the rat brain after repeated, extended doses of gadolinium-based contrast agents: comparison of linear and macrocyclic agents Investigative radiology. Wolters Kluwer Health **51**, 83 (2016)
2. Shiroishi, M.S., Castellazzi, G., Boxerman, J.L., D'amore, F., Essig, M., Nguyen, T.B., Provenzale, J.M., Enterline, D.S., Anzalone, N., Drfler, A., & others: Principles of T2*-weighted dynamic susceptibility contrast MRI technique in brain tumor imaging. J. Magn. Reson. Imaging **41**, 296–313 (2015)
3. Yi, Z., Li, X., Xue, Z., Liang, X., Lu, W., Peng, H., Liu, H., Zeng, S., Hao, J.: Remarkable NIR enhancement of multifunctional nanoprobes for in vivo trimodal bioimaging and upconversion optical/T2-weighted MRI-guided small tumor diagnosis. Adv. Funct. Mater. **25**, 7119–7129 (2015)
4. Yoshioka, H., Philipp, M., Schlechtweg, K.K.: Magnetic resonance imaging (MRI). Imaging Arthritis Metab. Bone Dis., 34–48 (2009)
5. Gravel, P., Beaudoin, G., De Guise, J.A.: A method for modeling noise in medical images. IEEE Transa. Med. Imaging **23**, 1221–1232 (2004)
6. Preza, M.G., Concib, A., Morenoc, A.B., Andaluza, Vi. H., Hernindezd, J.A.: Estimating the Rician noise level in brain MR image ANDESCON, 2014 IEEE, pp. 1–1 (2014)
7. Sijbers, J., den Dekker, A.J., Scheunders, P., Van Dyck, D.: Maximum-likelihood estimation of Rician distribution parameters. IEEE Trans. Med. Imaging **17**, 357–361 (1998)
8. Klosowski, J., Frahm, J.: Image denoising for real-time MRI. Magn. Reson. Med. **77**(3), 1340–1352 (2017)
9. Ali, H.M.: MRI medical image denoising by fundamental filters. In: High-Resolution Neuroimaging-Basic Physical Principles and Clinical Applications. InTech (2018)
10. Manjon, J.V.: MRI preprocessing. In: Imaging Biomarkers , pp. 53–63. Springer, Cham (2017)
11. Buades, A., Coll, B., Morel, J.-M.: A review of image denoising algorithms, with a new one. Multiscale Model. Simul. **4**, 490–530 (2005)
12. Rajan, J., Jeurissen, B., Verhoye, M., Van Audekerke, J., Sijbers, J.: Maximum likelihood estimation-based denoising of magnetic resonance images using restricted local neighborhoods. Phys. Med. Biol. **56**, 5221 (2011)
13. Coup, P., Yger, P., Prima, S., Hellier, P., Kervrann, C., Barillot, C.: An optimized blockwise nonlocal means denoising filter for 3-D magnetic resonance images. IEEE Trans. Med. Imaging **27**, 425–441 (2008)
14. Manjn, J.V., Coup, P., Marti-Bonmati, L., Collins, D.L., Robles, M.: Adaptive non-local means denoising of MR images with spatially varying noise levels. J. Magn. Reson. Imaging **31**, 192–203 (2010)
15. Coup, P., Manjn, J.V., Robles, M., Collins, D.L.: Adaptive multiresolution non-local means filter for three-dimensional magnetic resonance image denoising. IET Image Process. **6**, 558–568 (2012)
16. Manjn, J.V., Coup, P., Buades, A., Collins, D.L., Robles, M.: New methods for MRI denoising based on sparseness and self-similarity. Med. Image A.nal **16**, 18–27 (2012)
17. Vatsa, M., Singh, R., Noore, A.: Denoising and segmentation of 3D brain images. IPCV **9**, 561–567 (2009)

18. Sanches, J.M., Nascimento, J.C., Marques, J.S.: Medical image noise reduction using the Sylvester-Lyapunov equation. IEEE Trans. Image Process. **17**, 1522–1539 (2008)
19. Kwan, R.-S., Evans, A.C., Pike, G.B.: MRI simulation-based evaluation of image-processing and classification methods. IEEE Trans. Med. Imaging **18**, 1085–1097 (1999)

Dynamical Simulation of TT&C Based on STKX Components and MATLAB

Hu Mengzhong

1 Introduction

Dynamical Simulation of Measuring and Controlling Detector (DSMCD) software plays an important role in the range test, which can be used for test route planning, monitoring, and capability evaluation for Measuring and Controlling Detector (MCD) and dynamic track demo of MCD, and it can provide a good visual platform for the range designing test. However, the independent development of DSMCD software needs large amount of funds and time, and due to intellectual property, its source codes are not open to user. So it is not easy for users to modify and upgrade the software according to the actual work in the future.

STK software has a powerful graphics performance and data analysis, which is widely used in aerospace field and it can be used in DSMCD according to [1–4]. But its function is too complex, data interaction ability is poor. For domestic consumer, operation is inconvenient and difficult to master. To overcome these defects, more and more researchers try to integrate the STKX component technology and other technology to achieve special purposes with shielding the part function of the STK and establishing a friendly human–machine interface, this paper is just one of them, but it studies the seamless combination of STKX and MATLAB.

H. Mengzhong (✉)
Research Institute of CETC, No.38, Hefei, China
e-mail: weibohyb@163.com

H. Mengzhong
PLA, Unit 92941 Element 91, Huludao, China

© Springer Nature Singapore Pte Ltd. 2019
S. Tiwari et al. (eds.), *Smart Innovations in Communication and Computational Sciences*, Advances in Intelligent Systems and Computing 851,
https://doi.org/10.1007/978-981-13-2414-7_47

2   Background Details and Related Work

Generally speaking, STKX is often used in development platform of C and C#, such as [5–11]. The software developed by these methods can achieve the required software functions, but they are not easy for users to modify and upgrade the software quickly according to the actual work in the future, and modifying and upgrading often takes a lot of time and money. Through research, it is found that STKX which AGI company is introduced in the new version of STK is released in the form of ActiveX controls. The component allows developers to integrate the application program seamless into the spatial simulation environment and the data analysis engine, and provides technical support for the development of realistic, 3D visualized, powerful, precise spatial simulation application according to [3, 4]. As we know, MATLAB is programming easy and powerful software, which is widely used in the various disciplines, and provides a variety of specialized analysis tools. The GUI interface development of MATLAB is more and more simple, and the function is more and more complete. Therefore, the development of DSMCD software can be realized quickly by the combination of MATLAB and STKX.

However, research about the combination of MATLAB and STKX is rarely reported. The reason may be that interface for STKX and MATLAB is not perfect compared with other development platforms, and the interface reference is lack. To the problem, this paper will study the key technical problems of the combination of the MATLAB and STKX.

3   Development Framework for the Combination of MATLAB and STKX

STKX component (hereinafter referred to as the STKX) is a set of COM component provided by STK. It allows the developer to the STK in 2D and 3D visualization interface and a variety of data analysis of the ability to seamlessly integrated into the application; the component object model includes engine interface, 3D globe control, 2D map controls, and graphics analysis control components that are calculated and analyzed. Design MATLAB and STKX joint development software architecture as shown below (Fig. 1).

Interface engine component of STKX is a channel connecting STK analysis engine, which can be directly used in no GUI interface or GUI interface program. Using the component, developer can communicate with STK analysis engine and send commands to complete a given task simulation and calculation.

3D globe controls are used as the 3D graphics window for space environment simulation. The 3D space simulation interface can be integrated into a GUI application seamlessly, and the developers need only write event response code to respond to and control user's operations in the 3D interface.

2D map control can be integrated the 2D space simulation interface into the GUI application seamlessly. Developers only need to write the event response code to

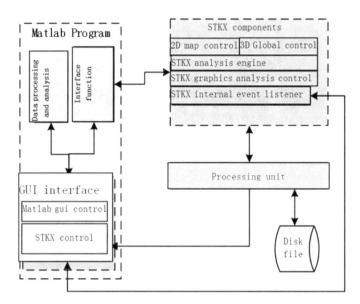

Fig. 1 Framework schematic for the combination of MATLAB and STKX

respond to and control the user's various operations in the 2D interface, which is similar to the globe 3D control.

Graphics analysis control component is a tool for space environment simulation analysis, which includes area tool, AzEl mask tool, obscuration tool, and solar panel tool.

The MATLAB program developed consists of three parts, which are GUI interface, data processing, and analysis module and interface command module. During development of MATLAB GUI interface, by adding an ActiveX control, the STKX control components such as AGI Globe Control 9, AGI Map Control 9, and AGI Graphics Analysis Control 9 can be added into our application visually. So it is easy to integrate the components into the MATLAB GUI for developer which is shown in Fig. 2.

Through the interface command module, the main MATLAB program can achieve a variety of interactive operation and analysis and calculation with the STKX control components. STKX internal event listener is the component developed by the user according to the mechanism of STKX.

The main interface contains the menu bar, toolbars, 2D and 3D control components, parameter setting module, and tracking analysis toolbar which include file manipulation tool button, deducing control button, display switch button for 2D and 3D window, and a 2D map aided tool for analysis and operation. 2D and 3D control components, parameter setting module, include tree-list control, location modification module, and table control for listing devices.

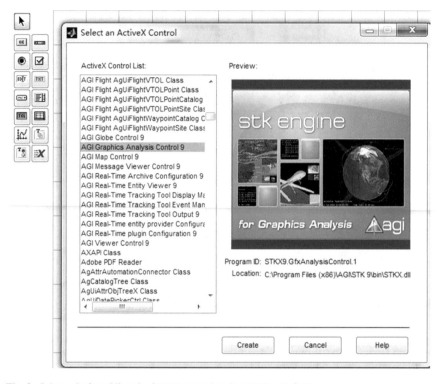

Fig. 2 Schematic for adding the STKX control to the MATLAB GUI

It can listen to the internal events of STKX, and feedback to the processing unit and the GUI interface, and then the application processes and responses of the feedback results. Based on the combination of MATLAB and STKX, the main interface of DSMCD software is shown in Fig. 3.

Tracking and analysis part mainly includes the track analysis button, which is used to change the tracking station and the target in real time. Overall, the main program has the following features:

(1) It has the function for visual demonstration of the dynamic tracking of the target effect in the 2D and 3D simulation windows for all kinds of test equipment.
(2) It has the 2D map zoom tool, real-time ranging tool, hand tools, location query tool, etc.
(3) User can change the location of stations to analyze equipment tracking in real time.
(4) Tracking data concerned by user can be displayed in the 3D simulation window in real time.
(5) 2D and 3D simulation windows can display zoom in, zoom out, and mutual switching.

Fig. 3 Main interface of the developed program

4 Research on the Interface Program for the Combination of MATLAB and STKX

To use the STKX control in MATLAB flexibly, the following key interface program needs to be solved.

4.1 Initialization of STKX Control in MATLAB

Modeling and analysis function in STK is very perfect, in order to save development time; some command codes can be added to automatically open STK program in the MATLAB GUI. Then user can edit, create initial models, and design the course of test conveniently and quickly. The key command for opening the STK program in MATLAB automatically is given as follows:

```
[FileName,PathName] = uigetfile('*.sc','Please select
the STK project file.');
winopen([PathName,FileName]);
```

The key command codes used for loading *.scfile and initializing the STKX-related control in MATLAB are given as follows.

```
fg=get(handles.activex1,'Application');
[FileName,PathName] = uigetfile('*.sc', 'Please select
the STK project file.');
cmd_str=sprintf('%s%s%s%s','Load / Scenario
"',PathName,FileName,'"');
fg.ExecuteCommand(cmd_str);
```

Among the codes above, "activex1" is the name of the STKX control that has been created in MATLAB. "ExecuteCommand" is the interface function for the STKX analysis engine executing user interface commands.

4.2   MATLAB Interface Program for 2D Map Control

2D map controls of STKX have less tool for map viewing and analysis; it only has zoom in tool and zoom out tool, which is inconvenient for the user. To the problem, the user can accord to the needs of the response action of the mouse in the 2D map control for rapid development. In order to meet the control inference process on the map of convenient operation, the author developed selective zoom in or out tool, narrow tools, map location query tools, and distance measurement tool. Let us take the selective zoom in tool as an example to introduce the key codes. The specific process of selective zoom in tool is that mouse location is recorded when in response to the left mouse button is pressed down, rectangular box is drawn from initial point to the current point in real time when the mouse is moved, and the region corresponding to the rectangle box is zoomed in when the mouse left button is bounced. At the same time, the rectangle box is deleted. The key codes are given as follows.

(1) Key codes for getting the current mouse position and saving the data in the "UserData" property of the uipushtool9 control.

```
    pickInfoData1 =
hObject.PickInfo(eventdata.X,eventdata.Y);
if pickInfoData1.isLatLonAltValid()
set(handles.uipushtool9,'UserData',[double(eventdata.X)
,double(eventdata.Y),...
double(pickInfoData1.Lat),double(pickInfoData1.Lon)]);
end
```

(2) Key codes for drawing or modifying the rectangular box in real time.

```
if (lin_plot)      % To judge whether to draw or modify
str2=sprintf('%s%d%s%d %d %d %d %d %d
%d%s%d','MapAnnotation * Add ',lin_num1+1,' Polygon
Coord Pixel Color white Position
4',double(xy0(1)),double(xy0(2)),
double(eventdata.X),double(xy0(2)),double(eventdata.X),
double(eventdata.Y),double(xy0(1)),double(eventdata.Y),
' Smooth yes WindowID ',wdn1);
dfg=fg.ExecuteCommand(str2);
else
str2=sprintf('%s%d%s%.1f %.1f %.1f %.1f %.1f %.1f %.1f
%.1f%s%d','MapAnnotation * Modify ',lin_num1,' Position
4
',double(xy0(1)),double(xy0(2)),double(eventdata.X),dou
ble(xy0(2)),double(eventdata.X),double(eventdata.Y),dou
ble(xy0(1)),double(eventdata.Y),' WindowID ',wdn1);
dfg=fg.ExecuteCommand(str2);
end
```

(3) Key codes for zooming the current map in the selected rectangle area.

```
str_cmd1=sprintf('%s%.5f %.5f %.5f %.5f %d','Zoom *
Region ',xy01(1),xy01(3),xy01(2),xy01(4),wdn1);
dfg=fg.ExecuteCommand(str_cmd1);
```

4.3 MATLAB Interface Program for 3D Globe Control

3D globe control of STKX can provide users with a three-dimensional map visualization window. Generally speaking, free view of 3D map can be realized through some keystrokes and mouse operations, but in order to achieve some special operation on 3D map, users can also develop some codes for responding mouse actions on the 3D globe control according to our special requirement. As DSMCD, the setting of 3D window visual angle can refer to the following commands.

```
fg=get(handles.activex1,'Application');
wd1=get(handles.activex1,'WinID');
hnd1_str='Satellite/Satellite1';
str_cmd1=sprintf('%s%s%s%s%s%d','VO * ViewFromTo Normal
From ',hnd1_str,' To ',hnd1_str,' WindowID ',wd1);
df=fg.ExecuteCommand(str_cmd1);
release(df);
```

4.4 MATLAB Interface Program for Analysis Engine Control

In the process of DSMCD, it is necessary to make a reasonable arrangement of TT&C equipment and analyze the performance of the equipment tracking moving target during the whole process. Key codes for analyzing the performance of any device tracking moving target are given as follows:

```
fg=get(handles.activex1,'Application');
str_sat='*/Facility/Facility1';
str_tar='Aircraft/Aircraft1';
    str_tx1=sprintf('%s%s%s','AER ',str_sat,'
',['*/',str_tar]);
df= fg.ExecuteCommand(str_tx1);
```

Through the above commands, the tracking route for target will be shown in the 2D map, which is displayed in a rough line. If you need to add the tracking data in the 3D map window during the process of deduction, you can refer to the following codes:

```
str_cmd=sprintf('%s%s%s%s%s%d','VO ',str_sat,'
DynDataText Access "AER" ',str_tar,' ShowOn Color
yellow Format Horizontal ',wdn1);
dfg = fg.ExecuteCommand(str_cmd);
```

In order to change the location of the tracking station for optimization of the station location distribution, the following commands can be referred to change the site location by mouse or change the site location by text inputting:

```
%Get the station position
fg=get(handles.activex1,'Application');
str_sat='*/Facility/Facility1';
str22=sprintf('%s %s','Position', str_sat);
dfg = fg.ExecuteCommand(str22);
%modify the station position
str22=sprintf('%s %s %s %s %s %s %s','SetPosition',
str_sat,'Geodetic',41.00,120.44,210,'MSL');
fg=get(handles.activex1,'Application');
dfg = fg.ExecuteCommand(str22);
```

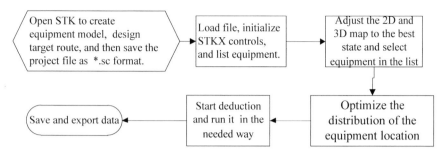

Fig. 4 Process flow

4.5 MATLAB Interface Program for Execution Control of DSMCD

In DSMCD, it is needed to begin, end, reverse, reset, step, or accelerate deduction, so the relevant codes are given as follows:

```
fg=get(handles.activex1,'Application');
fg.ExecuteCommand('Animate * Start ');
fg.ExecuteCommand('Animate * pause');
fg.ExecuteCommand('Animate * Start Reverse');
fg.ExecuteCommand('Animate * reset ');
fg.ExecuteCommand('Animate * Slower');
fg.ExecuteCommand('Animate * Faster');
fg.ExecuteCommand('Animate * Step Reverse');
fg.ExecuteCommand('Animate * Step Forward');
```

5 Process Design for DSMCD

Through the research and development of the interface program, the design of the specific work process for DSMCD is shown in Fig. 4.

(1) Open STK in MATLAB to create equipment model and design target route first, and then save the project file as needed *.sc format.
(2) Load the *.sc project file, initialize the STKX-related controls, and list-related deduction equipment.
(3) Through the map tools developed in the paper, adjust the 2D and 3D map window to the best state, select the tracking targets, and initial tracking equipment in the list of equipment.
(4) By changing the location of tracking equipment, repeat to analyze and view the current equipment tracking situation until the appropriate location is found.

(5)   Through the deduction control button, start or stop the dynamic tracking of TT&C equipment in 2D and 3D map window, and control the speed of deduction according to the need.
(6)   Save or export the required data, and then end the deduction program.

6 Conclusions

STKX components have beautiful interface and can provide realistic 2D and 3D simulation window. It can be combined with user GUI interface development through ActiveX controls. In this paper, as the current status that interface STKX and MATLAB is not perfect compared with other development platforms, and the interface reference is lack, the software architecture for integrating STKX into MATLAB is built according to the characteristics of STKX component and MATLAB software. The key problems of the interface between the two software are researched and solved, and then, according to the requirement for the dynamic tracking of TT&C equipment and the optimization of the station distribution, a kind of workflow is designed. In the end, the rapid development of the application program for dynamical simulation of measuring and controlling detector is successfully realized based on MATLAB and STKX. The research results in this paper can be applied to the dynamic simulation and analysis of all kinds of measurement and control, communication, radar, electronic countermeasure, and space mission, so it has broad application prospects.

References

1. Yin, Y., Qi, W.: STK Application in Computer Simulation. National Defense Industry Press, Beijing (2005)
2. STK8.0 Random Documents
3. Suquan, D., Bo, Z., Shiyong, L., et al.: SAT Simulation Analysis in Space Missions Application in China. Beijing: National Defense Industry Press, pp. 1–8, (2011)
4. Yasheng, Z., Pengshan, F., Haiyang, L. et al.: Master and Master of Satellite ToolsBox. Beijing: National Defense Industry Press, pp. 1 –20 (2011)
5. Bo, C., Gang, Z., Na, W., Yang, Z.: Key technology of visual simulation based on STKX component. Comput. Eng. 3(19), 261–263 (2011)
6. Xiue, G., Xiucan, W.: Satellite to facility link simulation model based on STKX. Inf. Technol. 31(2), 13–17 (2012)
7. Zhiwei, Z., Dengdi, L., Jinhui, Z., et al.: Research on aerospace simulation system based on STKX and HLA. Comput. Eng. 38(6), 259–261 (2012)
8. Ying, Li, Yang, L.: Visualization of TDRSS task operation based on STK. Inf. Electron. Eng. 10(4), 465–469 (2012)

 9. Zhefeng, D., Chuanyu, Z.: Spatial simulation model based on STKX. Sichuan Ordnance J. **30**(10), 141–143 (2009)
10. Junkang, C., Xinhong, L., Shixuan, L.: STKX and QT technology integration in the application of satellite visual simulation. Foreign Electron. Meas. Technol. **33**(1), 74–77 (2014)
11. Jie, H., Tongxin, D., Yongjun, Z.: An approach for integrating VC with STK and its application in simulation. Comput. Simul. **24**(1), 291–294 (2007)

Author Index

A
Abhay Agarwal, 105
Abhilash Panda, 117
Abhishek Burse, 55
Abhishek Mahajani, 465
Adrija Sarkar, 105
Ajitkumar Shitole, 373
Akshay Talke, 373
Alind, 247
Ameya Jawalgekar, 373
Ananiah Durai S., 209
Ankit Shah, 283
Ankur Rai, 485
Apurva Goel, 257
Arvind Yadav, 25
Asha G. R., 189
Ashok Gupta, 335
Ashwani Kumar Dubey, 105, 475
Astitva Narayan Pandey, 443

B
Balaji M., 179
Bankar Rushikesh T., 3
Biswas T., 45
Bose A., 45

C
Chen-Fei Qu, 147, 137
Chodnekar S. P., 97

D
Darshan Mehta, 77
Das S., 361
Deepak Mishra, 161
Deepak Sharma, 465

Deshpande P., 413
Dhabu M., 413, 433
Dhanani J., 381
Dipti Durgesh Patil, 11
Dong-Lin Wang, 147, 137
Dutta S., 127

F
Fraol Gelana, 25
Funde N., 413, 433

G
Ghatak S., 127
Ginni Arora, 475
Goel P., 271
Gowrishankar Subrahmanyam, 189
Gupta A., 361
Gupta P., 271
Gurjar D., 495

H
Harish T., 97
Harsh Vikram Singh, 485
Hetal Patel, 161
Himanshu Gupta, 443
Himanshu Sahu, 247
Hu Mengzhong, 507

I
Isaac Maria, 465

J
Jagannath Das, 179
Jay Shankar Prasad, 423
Jyotyana M., 495

© Springer Nature Singapore Pte Ltd. 2019
S. Tiwari et al. (eds.), *Smart Innovations in Communication and Computational Sciences*, Advances in Intelligent Systems and Computing 851,
https://doi.org/10.1007/978-981-13-2414-7

K

Kanagachidambaresan G. R., 179
Kakulapati V., 323
Kausar Parveen, 173
Kavita Burse, 55
Krishna Prakash N., 67
Kuila P., 45
Kumaresan P., 77
Kumar Sharma K., 495
Kumari V., 495

M

Mahender Reddy S., 323
Maheswari R., 401
Mahima V., 179
Mainak Adhikari, 347
Manju Kumari, 173
Manpreet Singh Sehgal, 423
Margi Engineer, 283
Mehta R., 381
Mohammad Sabir, 173
Mohapatra H., 313

N

Neeraj Mishra, 161
Niveditha Devarajan, 227

P

Pal R., 127
Paranjape P., 413, 433
Pathak R., 413
Patial Arthav S., 201
Pengcheng Zhao, 137
Polamarasetty Anudeep, 67
Prachi Arora, 257
Preetham Kumar, 35
Pulkit Pandey, 201

R

Rana D., 381
Rachna Somkunwar, 237
Rajarao S., 401
Rajesh Ingle, 455
Raj S., 97
Rashmi Burse, 55
Rath A. K., 313
Rohit Kr. Singh, 373
Rohit Pal, 127
Roy R., 127

S

Sakshi Arora, 295
Sanyam Raj, 373

Sarkar A., 127
Seeja K. R., 393
Shadab Alam, 335
Shams Tabrez Siddiqui, 335
Sharmila P., 401
Sheeba Rani S., 401
Shreenidhi H. Bhat, 35
Sindhu Hak Gupta, 201, 227
Sivagami Periannan, 85
Sravani M. M., 209
Sriraman H., 97
Sudeshna Bhakat, 85
Suresh S. Salankar, 3
Swati Nikam, 455
Sweta Jain, 257
Sweta Swain, 393

T

Tanay Deshmukh, 77
Tarasha Khurana, 361
Tarachand Amgoth, 347
Tidke B., 381
Tiwari D., 361
Tusar Kanti Mishra, 117
Tyagi N., 271

V

Varshney P., 271
Vaze Vinod M., 237
Vinay Pandya, 465
Vippon Preet Kour, 295
Virendra Patil, 373
Vishal Goar, 173
Vishnu Ganesh Phaniharam, 117
Vishnu Pratap Singh Kirar, 55

W

Wadhai Vijay M., 11
Wadiwala Krishna K., 161
Weiyan Li, 137
Wen-Jing Wu, 137, 147
Wen-Xia Liu, 147

X

Xuantao Zhang, 137

Y

Yokesh Babu Sundaresan, 77

Z

Zainul Abdin Jaffery, 475
Zaki Ahmad Khan, 335

Printed in the United States
By Bookmasters